The effective management of plants is fu
to all agricultural enterprise, making botany a key
science for all growers. This book provides an
integrated explanation of all aspects of plant
structure and function for students of agriculture,
horticulture and applied biology, with the aim of
highlighting the practical relevance of botany to
agriculture. Each chapter is self-contained and self-
explanatory, with specific chapters covering water,
minerals, structure, growth and development from
sowing to harvest, environmental effects and
controls, breeding, vegetative propagation, field
production and yield, and the nutritional content of
produce. Taken as a whole, *Plants in Agriculture*
fulfils the need for a single text which promotes a
comprehensive understanding of how plants operate
in agriculture.

PLANTS IN AGRICULTURE

PLANTS IN AGRICULTURE

J.C. FORBES

*Monsanto Agricultural Co.,
St Louis, Missouri, USA*

R.D. WATSON

*Environmental and Landuse Consultant,
Landwise Scotland, Alford, Scotland*

CAMBRIDGE
UNIVERSITY PRESS

Published by the Press Syndicate of the University of Cambridge
The Pitt Building, Trumpington Street, Cambridge CB2 1RP
40 West 20th Street, New York, NY 10011–4211, USA
10 Stamford Road, Oakleigh, Victoria 3166, Australia

First published 1992

Printed in Great Britain at the University Press, Cambridge

A catalogue record for this book is available from the British Library

Library of Congress cataloguing in publication data available

ISBN 0 521 417554   hardback
ISBN 0 521 427916   paperback

CS

Contents

Contents

Contents

Contents

Preface

Our aim in writing this book has been to bring out the meaning of botany for modern agriculture. It is directed primarily at students of agriculture, horticulture and applied biology. Since nearly everything that can be said about crop plants is, in some measure, applicable to all angiosperm plants, we believe the book will also be found useful by those studying other aspects of biology.

In 1946, Nelson published his wide-ranging text 'The Principles of Agricultural Botany' to meet the needs of his day. Since then, authors such as Gill and Vear in 'Agricultural Botany', Berrie in 'An Introduction to the Botany of Major Crop Plants' and Langer and Hill in 'Agricultural Botany' have provided fuller treatments of the form and classification of crop plants. However, in the years since Nelson published his book, more research has been carried out into all aspects of plants than in the entire earlier history of the subject, and our knowledge of how plants work and our mastery of new techniques of handling them have advanced apace. Much of this information has been published in specialist volumes, but the need for a single text that would bring together an understanding of plants as they operate in agriculture has grown accordingly, till the lack of it is now a serious impediment to the teaching of agriculture. The excellent and well illustrated text by Janick and his co-authors 'Plant Science', attempts to meet this need, though from a somewhat different stance from our own.

Agriculture is now a series of highly complex technologies. The technology of crop husbandry, for example, involves well timed cultivations, the use of sophisticated machinery to handle the crop, the choosing of appropriate crop varieties, and the application of chemicals to control weeds, pests and diseases and in some cases to modify the growth pattern of the crop. The demands of the marketplace have become much more stringent; the processors of food and other agricultural products have detailed quality requirements that must be met, and people in many parts of the world can now afford to be more discriminating in what they eat. The costs of growing a crop have escalated so that higher yields are required to cover these costs and leave a margin of profit. The environmental costs of intensive agriculture, such as degradation or erosion of soils, depletion of aquifers, and nitrate pollution of rivers and lakes, are more widely recognized and less likely to be tolerated. This provides another major constraint within which the farmer must operate.

At the root of all this is the management of the plant, which underpins all agricultural enterprises, making botany the key science for growers. It is also necessary to relate botany to the many other disciplines in agriculture such as animal husbandry and soil science. Thus our colleagues in such disciplines may be surprised to find us frequently dipping into 'their' subjects. It is an abiding evil of most systems of education that the student's knowledge tends to become fragmented, pigeon-holed according to the disciplines under which it was taught, though in reality knowledge does not exist in discrete compartments. This fragmentation is a serious impediment to persons attempting to make such knowledge work for them, and as knowledge expands and the gulfs between specialisms widen, the problem worsens. For the student's sake, we have tried to put our subject back together again, and relate it to other relevant disciplines.

In writing this text, we were aided by the unusual system found in the School of Agriculture, Aberdeen while we lectured there. This combined teaching at many levels, extension and advisory work, and research and development, all in one institute. Both authors found themselves, in a typical week, not only teaching aspects of agricultural botany to students at various levels of academic ability, but also involved as specialist extension officers, helping to solve problems on farms, and doing research and development. These varied responsibilities required us to remain abreast of scientific developments in our subjects, while retaining a feel for the underlying commercial realities of their application. From this stance, we have tried in our book to bear in mind three needs of our readers, which, we feel sure, are not confined to agriculture, though they may be more acute within it.

First, in an age in which scientific knowledge advances at an accelerating rate there is a growing need to explain, in as simple terms as possible, how that science is applied. Too many academic authors, we feel, fall into the trap of writing textbooks as if all their readers were destined to be fellow scientists, even fellow specialists in those authors' own chosen fields. Where, for instance, the stated target readership is students of agriculture, there is a tendency to dwell on information of interest only to the botanist, or biochemist, or veterinary scientist, or economist. In this book we, as botanists, are primarily talking not to future botanists, or even future agricultural botanists, but are mentally standing in a field talking to future growers and suppliers of food and to the men and women in ancillary occupations who will support them.

Second, there is a need to cope with a variety of problems of teaching any industrial vocational course, which are particularly chronic in agriculture. Agriculture is an industry mainly of small production units, so that a relatively high proportion of its workforce is in management positions. Further, because it is in small units, it utilizes many service industries. The modern grower's management decisions are strongly influenced by streams of information from such diverse sources as salespeople, government advisers, feed manufacturers and bankers. Moreover, agriculture is intensely variable. Weather, soil and farm – they never repeat themselves in quite the same combination. All this is in an industry of rapid change. Bewilderment is the lot of any grower not trained to comprehend. The educational principle of flexible thinking through sound comprehension could not be more relevant than here. That comprehension it is our aim to supply in this book.

Third, the division of labour in any modern production system often obscures the significance of one operation for the whole system. Rarely nowadays is one person involved throughout the growing, processing, packaging, transporting and marketing of the product that provides his or her

living, whether it be eggs or oats. Nevertheless the grower should appreciate the impact of his decisions at an early stage of the production process – decisions, say, related to fertilizer usage – on the quality of the final product received by the consumer, and the influence of this in turn on the diet and health of the community. Similarly, the processor, wholesaler and consumer must understand the restraints that biology imposes on the primary producer. We aim, therefore, to give the reader an overview of the agricultural 'production line'.

We have assumed in our readers a knowledge of elementary biology, chemistry and physics. To aid students with minimal knowledge of plant form and function, we have written each chapter in such a way that they will find the simpler as well as the more advanced terms and concepts explained as they arise, even at the risk of boring our more knowledgeable readers. In any case, because the field we have tried to cover is so large, our intention is that each chapter can largely be read on its own and understood on its own. As far as possible, where concepts or terms from other chapters have been used, we have provided cross-references to the appropriate chapter and section.

Our approach within each topic has been to set the plant in its context, rather than as that somewhat abstract organism that inhabits the pages of so many textbooks but nowhere else in reality, apparently germinating in space, and pursuing its development unharried by the adversities of climate and soil conditions. For example, in dealing with plants and water, we outline the basic functioning of water in the plant, but then relate this to the behaviour of water in soils, and the impact of climate and husbandry on water relations within the plant and in the crop as a whole. Similarly, with plants and minerals, we attempt to relate the behaviour of minerals in the plant and in the soil to the resultant impact on crop growth and yield.

To avoid interrupting the flow of the text, we have not used full botanical names beside the common names of plants. Readers wishing to check botanical names will find them in the index alongside the names used in the text. Also in the interest of readability, we have used direct personal wording in the text in preference to the more indirect form (for example, 'We shall consider X . . .' instead of 'X will be considered . . .'). This is unusual in scientific texts, but we feel the personal style gives more relaxed and simpler reading and helps smooth the student's path.

Certain aspects of agricultural botany we have left largely untreated. The systematic structure and classification of crop plants are well covered in other texts, and we have seen no advantage in repeating them. Two large areas, agricultural ecology and crop protection, are now so substantial in their subject matter and their application in the field that they can be properly covered, we feel, only as separate texts, and can not be done justice to in single chapters. The reader will, however, find frequent reference in our pages to aspects of these subjects where they have relevance to the topic in question.

Though our book is meant to be usable all over the world, in our selection of examples we have consciously drawn more heavily on cool temperate agriculture, of which we have extensive practical experience. We apologize for this bias, but notwithstanding we have tried to draw out our points in such a way that their universal significance is made plain, and we have cited, where appropriate, illustrative examples from warm temperate, tropical and arid climatic regions.

The text can be divided into three sections. In Chapters 1 to 4, we examine how the plant uses the basic resources of energy, water and minerals respectively, and how it derives its structural strength and preserves its integrity. Then, in Chapters 5 to 10 we deal with the growth and development of the plant from seed to maturity, including its reproduction as managed by man in propagation and breeding. Finally, in Chapters 11 and 12 we focus on plants as crops, looking in turn at the two most important attributes of a crop for the grower – its yield and the quality of its produce as food.

Any author of a basic text such as this draws on the work of literally thousands of his fellow scientists in the laboratory and the field, and on the hard-won experience of unnumbered growers. To them all we acknowledge our overwhelming debt, while retaining responsibility for any misinterpretation in the text. We thank also our colleagues in the School of Agriculture at Aberdeen who have patiently answered our questions and read parts of our work, and the office staff who carefully typed successive drafts. We hope the results justify their patience. To two people in particular, Mr Alastair McKelvie and Dr Stanley Matthews, successive Heads of the Agricultural Botany Division, we extend our thanks, for having provided time and encouragement to undertake the work. We are also greatly indebted to Dr Hazel Carnegie who prepared the botanical illustrations, and Mr Stephen Kinnaird who drew the technical diagrams. Finally, our thanks are due to Monsanto Company for its support in completion of the work, and to Mrs Elinor Forbes for assistance in proofreading and indexing.

Jim Forbes
St Louis, Missouri

Drennan Watson
Alford, Scotland

1

Plants as the Basis of Agriculture

Oats, peas, beans and barley grow –
Can you or I or anyone know
How oats, peas, beans and barley grow?

Traditional children's rhyme.

It has probably never been sufficient for farmers to know simply that oats, peas, beans and barley grow. Even in agricultural systems untouched by modern technology, farmers possess a great store of indigenous knowledge which has accumulated over generations of local experience. In the tough commercial environment of developed agriculture today, an understanding of *how* crops grow is essential to the farmer who has to adopt and adapt the latest techniques in order to optimize quality and yield and thereby maximize profit. Equally essential is an understanding of the limitations of these techniques and their long term implications for the environment on which he and all the rest of us so totally depend.

Crops have been described as 'plants in the service of man'. They do not differ in any fundamental way from other plants. The basis for an understanding of crop technology is therefore a broad knowledge of the science of plants. Let us begin by reviewing the ways in which humans are dependent on plants for their very existence, for their comfort, and for the general quality of their life.

1.1 The importance of plants for mankind

Plants provide all our *food*, either directly (oats, peas, beans and barley) or indirectly by nourishing the animals whose meat, milk or eggs we in turn consume. An overwhelming proportion of our food is produced by agriculture (including horticulture) with a small proportion contributed by fisheries. The amount of food provided by the

hunting of wild animals and the gathering of wild plants is now negligible.

Plants also provide *fibres* such as cotton and flax, from which much of our clothing and other woven fabrics are made. They give *wood*, still one of our most important construction and furnishing materials and still the main fuel supply of most of the world's population. *Coal* and *oil*, the principal fuels of the developed world, are the remains of plants which lived millions of years ago. From them are derived plastics, manufactured fibres and thousands of other products. As reserves of oil dwindle and become harder to extract, alternative fuels based on ethanol or methanol are being developed, derived from the produce of present-day plants. As if all this were not enough, plants supply a vast range of industrial and domestic materials including paper, rubber, scents, oils, dyes and drugs.

Let us not forget the importance of plants to the quality of our environment. The great natural and semi-natural forests, scrublands, moorlands and grasslands of the world give us 'breathing space' and provide habitats for wildlife without which this planet would be immeasurably poorer. On a smaller scale, copses and thickets, hedgerows and riverbanks in farmland provide oases of plant and animal diversity which most thinking farmers will seek to preserve even at the cost of some yield.

Plants enormously enrich our lives with their beauty, their perfume and their limitless range of form. Many crops are grown on a horticultural and increasingly on an agricultural scale as *ornamentals*. Plants in the form of *turf* provide an ideal surface on which to play golf, football or cricket, and an attractive and easily maintained ground cover in our parks and gardens.

In short, plants are as important to us as the air that we breathe. Indeed the very *oxygen* in the air, on which all animal life depends, would not be there without plants to release it.

Weeds are plants that interfere with human activities, especially in agriculture. They are not in themselves undesirable; in certain situations, however, such as in a wheat field or on a gravel path, they become a nuisance and may need to be controlled. A weed is not a *kind* of plant but a plant looked at in a particular way. Our *judgement* of it makes it a weed, not some inherent characteristic of the plant itself.

Our main concern in this book is with plants as crops. Much of what we have to say, however, applies equally well to any of the other plants that impinge upon agriculture, including weeds.

1.2 What is a plant?

A plant is, first of all, a living thing and shares the characteristics which we recognize as 'life' with other living things.

Traditionally, living things were divided into two great 'kingdoms': the animal kingdom and the plant kingdom. Zoologists defined what they meant by animals, and left everything else to the botanists. Thus it was that most *microbes*, in particular *fungi* and *bacteria*, came to be classified as 'plants'.

We now recognize that fungi and bacteria are as different from each other, and from plants, as plants are from animals, and therefore we accord them the status of kingdoms. A fifth category, the *viruses*, occupies the twilight zone between the living and the non-living. The chief characteristics of these groups are summarized in Table 1.1 and typical examples of microbes are illustrated in Fig. 1.1.

True plants differ from all other organisms (except a few bacteria) in that they build up the materials of which they are constructed from the simplest of raw materials: water and carbon dioxide gas. Other organisms require an external source of carbohydrates (see Section 1.6.1) or organic materials that can be readily metabolized to carbohydrates. Animals, for example, obtain these materials from their food.

Carbohydrates store a lot of energy in their chemical bonds. This energy can be released, in a

Table 1.1 The five main categories (kingdoms) of organisms recognized by most biologists

Viruses	Consist only of nucleic acid and protein, having no protoplasm of their own. Can be reproduced only inside living protoplasm of other organisms. (Usually not categorized as living organisms.)
Bacteria	Microscopic single-celled organisms; nucleic acid rarely localized in a nucleus. Some photosynthetic but most derive energy from preformed organic matter in other living organisms or their dead remains.
Fungi	Microscopic to large (e.g. mushrooms); nucleic acid localized in nuclei but protoplasm not truly compartmentalized into cells. Not photosynthetic, deriving energy from preformed organic matter.
Animals	Single- or many-celled organisms with a nucleus in each cell; cells bounded by membrane. Not photosynthetic, deriving energy from preformed organic matter (food). Most species capable of locomotion.
Plants	Single- or many-celled organisms with a nucleus in each cell; cells bounded by membrane and cellulose cell wall. Most photosynthetic, requiring no external source of organic food. Not capable of locomotion.

form that can be used to support all the functions of growth and life, by a metabolic process called *respiration*, which we deal with in Section 1.8. It takes a lot of energy to build carbohydrates from simple raw materials. How do plants achieve it? Their source of energy is the sun. They trap energy from sunlight with *chlorophyll*, which is the green pigment characteristic of virtually all plants. The trapped energy is then stored in the molecular bonds of sugar and other carbohydrates and can subsequently be used in respiration or metabolized to any other substance required by the plant. Oxygen gas is released as a waste product. This unique process of fixing the energy of the sun in the chemical bonds of sugar – *photosynthesis* – is the basis of agriculture, indeed of all life on earth, and will be dealt with in greater detail in Section 1.6.

Having no need to move from place to place in search of food, plants, unlike most animals, have no powers of locomotion. The few simple raw materials they require, carbon dioxide, water, nitrate ions as a source of nitrogen, and certain other mineral ions, are all around them. Marine

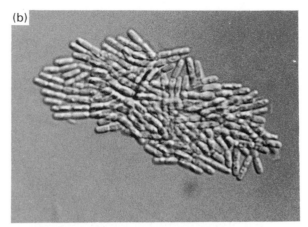

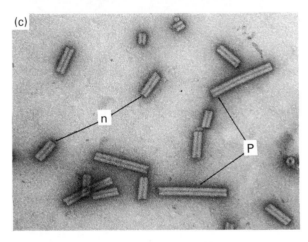

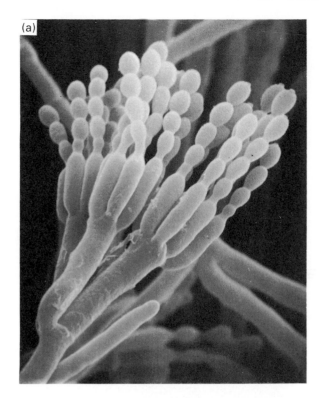

Fig. 1.1 Examples of (a) a fungus, (b) bacteria and (c) viruses. (a) Spore-bearing bodies of *Penicillium*, a mould fungus, × 1600 (courtesy of D Jones, Macaulay Land Use Research Institute). (b) A colony of a species of *Bacillus*, a type of rod-shaped bacterium, × 1800 (courtesy of A M Paton, University of Aberdeen). (c) Particles of tobacco rattle virus, consisting of a nucleic acid core (n) and a protein coat (p), × 320 000 (courtesy of G H Duncan, Scottish Crop Research Institute).

and freshwater plants are constantly bathed in a dilute solution of the raw materials they need, while terrestrial plants draw their carbon dioxide from the air and their other requirements from the soil.

When any plant tissue is examined under a microscope, it can be seen to be composed of fluid-filled compartments called *cells*. The cells of plants differ in a few notable ways from those of animals, and also from the bodies of bacteria and fungi which in some respects can be likened to single cells. As shown in Fig. 1.2, the centre of each plant cell, and indeed most of its volume, is occupied by an apparently featureless, water-filled area called the *vacuole*. Around the outside of the vacuole, in a thin layer, is a fluid with a granular appearance and containing various bodies. This is called the *cytoplasm*, and the most prominent body immersed in it is the *nucleus*, only one of which is normally present in each cell. Each plant cell is bounded by a *cell membrane* and, outside the membrane, a *cell wall*, which is not part of the living system. It is constructed primarily of *cellulose*, a carbohydrate consisting of long chains of sugars joined end to end. Whereas the membrane of a plant or animal cell is so thin that it cannot be seen with a conventional light microscope, the cell walls of plants are readily visible at fairly low magnification (Fig. 1.3). Walls of adjacent cells are cemented together by another important carbohydrate, *pectin*.

A cell can come into existence only by the division of a pre-existing cell into two. When this happens, in plants as in animals, the nucleus also divides. Certain groups of cells in a plant retain the capacity for cell division throughout the life of the plant. These are the *meristems* which continually

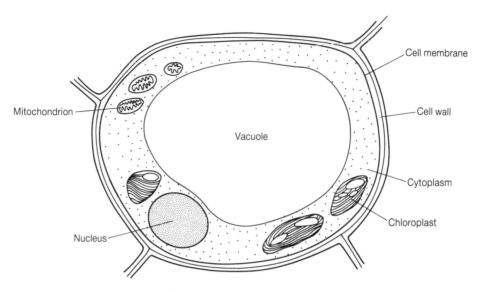

Fig. 1.2 (Above) Generalized diagram of a plant cell. Cell membrane, cytoplasm, nucleus and mitochondria are found in both animal and plant cells, but only plant cells have cell walls, chloroplasts and a large vacuole occupying most of the volume of the cell.

Fig. 1.3 (Below) Cells of a maize root, cut in tranverse section. Plant cell walls are readily visible at low magnification with a light microscope.

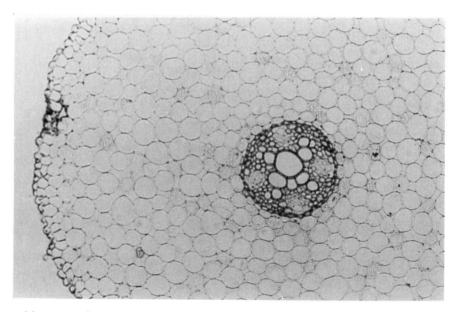

produce or are able to produce new growth. The final number of parts – say, leaves – is not determined at an early stage, as is the case in most animals. Plants are therefore said to show an *indeterminate* habit of growth. One consequence of this is that an individual plant will grow to a size dictated largely by the availability of raw materials and sunlight and governed only to a limited extent by its genetic blueprint (Fig. 1.4).

As in the cells of animals, the cell membrane surrounding the cytoplasm is an integral part of the living system and controls the transfer of materials into and out of the cell. The vacuole and various bodies visible within the cytoplasm, including the nucleus, are also surrounded by membranes.

Fig. 1.4 Shepherd's purse plants from a poor soil with inadequate supply of mineral nutrients and water (right) and from a rich, well-watered soil (left). Because plants have an indeterminate habit of growth, they grow to a size dictated by the availability of raw materials and sunlight.

1.3   The kinds of plants

Plant life, like animal life, originated in the sea. The aquatic environment offers far fewer problems to an organism than the terrestrial environment, with its constant threat of desiccation and exposure to extremes of temperature. As plants evolved to take over the land surfaces of the earth they became less and less dependent on free water, and developed increasing structural complexity to cope with the problems of life on land. Those plants which remained in the sea developed much less complexity. As we survey the main types of plants alive today, it can be seen that in some respects the evolutionary history of the most advanced land plants is reflected in the less advanced types.

It is useful to recognize five levels of evolutionary advancement in the plant kingdom, as shown in Fig. 1.5. Some examples of plants at the first four of these levels are illustrated in Fig. 1.6.

1.3.1   Algae

The *algae* are a highly diverse group of mainly aquatic plants, the simplest and most primitive of all plants. Their cells show little specialization of structure or function except for those concerned with reproduction. Most reproduce sexually, that is new individuals arise by the fusion of two special sex cells called *gametes*. In most algae the gametes, both male and female, are released from the parent plant to fuse in the open water. Algae are either microscopic and single- or few-celled, floating near the surface (*phytoplankton*), or larger and normally attached to rocks (*seaweeds*).

The importance of the planktonic algae is immense. They nourish, directly or indirectly, all marine animal life, including commercially important fish. It has been estimated that about 90% of all the photosynthesis that goes on in the world is performed by these minute plants. They therefore account for about 90% of all the oxygen in the air.

The direct economic importance of the algae is, however, limited. Certain red seaweeds are occasionally harvested for human consumption and others provide industrial colloids such as agar. Brown seaweeds yield *alginates* which find many industrial uses, and the same seaweeds are occasionally used as organic manure in agriculture, being rich in some of the mineral elements required by crops.

Some microscopic algae have a role in sewage treatment. Others can become a problem in water bodies, forming blooms or scums which sometimes release toxic substances into the water, killing fish and presenting a serious health hazard to man and livestock.

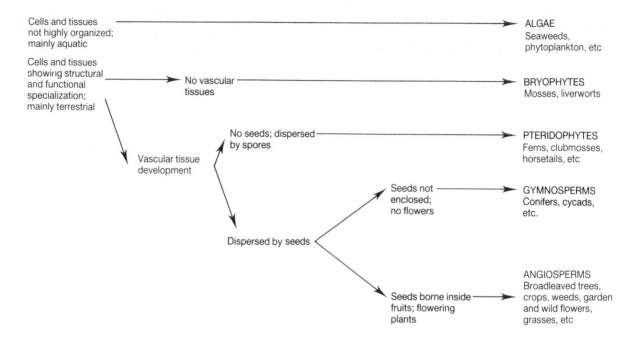

Fig. 1.5 The five levels of evolutionary advancement in the plant kingdom.

1.3.2 Bryophytes

The second level of evolutionary advancement is represented by the *bryophytes*, which include liverworts and mosses. These are small, mainly terrestrial plants showing some degree of specialization of cell structure and function. Cells are organized into *tissues* and tissues into *organs* except in the simplest types which are alga-like in their organization. The more advanced bryophyte plant consists of a stem bearing small leaves, and is a more efficient photosynthetic structure than an algal plant. Bryophyte gametes require to be virtually immersed in free water for their transference in sexual reproduction, but the species are dispersed by tiny airborne *spores*, any one of which can give rise to a new plant if it falls in a suitable location.

Liverworts are the more primitive of the bryophytes. They occur only in damp, shady places and occasionally become weeds in glasshouses. Mosses are more advanced and more common than liverworts but are still largely restricted to damp places such as woodlands and bogs. Some of these may be problem weeds of ornamental or sports turf but they are of little importance in agriculture. Sphagnum, or bog moss, is largely responsible for the accumulation of peat, a valuable horticultural growing medium and still used in some parts of the world as a fuel.

1.3.3 Pteridophytes

The *pteridophytes* are terrestrial plants of the third level of evolutionary advancement. They include clubmosses, horsetails and ferns. They show a high degree of cell and tissue specialization, and, in particular, have well-developed *vascular tissues* for the transport of water and other materials about the plant. This has enabled them to grow to a much greater size than the bryophytes. They require only a film of free water for the transference of gametes, and like the bryophytes are dispersed by airborne spores.

Only among the ferns do we see the attainment of great size, this being permitted by the development of *wood* as a supporting tissue. Three hundred million years ago pteridophytes were the dominant land plants and clubmosses, horsetails and ferns all attained the size of great trees. Their remains form the coal deposits of the world. Pteridophytes are now of only minor and local importance as components of the vegetation. They do, however, include several weeds of economic importance in agriculture. Examples are bracken (a fern), and several species of horsetail. These weeds are poisonous to grazing animals.

Fig. 1.6 Some examples of (a) algae, (b) bryophytes, (c) pteridophytes and (d) gymnosperms.

1.3.4 Gymnosperms

The fourth level of evolutionary advancement is represented by a small group of trees and shrubs with a highly developed system of vascular tissues – the *gymnosperms*. They include the conifers which have needle-like leaves and reproductive organs borne in cones, which are of two types, male and female. The male cones produce airborne spores, or *pollen*, which carry the male gametes to the female cones, where the female gametes or *egg cells* are borne in small structures called *ovules*. Here fusion of male and female gametes, the process of *fertilization*, takes place. The fertilized egg cell grows into the *embryo* of a new plant and the ovule, swelling to accommodate the embryo and a reserve of carbohydrate for its subsequent growth, becomes a *seed*. All this takes place in the female cone while it is still attached to the parent plant. The seed is the unit by which gymnosperms are dispersed. It can subsequently *germinate* to produce a seedling which in the course of time becomes a new plant.

Coniferous forests cover large tracts of northern Europe, Asia and North America. The spruces, firs, pines and larches are of great commercial importance as timber-producing trees. Their wood is relatively soft and easily worked for construction purposes and is ideal for pulping to make paper.

1.3.5 Angiosperms

The *angiosperms* are by far the largest group of plants on earth and are generally agreed to represent the highest level of evolutionary advancement in plants at the present day. Like gymnosperms they are seed plants. The pollen- and ovule-producing organs are borne, however, not in cones but in *flowers*, and the ovules are enclosed in a hollow structure called an *ovary*, one or several to each flower. After fertilization, as each ovule develops into a seed, the ovary also expands to become a *fruit*.

The flower evolved largely as a device to reduce the randomness of pollen dispersal by attracting insects which would carry the pollen from flower to flower. Many angiosperms have, however, returned to a system of wind pollination; these usually have smaller, less conspicuous flowers than those which continue to rely on insects.

Angiosperms, gymnosperms and pteridophytes are all *vascular plants*, that is, they possess vascular tissues, and they undoubtedly share common ancestry. The angiosperms have now displaced the more primitive kinds of vascular plants as the dominant land vegetation of the earth. They are of supreme economic importance, including almost all agricultural and horticultural crops with the exception of a few ornamentals, and all hardwood trees. Almost all weeds, with the exception of the few mentioned under bryophytes and pteridophytes, are also angiosperms.

Virtually the whole of the rest of this book, therefore, concerns the angiosperms alone. When we talk of a plant in general terms, we shall mean an angiosperm plant; in many cases, however, what is true of an angiosperm is also true of other kinds of plant.

1.4 Organization of the angiosperm plant

1.4.1 Structural organization

An angiosperm begins life as a fertilized egg cell which grows into an embryo, enclosed in a seed. On germination of the seed, the embryo becomes a seedling, which by further growth over a period of time becomes a mature plant. This growth takes place in two phases: a *vegetative* phase, when no reproductive structures are produced, and a later *reproductive* phase, characterized by the formation of flowers and later of fruits bearing seeds, to complete the life cycle of the plant. In agriculture, this sequence of events may be curtailed. Many crops, such as sugar beet or cabbage, are harvested while still in the vegetative phase of growth; a few, such as cauliflower, are allowed to enter the reproductive phase but go no further; and some, such as peas or broad beans, are harvested before their seeds are fully ripe (Fig. 1.7).

The plant in the vegetative phase is composed of essentially only three types of organ: the *leaf*, the *stem* and the *root*. Most plants have a short transition zone, the *hypocotyl*, between stem and root (Fig. 1.8). Basically, the leaf is the organ of photosynthesis, the stem is the supportive structure that bears the leaves, and the root anchors the plant in the soil and absorbs water and mineral ions. A continuous system of vascular tissues provides the necessary pathways for movement of materials between roots, stems and leaves.

Many plants show considerable modification of these basic organs to serve other purposes. Couch grass, for example, a troublesome weed of arable and horticultural crops, has long, horizontally

1.4 Organization of the angiosperm plant

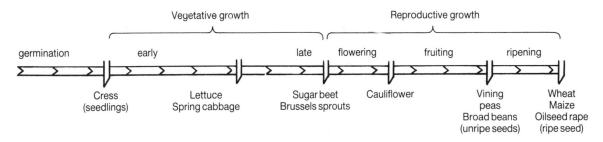

Fig. 1.7 The phases of growth of plants, showing the stages at which various crops are normally harvested.

Fig. 1.8 The structural organization of a plant in the vegetative phase of growth.

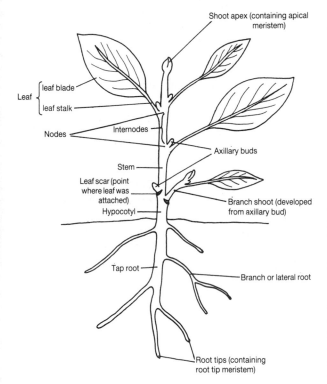

exceptions, leaves do not retain zones of meristematic tissue throughout their lives.

The stem, however, is an organ of indeterminate growth. At its tip is the *apical meristem* which remains functional throughout the vegetative stage of the life cycle of the plant. This meristem gives rise not only to new stem tissue but also to leaves. The whole system of organs produced by the apical meristem is known as the *shoot*; essentially it is a stem with leaves arranged on it.

Small zones of meristematic tissue are left behind by the advancing apical meristem and are to be found at intervals along the maturing stem. Their distribution is not random. They are always located just above the point of attachment of a leaf to the stem and are known as *axillary buds*. Without exception, every leaf has an axillary bud (the *axil* is the angle formed by the leaf and the stem above it) and every axillary bud is subtended by a leaf (Fig. 1.8). An axillary bud may grow to produce a branch shoot. The branching pattern of the plant is determined by the positioning of the axillary buds and by which do and do not grow into branches.

The point on a stem at which a leaf and its associated axillary bud are borne is called a *node*, and the portion of stem between any two adjacent nodes is an *internode*. Most commonly there is only one leaf at each node, in which case the leaf arrangement is said to be *alternate*. In some plants there are two leaves at each node, the so-called *opposite* leaf arrangement. Where there are more than two leaves per node the arrangement is said to be *whorled*. The arrangement may differ in different parts of the same plant, for example in runner beans, in which the first two leaves are opposite and subsequent leaves are alternately arranged.

Many plants (e.g. pea, broad bean) have *compound* leaves which at first sight resemble stems bearing leaves at nodes (Fig. 1.9). The individual leaf-like portions of these leaves are correctly

creeping underground stems acting as a means of spread. The familiar tubers of the potato are highly modified portions of stem acting as carbohydrate storage organs, again borne underground. The spines of gorse and many other plants are modified leaves giving some defence against grazing animals.

The leaf is an organ of determinate growth. Its final shape and to some extent its final size are determined at an early stage in its life. With few

9

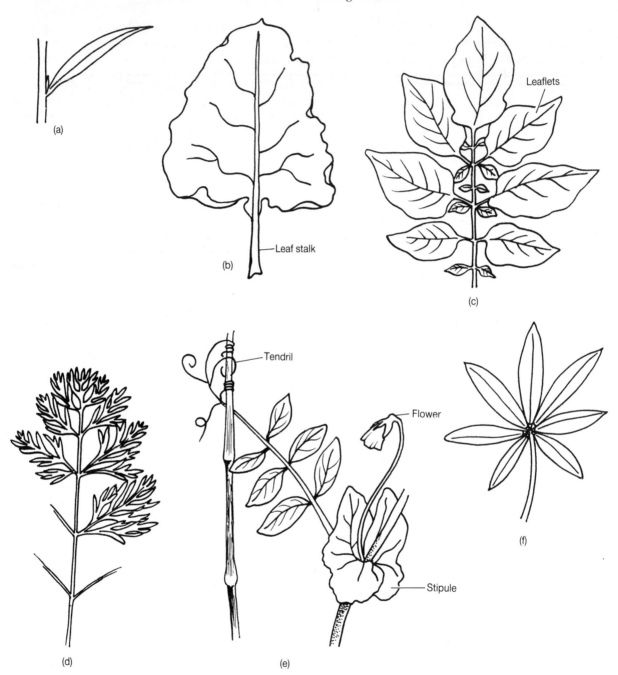

Fig. 1.9 Leaves of (a) flax, (b) beet, (c) potato, (d) carrot, (e) pea and (f) lupin. Leaves may be simple (a,b), with the blade attached directly to the stem (a) or with a leaf stalk at the base (b). Alternatively, they may be compound (c,d,e,f) with individual leaflets attached to the leaf stalk. In pinnate leaves (c,d,e) the leaflets are arranged in rows on opposite sides of the leaf stalk; in palmate leaves (f) they arise from a single point at the end of the leaf stalk.

Some compound leaves have their leaflets further sub-divided (d). In pea leaves, leaflets at the tip of the leaf stalk are modified as tendrils which wrap round the stems of other plants to give the pea plant support, and wing-like structures (stipules) are borne at the base of the leaf stalk. The flower can be seen arising in the axil of the leaf (e).

10

called *leaflets* and are arranged not on a stem but on a *leaf stalk*. A leaf stalk can always be distinguished from a stem by the fact that it does not have an apical meristem at its tip, nor does it have axillary buds associated with the leaflets. In internal structure, too, leaf stalks and stems are quite different.

The root, like the stem, is an organ of indeterminate growth, having a meristem at its tip which remains functional throughout its life. Branch roots arise quite differently from branch stems, as will be described in Section 6.1.2.

In the reproductive phase of growth, vegetative organs may continue to be produced but other organs specialized for reproduction also appear. The flower is an assemblage of organs, some concerned with the production of ovules, some with the formation and release of pollen, some with the attraction of insects and some with a protective role. The fruit, which as we have seen is formed from part of the flower, gives some protection to the seed and may aid in its dispersal or survival after it has been shed from the plant.

1.4.2 Functional organization

Just as in a motor car there are various *functional systems* (for example electrical system, braking system, cooling system), each concerned with one aspect of the operation of the whole machine, or similarly in the human body we have a digestive system, skeletal system, blood circulatory system and so on, in the same way the plant is organized into various interacting functional systems. The organization of much of this book reflects the functional organization of the plant.

Metabolic system. Each cell of a plant is equipped with a battery of highly specialized protein molecules known as *enzymes* which work together to assist in the production of all the complex organic molecules of which the plant is built and with which it functions. Two major processes of the metabolic system are photosynthesis and respiration, which supply the energy needed to operate the rest of this system, and all the other functional systems of the plant. These processes are basic to plant life and therefore to agriculture and are dealt with later in this first chapter.

Transport systems. The plant has two parallel transport systems, one a living tissue for the transport of organic materials manufactured in the plant and one a non-living tissue for the conduction of water and dissolved salts. These are the vascular tissues of the plant, known respectively as phloem and xylem, both ramifying throughout the plant. The phloem system, because of its role in distributing the carbohydrates made by photosynthesis in the leaves, is dealt with in this chapter, and the xylem system, as part of the subject of plant-water relationships, is described in Chapter 2.

Absorption system. The absorption of water and mineral ions by plant roots is covered in Chapters 2 and 3 respectively. Chapter 3, on plant-mineral relationships, also has more to say on transport systems, which are intimately linked to the absorption system.

Support system. The plant has various strengthening tissues to carry the weight of its organs and to resist breakage, especially under the extra loads imposed by wind and rain. This support system is the subject of part of Chapter 4.

Protective system. Also dealt with in Chapter 4 are the outer tissues of the plant, which protect its delicate inner workings from the hostile world outside. Other defence mechanisms are also considered.

Meristematic system. The meristems, as has been mentioned, are the basis of all growth in the plant. The change of form accompanying growth, *development* as it is called, arises through differential meristematic activity in different zones of the plant. Growth and development are a major theme of Chapters 5–8.

Reproductive system. Flowering, fruiting and the production of seeds are examined in Chapter 7. Other systems of reproduction which do not involve flowers, fruits and seeds are often called *vegetative propagation*. They and their use by man are dealt with in Chapter 10.

Sensory systems. Plants have various systems for sensing changes in external conditions, including atmospheric humidity, light intensity, daylength and gravity. These systems are not, however, linked to a central nervous system as in animals, and plants respond quite automatically to environmental changes detected by their sensory systems. The suggestion, occasionally made, that plants are capable of feeling pain or pleasure, has little evidence to support it. The impact of the sensory systems on plant growth and development is dealt with in Chapter 8.

Control systems. The operation of most of these systems is regulated by chemicals called *hormones*, produced in minute amounts in the plant, but having influences out of all proportion to their concentration. The operation of this control system is also covered in Chapter 8.

Genetic system. Molecules called *nucleic acids*, found mainly in the nucleus of each cell, provide the basis for the genetic blueprint of the plant. Organic subgroups, known as bases, on the nucleic acid molecule are ordered in such a way that their sequence can be translated by the chemical machinery of the cell into a sequence of amino acids, the building blocks of protein molecules. A single molecule of nucleic acid can therefore act as a blueprint for any number of identical protein molecules (Fig. 1.10). As proteins are a major structural component of plants, and many operate as enzymes, controlling the entire metabolic system, a blueprint for proteins essentially provides a blueprint for the entire structure and functioning of the plant. An account of *heredity*, the means by which the blueprint is passed on from generation to generation, appears in Chapter 9.

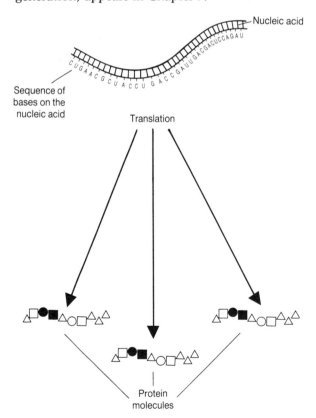

Fig. 1.10 Protein molecules with identical sequence of amino acids, here represented by the symbols △ □ ■ ○ ● being produced by translation from a nucleic acid carrying a sequence of bases here represented by the letters C, U, G and A.

The genetic system dictates the organization of all the other functional systems of the plant and is the only one whose operation is not influenced by the environment in which the plant grows. Throughout the first eight chapters of the book there will be an examination of the ways in which the environment affects the functioning of the plant, and particular attention will be paid to those aspects of the environment that can be controlled by the farmer.

1.5 What is a crop?

Earlier in this chapter we asked the question, what is a plant? Let us now ask, what is a *crop*?

A crop is, first of all, not one plant but many growing together, interacting with one another as well as with their environment. This environment is normally highly modified by the activities of the farmer, gardener or forester. A crop consists of plants which have been selected and changed to serve certain purposes; therefore the crop as well as its environment is much modified by human intervention.

Crops are grown for a *purpose*. The produce of a crop must be suited to the purpose for which it is intended, otherwise it may be unmarketable. The grower must therefore seek to maximize the *quality* of his produce, as we discuss in Chapter 12.

The *quantity* of material, that is, the *yield*, produced by the crops must also be optimized. Very often yield and quality do not go together, and it is necessary to compromise between them. Crop productivity and yield, the subject of Chapter 11, depend to a great extent on the process of photosynthesis, and it is to this process that we must now turn.

1.6 Harnessing of energy by plants: photosynthesis

Photosynthesis is the manufacture of organic matter from carbon dioxide, using hydrogen derived from water and energy derived from sunlight. Its raw materials are thus very simple. The first products of photosynthesis are carbohydrates, which can then be metabolized to other types of organic substance.

1.6.1 Carbohydrates

There is no simple precise definition of a carbohydrate. They contain carbon, hydrogen and oxygen atoms in roughly a 1:2:1 ratio, and they may also contain small proportions of other elements. They store large amounts of energy in their chemical bonds; this energy is released when the bonds are broken by oxidation back to carbon dioxide and water.

The simplest carbohydrates are the *monosaccharides* with three or more carbon atoms in the molecule. In plants the most important of these are *trioses* ($C_3H_6O_3$), for example glyceraldehyde, *pentoses* ($C_5H_{10}O_5$), for example ribose, and *hexoses* ($C_6H_{12}O_6$), for example glucose, fructose.

Most of the carbohydrate content of plants is made up of hexose units, although only a very small proportion of these are in the monosaccharide form. The hexose units are usually linked together, mainly in pairs to form *disaccharides* or in large numbers to form *polysaccharides*. Cane or beet sugar, one of the world's major agricultural products, is the disaccharide *sucrose*, which consists of a glucose and a fructose molecule joined together.

$$C_6H_{12}O_6 + C_6H_{12}O_6 \rightarrow C_{12}H_{22}O_{11} + H_{22}O$$
glucose   fructose   sucrose

This equation does not describe a single enzyme reaction but the effect of several enzyme reactions involved in the synthesis of sucrose. Sucrose is the main form in which carbohydrates are transported in the plant.

Starch, the principal energy reserve of cereals and potatoes, and therefore the main source of dietary energy for most of the human population of the world, is a large polysaccharide consisting of branched chains of hexose units. Plant cell walls are, as we have seen, constructed mainly of *cellulose*, a polysaccharide made of unbranched chains of hexose units. It is the main strengthening and supporting material of the plant.

Although the human digestive system can use starch as a food, it does not have enzymes capable of breaking down cellulose. We cannot therefore directly utilize cellulose as a dietary energy source. It is nevertheless an important constituent of our diet, providing the 'roughage' necessary for proper intestinal function. Ruminants such as cattle and sheep have microbes in their digestive tract which break down cellulose for them and thereby enable them to thrive on a diet of grass. Cellulose is economically important not only as a food for ruminants but also in the form of fibres for the textile industry (e.g. cotton, flax) and as the main constituent of timber and paper.

It is customary to treat hexose ($C_6H_{12}O_6$) as the end-product of photosynthesis, although virtually all the hexose manufactured is further metabolized to sucrose, starch, cellulose and non-carbohydrate materials. Photosynthesis takes place in two stages. The first of these is the capture of energy from sunlight and its conversion to chemical energy. The second is the building of carbohydrate from carbon dioxide in a series of enzyme reactions driven by this chemical energy.

1.6.2   Capture and fixation of light energy

There are a number of pigments in the plant which absorb light energy for photosynthesis, but the most important are the *chlorophylls*, green pigments containing a magnesium atom in an otherwise organic molecule. In this book we shall refer to the whole complex of photosynthetic light-absorbing pigments as 'chlorophyll'.

The chlorophyll molecules are aggregated in groups of several hundred, each group acting as a light trap. Some of the energy they absorb is transferred to special chlorophyll molecules in the light trap which respond by ejecting a high-energy electron; they maintain electrical neutrality by snatching a low-energy electron from hydroxyl (OH^-) ions which are always present in water. The net effect is that a low-energy electron from water is transformed into a high-energy electron by the absorption of energy from sunlight (Fig. 1.11). The high-energy electron is accepted by another specialized pigment which emits an electron at a slightly lower energy level. This is then intercepted by yet another pigment which again releases a lower-energy electron. The process is repeated through several such pigments, the net effect being *electron transport* down a potential energy gradient.

The importance of these pigments is that they allow some of the energy lost by the electron travelling down the gradient to be fixed in the form of an energy-carrying molecule called *adenosine triphosphate (ATP)*.

At the end of the electron transport chain the electron is given another energy boost by a second chlorophyll light trap and is then used to reduce another energy-carrying molecule, nicotinamide adenine dinucleotide phosphate (*NADP*) to a form called *NADPH*. This requires the withdrawal of hydrogen ions (H^+) from water, which balances the withdrawal of hydroxyl ions (OH^-) by chlorophyll as it makes good its electron deficit (Fig. 1.11).

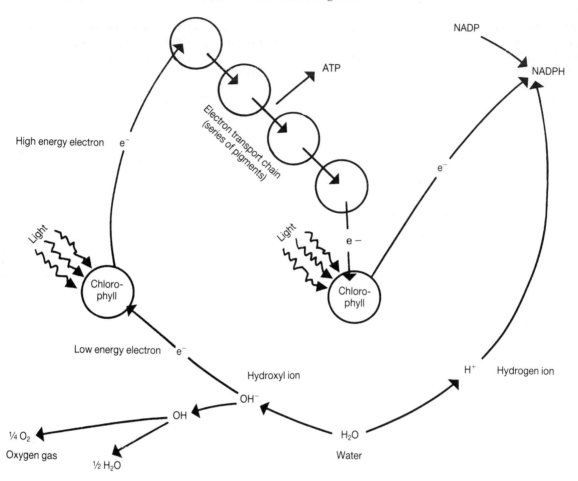

Fig. 1.11 The conversion of light energy to chemical energy in photosynthesis. Chlorophyll absorbs light energy and emits high-energy electrons which are used to make ATP and NADPH. In the process, oxygen gas is released.

The hydroxyl groups, having lost electrons to chlorophyll, combine in pairs to regenerate water and produce oxygen, a waste product which is eventually released as a gas:

$$4OH^- \rightarrow 4[OH] + 4e^-$$
$$4[OH] \rightarrow 2H_2O + 2[O]$$
$$2[O] \rightarrow O_2$$

Photosynthesis, then, is the ultimate source not only of all human food but of all the oxygen in the air. However, the abundance of free oxygen around the photosynthetic system is something of a disadvantage to the plant, as it encourages a metabolic process called photorespiration, of which more will be said later (Section 1.6.5).

Not all the high-energy electrons resulting from the second energy boost shown in Fig. 1.11 are used to reduce NADP. Some are returned to the electron transport chain where they are used to make ATP. Overall, about three molecules of ATP are made for every two molecules of NADP reduced.

1.6.3 Assimilation of carbon dioxide

ATP and NADPH provide the energy for the building of carbon dioxide into carbohydrate molecules, a process known as carbon dioxide *assimilation*. This takes place by a metabolic pathway known as the *Calvin cycle*. One of the key reactions in this pathway is mediated by the enzyme ribulose diphosphate carboxylase. In this reaction carbon dioxide is added to the five-carbon sugar ribulose diphosphate (RuDP) to make a three-carbon product, phosphoglyceric acid (PGA):

$$RuDP + CO_2 + H_2O \rightarrow 2\,PGA$$
$$(C_5) \qquad\qquad\qquad (2C_3)$$

The main energy input to the Calvin cycle comes with the reduction of PGA to another three-carbon sugar, triose phosphate, which requires two molecules of NADPH and two of ATP for every molecule of carbon dioxide assimilated (Fig. 1.12).

On average, for every twelve molecules of triose phosphate produced, and therefore for every six molecules of carbon dioxide assimilated, two combine to make hexose diphosphate, from which all disaccharides and polysaccharides can be synhthesized:

$$2\text{ triose phosphate} \rightarrow \text{hexose diphosphate}$$
$$(2C_3) \qquad\qquad\qquad (C_6)$$

The other ten molecules of triose phosphate enter a complex series of reactions resulting in the formation of six molecules of RuDP, thus completing the cycle. One molecule of ATP is needed for the formation of each molecule of RuDP:

$$10\text{ triose phosphate} \rightarrow 6\,RuDP$$
$$(10C_3) \qquad\qquad\qquad (6C_5)$$

The Calvin cycle is summarized in Fig. 1.12. Overall, to make one molecule of hexose requires the assimilation of six molecules of carbon dioxide,

the reducing power of twelve molecules of NADPH and the additional energy of eighteen molecules of ATP. Six molecules of water are generated in the process.

1.6.4 Linking light fixation to carbon dioxide assimilation

We know that for every four electrons extracted from hydroxyl ions by chlorophyll, one molecule of oxygen is produced as a waste product.

$$4OH^- \rightarrow O_2 + 2H_2O + 4e^-$$

These four electrons reduce two molecules of NADP to NADPH and, on average, allow three molecules of ATP to be made. This is exactly enough ATP and NADPH for the assimilation of one molecule of carbon dioxide. The assimilation of six molecules of carbon dioxide to make one hexose molecule therefore requires the extraction of twenty-four electrons from hydroxyl ions, which results in six molecules of oxygen being released.

For every hydroxyl ion attacked by chlorophyll and removed from water, a molecule of water dissociates to keep the concentration of hydroxyl ions constant:

$$H_2O \rightarrow H^+ + OH^-$$

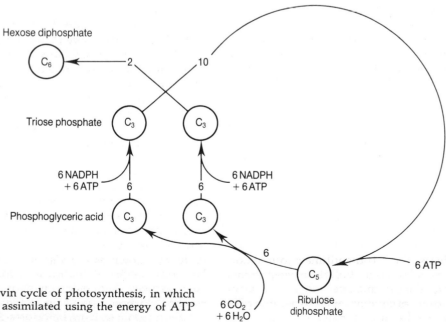

Fig. 1.12 The Calvin cycle of photosynthesis, in which carbon dioxide is assimilated using the energy of ATP and NADPH.

The excess hydrogen ions are, as we have seen, used in the reduction of NADP to NADPH. A total of twenty-four molecules of water are dissociated in the synthesis of one molecule of hexose. Twelve of these are regenerated immediately in the formation of oxygen and a further six are regenerated in the Calvin cycle. There is thus a net consumption of six molecules of water in photosynthesis.

The overall equation for photosynthesis can now be written:

$$6CO_2 + 6H_2O \rightarrow C_6H_{12}O_6 + 6O_2$$

It has to be remembered, however, that this simple equation does no more than indicate the relative quantities of materials required or produced, and does not describe the complex way in which carbon dioxide is assimilated.

1.6.5 Photorespiration

Ribulose diphosphate carboxylase is an unusual enzyme. As we have seen, it is responsible for the assimilation of carbon dioxide into RuDP to make two molecules of PGA. If oxygen is present, however, the same enzyme facilitates the addition of oxygen to RuDP to make two molecules of another substance, phosphoglycolic acid, which does not enter the Calvin cycle. Oxygen and carbon dioxide compete for RuDP; the more oxygen there is around, the more RuDP will be diverted from the Calvin cycle. We have seen that oxygen is liberated as a result of the activity of chlorophyll; there is therefore no shortage of oxygen to compete with carbon dioxide.

The subsequent metabolism of phosphoglycolic acid involves the release of one molecule of carbon dioxide for every molecule of oxygen taken up. This process is known as *photorespiration*. Its similarity to respiration is a superficial one; although it breaks down carbohydrate with the uptake of oxygen and the release of carbon dioxide, it yields no useful energy. It takes place only when photosynthesis is also going on, since the Calvin cycle is necessary to generate RuDP. It therefore occurs only in light.

Photorespiration is generally thought of as a wasteful process, immediately consuming 10–25% of the carbohydrate formed in photosynthesis, and giving no return. It may, however, perform a useful service in mopping up hydrogen peroxide which is sometimes generated in the electron transport system and which would cause serious damage if allowed to accumulate.

1.6.6 Other pathways of carbon dioxide assimilation

Most plants of temperate climates are C_3 *plants*, that is to say, plants in which the first product after carbon dioxide assimilation has three carbon atoms per molecule. This product is phosphoglyceric acid (PGA), as we have seen. Many plants of warmer climates, including major crops like maize and sugar cane (but, interestingly, not rice) are C_4 *plants*. These plants assimilate carbon dioxide into oxaloacetic acid, a product with four carbon atoms per molecule. The details and significance of C_4 photosynthesis will be dealt with in Section 1.6.10.

A third pathway, known as Crassulacean acid metabolism (*CAM*), occurs in a wide range of plants but in few of agricultural importance. CAM plants assimilate carbon dioxide into organic acids in the dark, no ATP or NADPH being necessary for this part of the process. The organic acids later release their carbon dioxide during the day, when it can be reassimilated into carbohydrates by the Calvin cycle using ATP and NADPH formed in the light. The CAM mechanism is an adaptation to reduce water loss in plants of arid climates and its significance as such will be explained in Section 2.6.2.

1.6.7 The chloroplast

The chlorophyll which gives plants their green colour is green because it absorbs light of green wavelengths rather less efficiently than that of red or blue wavelengths. Chlorophyll is not, however, distributed randomly in green tissues or even in a single photosynthesizing cell. Instead it is amassed in small, usually oval, bodies called *chloroplasts* which are immersed in the cytoplasm of photosynthesizing cells. Chloroplasts can readily be seen with a conventional light microscope, often moving slowly around the cell.

At high magnification a chloroplast can be seen to have dark green bodies called *grana* inside it; these contain most of the chlorophyll while the surrounding fluid or *stroma* is only pale green. The grana carry not only the chlorophyll light traps but also all the apparatus for electron transport, ATP synthesis and NADP reduction. The enzymes for carbon dioxide assimilation are probably located in the stroma. The chloroplast is enclosed in a membrane which prevents intermediate products of photosynthesis from being attacked by enzymes other than those concerned with photosynthesis itself.

Chloroplasts can manufacture starch from hexose; sometimes starch grains are visible inside chloroplasts. Surprisingly, however, they appear unable to make sucrose, which is synthesized from glucose and fructose out in the cytoplasm.

Chloroplast-containing cells are generally thin-walled, with large vacuoles. They form soft, rather delicate tissues of the type known as *parenchyma*. Photosynthetic parenchyma occurs in all leaves, in the outer layers of young stems and indeed in all green parts of plants.

1.6.8   Transport of the products of photosynthesis

Carbohydrates are carried, mainly in the form of sucrose, from the photosynthetic tissues to all parts of the plant. A mass flow of sucrose solution takes place along a system of tubes that forms a complete distribution network throughout the entire plant. These tubes are formed from long series of highly specialized living cells. They are called *sieve tubes* because the cells are joined end to end with perforated walls which, when seen under a

Fig. 1.13 Sieve tubes in the phloem. Each sieve tube is a series of living cells joined by perforated end walls. In this photograph one end wall is seen face on, showing the perforations through which sucrose and other materials pass.

microscope, resemble sieves (Fig. 1.13). Their cytoplasm is continuous from cell to cell through the perforations in the end walls, making an unbroken pathway for the transport of sucrose along the whole length of the sieve tube.

Sieve tube cells are remarkable in not possessing a nucleus. Each sieve tube cell is, however, associated with a small parenchyma cell; the nucleus of this *companion cell* seems to take over the control function of the missing sieve tube cell nucleus.

Sieve tubes and companion cells form the *phloem*, one of the two conducting or *vascular tissues* of the plant. (The other is the water-conducting tissue or *xylem*, dealt with in Section 2.1.2). Movement of any substance from one part of a plant to another is termed *translocation*.

Sucrose is not the only material to be translocated in the phloem. Any substance present or being produced in one part of a plant and required elsewhere in the plant may move in this tissue. Phloem translocation is often described as movement of materials from *source* to *sink*. In the case of sucrose, the source is, first and foremost, the leaves. There are two distinct categories of sink for sucrose: actively growing regions of the plant where the carbohydrate is needed immediately as a supply of energy, and regions where carbohydrate is stored for drawing on later when the need for a ready-made energy supply arises.

Sieve tube translocation is an active rather than a passive process such as diffusion; that is, it requires

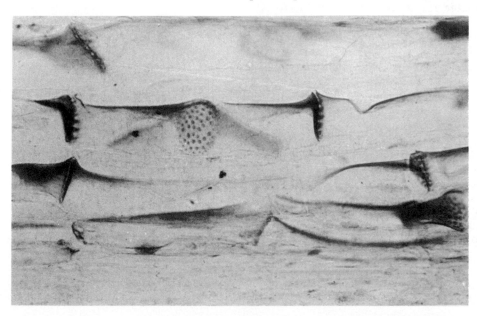

a supply of energy in the form of ATP to drive the sucrose solution along. Typical rates of flow are from 100 to 200 cm h^{-1}, but much faster rates have been recorded. The rate and direction of flow are controlled in a complex way by source supply and sink demand. It is quite common for adjacent sieve tubes to be translocating in opposite directions.

The sieve-like end walls of sieve tube cells act as bulkheads. When a sieve tube is cut, the sieves rapidly become clogged with a deposit of carbohydrate. This helps to localize the damage and prevent excessive loss of translocating materials.

An important category of crop pests – aphids, which include greenfly and blackfly – obtain their food directly from the sucrose solution in sieve tubes. The aphid inserts its fine tubular mouthpiece, or stylet, through the outer layers of stem or leaf tissue into a single sieve tube. It does not need to suck the sucrose solution out – the pressure which drives the translocation process inside the sieve tube pushes the material out through the aphid stylet. Indeed, the sucrose is usually pushed out faster than the aphids can use it and the surplus passes through the aphids and forms a sweet sticky deposit known as honey dew around where they are feeding.

1.6.9 *The leaf as the organ of photosynthesis*

All green parts of a plant owe their colour to the presence of chlorophyll and are capable of photosynthesis. But in most plants only the leaves are specially adapted for efficient photosynthesis. As we consider the structure of a typical leaf we will note how well it is designed for the job it has to do.

Plant leaves exhibit an enormous range of variation. Contrast, for example, the small, unstalked leaves of flax with the large, stalked leaves of sugar beet, or contrast either of these *simple* leaves with the *compound* leaves of potato or carrot (see Fig. 1.9). But the typical leaf, if there is such a thing, is broad, flat and thin, intercepting as much light while carrying as little weight as possible, and having a large surface area in relation to its volume to ensure efficient exchange of gases with the atmosphere. Leaves are arranged on the stem in such a way that light passing through spaces between upper leaves is intercepted by lower leaves. In a well-grown crop the soil surface is totally obscured by leaves.

Most *dicotyledons* (in general the broad-leaved plants such as beans, beets and beech trees) have a flexible *leaf stalk* joining the expanded *leaf blade* to

the stem, but in some, including flax, the leaf blade is attached directly to the stem. In most *monocotyledons* (including cereals and grasses) the long narrow leaf blade is joined to the stem by a *leaf sheath* which forms a cylinder around the shoot (Fig. 1.14). The leaves of dicotyledons are characterized by a net-like pattern of *veins* (bundles of vascular tissue) in the blade, often with a single main vein or *midrib*, continuous with the stalk, running the full length of the blade. In monocotyledon leaves, on the other hand, the veins do not branch but run parallel to one another through the entire leaf sheath and blade.

The photosynthetic tissue of the leaf is called the *mesophyll* and is a parenchyma tissue characterized by a high density of chloroplasts in the cells, to make full use of the intercepted light. Permeating the entire mesophyll tissue is an extensive air space between the cells. This air space represents the internal ventilation system of the leaf and is continuous, as will be shown later, with the atmosphere outside. When carbon dioxide is taken in and oxygen given out in photosynthesis, it is the internal air space that exchanges gases with the mesophyll cells.

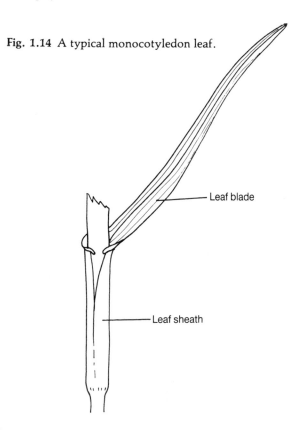

Fig. 1.14 A typical monocotyledon leaf.

Leaf blade

Leaf sheath

In dicotyledons, but not in monocotyledons, the mesophyll is usually differentiated into an upper and lower layer. The upper, or *palisade layer*, consists of more or less cylindrical cells very rich in chloroplasts, but with relatively little air space between the cells. The lower, or *spongy layer*, has more irregularly shaped cells with fewer chloroplasts but with a large air space (Fig. 1.15). This differentiation within the mesophyll tissue reflects the fact that dicotyledons usually hold their leaves more or less horizontally, so that the upper surface receives much more light than the lower surface. Monocotyledon leaves are held more upright, with both surfaces receiving similar amounts of light.

The mesophyll is a delicate tissue, and is protected and to some extent held together by a single layer of more robust cells on both upper and lower surfaces of the leaf. This layer is the *epidermis*, and forms part of a protective skin covering the entire plant (see Section 4.5.1). There are no intercellular spaces in the epidermis, except for tiny pores called *stomata* (singular: stoma) which provide a connecting pathway between the atmosphere and the internal air space. Each stoma is flanked by two specialized cells called *guard cells* (Fig. 1.15).

The function of the guard cells is to open and close the stomatal pore. If the stomata are open, carbon dioxide readily enters the internal air space of the leaf to make good the carbon dioxide depletion cause by photosynthesis. If the stomata are closed, the mesophyll cells quickly run out of carbon dioxide and photosynthesis stops. What, then, is the point of having a system for closing the stomata?

The answer is to prevent excessive loss of water from the plant by evaporation from the leaves. The leaf, well designed as it is for efficient light interception and gas exchange, also provides an ideal surface for the evaporation of water. Loss of water through the epidermis is minimized by a waxy waterproof coating, or *cuticle*, which covers both surfaces of the leaf. There are, however, so many stomata in the epidermis that when these are open, evaporation may be almost as rapid as if the cuticle did not exist. By having pores which can open and close, the plant is able to photosynthesize at full speed when water is in plentiful supply, and temporarily 'close down' during periods of water shortage. The role of the stomata is dealt with further in Section 2.4.1.

The veins of the leaf blade are embedded in the

Fig. 1.15 Internal structure of a typical dicotyledon leaf in cross-section.

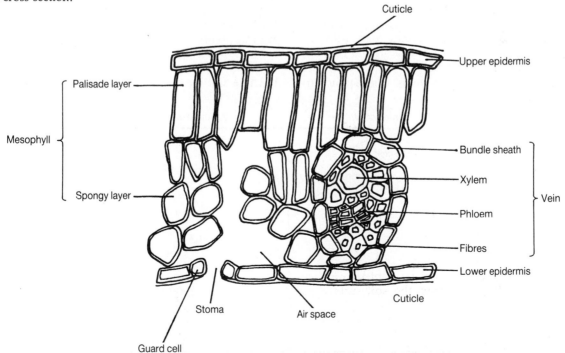

mesophyll. They combine three important functions: support, distribution of water to all parts of the leaf blade, and transport out of the leaf of sucrose and other materials manufactured there. As we have seen the sucrose is carried in the living tissue called phloem, and the water moves in the xylem. Some of the water is needed as a raw material for photosynthesis, but most of the water brought to the leaf blade in the xylem is lost by evaporation through the stomata.

The importance of the veins in providing a supporting skeleton for the leaf blade is frequently understated. The main reinforcing material in the veins is a complex carbohydrate called *lignin*. This substance impregnates the considerably thickened walls of certain types of cell, which are then said to be *lignified*. Heavily lignified cells are almost always dead. The water-conducting cells of the xylem come into this category; they have the dual function of water transport and structural support. Many leaves also have bundles of *fibres* in the veins. These are greatly elongated, very thick-walled, heavily lignified cells whose sole function is to help support the leaf. Above and below the veins there is frequently yet another support tissue, this time not lignified, but with extra cellulose deposition, and more flexible. This is known as *collenchyma*. It is often particularly well developed around the midrib and in the leaf stalk.

Each vein thus consists of a bundle of tissues: xylem, phloem and fibres. The xylem always lies towards the upper surface of the leaf, with the phloem just below it. If fibres are present they are most commonly found below the phloem (see Fig. 1.15). The whole bundle is completely surrounded by a *bundle sheath* of parenchyma cells which are elongated in the direction of the veins and which lack intercellular spaces.

The function of the bundle sheath in C_3 plants is not known with certainty, but it is clear that all carbohydrate passing from the mesophyll into the phloem has to pass through it. In C_4 plants the bundle sheath is particularly well endowed with chloroplasts, which are larger and paler than those of the mesophyll and seem to lack properly organized grana (see also Section 1.6.10 below).

1.6.10 Photosynthesis in C_4 plants

The mesophyll cells of C_4 plants are surrounded, as in C_3 plants, by the internal air space of the leaf. It is these cells that fix atmospheric carbon dioxide. In C_4 plants the carbon dioxide is assimilated in the cytoplasm, not in the chloroplasts, of mesophyll cells. It combines with a three-carbon substance, phosphoenolpyruvic acid, to form a four-carbon substance, oxaloacetic acid. This is then metabolized to malic acid or, in some species, aspartic acid, which is transported to the bundle sheath.

Here malic acid (or aspartic acid) gives up its carbon dioxide, which then enters the bundle sheath chloroplasts. The three-carbon product left is carried back to the mesophyll where it is used to regenerate phosphoenolpyruvic acid. In the bundle sheath chloroplasts carbon dioxide is reassimilated by the Calvin cycle with the formation of hexose, as in C_3 plants (Fig. 1.16).

Most of the light energy fixation is done by the

Fig. 1.16 Carbon dioxide assimilation in a C_4 plant.

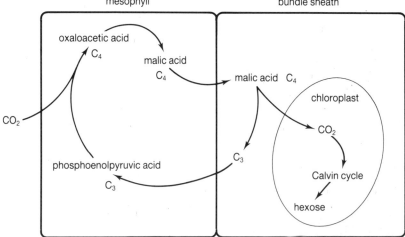

mesophyll chloroplasts. Very little oxygen is therefore generated in the bundle sheath chloroplasts. Thus ribulose diphosphate carboxylase in the bundle sheath is less likely to be attacked by oxygen and the rate of photorespiration is very much lower than in C_3 plants. Furthermore, any carbon dioxide produced by photorespiration has to pass out through the mesophyll, where it is likely to be fixed again into oxaloacetic acid. This makes photorespiration undetectable in C_4 plants.

The much more complete assimilation of carbon dioxide made possible by the C_4 pathway confers certain advantages on plants living in a warm, dry climate, as we shall see later. The pathway is, however, less efficient in its use of light energy, because of the additional energy (three molecules of ATP per molecule of carbon dioxide) needed for regeneration of phosphoenolpyruvic acid.

1.7   Storage of energy in the plant

A proportion of the carbohydrate manufactured by photosynthesis during the day has to be stored to sustain the plant at night. In addition to this short-term energy reserve, plants also lay down longer-term stores of fuel for sustenance at times when photosynthesis is limited, even during the day. There are three situations of great agricultural significance where this energy stockpile is necessary.

1.   A seed planted in the soil cannot derive the energy it needs for growth from photosynthesis. It is therefore totally dependent on the reserve of fuel laid down in it by the parent plant. The more fuel the seed contains, the longer the seedling can continue to grow without becoming photosynthetically self-sufficient. This means in practice that a large seed with a considerable energy store can generally be sown at a greater depth than a small seed with a limited energy store (Fig. 1.17).

2. In cool temperate climates, plants have to survive the winter when conditions are not favourable to growth. In many warmer climates there is a dry season which is similarly inimical to growth. Most plants become dormant during the cold or dry season. In many cases they survive only as seed, the entire parent plant dying at the onset of the unfavourable conditions. In other cases the plant lays down a reserve of fuel in some vegetative organ resistant to or protected from frost or drought. Whether the reserve is in the seed or in a part of the plant itself it has to do two things: sustain the seed or plant during its dormant period,

and provide the energy to re-establish the photosynthetic apparatus in the form of leaves as soon as favourable conditions return. In virtually all our important food crops, what is harvested is the energy reserve accumulated at the end of the growing season, for example the grain of cereals, the roots of sugar beet, carrots and swedes, and the tubers of potatoes. Crop plants produce harvestable energy reserves far in excess of what they themselves require. This is the result of both unintentional and purposeful selection by humans over the centuries.

3.   A number of plants, notably the grasses, have great powers of regeneration after almost complete defoliation. Removal of the photosynthetic structure by grazing animals or by cutting for hay or silage means that the grass plant must call on stored energy to enable it to build up its leaf area again. Cutting too frequently or grazing too heavily, however, exhausts the energy supply and the grass dies out. Underground energy reserves also permit rapid regeneration of weeds such as bracken or docks after cutting. Unless cutting is repeated frequently the energy reserves accumulate again and the weed remains as vigorous as ever.

Plants use a number of chemical substances to store energy. In stems, roots and other vegetative storage organs the commonest substances are sucrose and starch. Sucrose is the major energy store of carrots, swedes and of course sugar beet and sugar cane. Starch is the form in which energy is stored in potato tubers. Jerusalem artichoke tubers and chicory roots are unusual in that their main energy storage substance is *inulin*, a short-chain polysaccharide made up of fructose units. Inulin is commercially valuable as a source of fructose and as a food for diabetics.

In seeds, energy is stored mainly as starch and oils. The proportion of oils to carbohydrates is low in wheat and barley, higher in oats and maize, and very high in the seeds of sunflower, linseed, rape and peanut, all of which are grown for oil extraction. Chemically, vegetable oils are esters of glycerol and three fatty acids. In animal fats, most of the fatty acids are said to be saturated – they lack double bonds between carbon atoms. In some vegetable oils, on the other hand, there are frequent double bonds in the fatty acid component. The 'polyunsaturated' nature of these vegetable oils apparently makes them a more suitable constituent of the human diet than animal fats, in circumstances in which the latter are thought to increase the risk of heart disease.

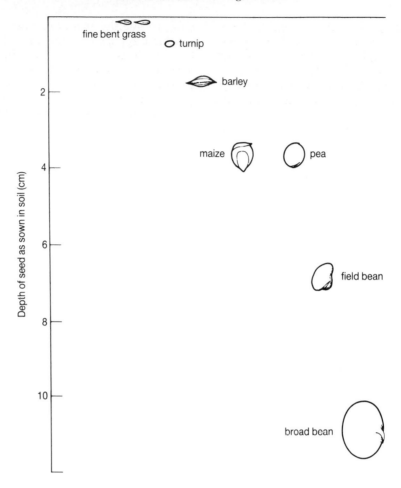

Oil contains more stored energy per unit weight than carbohydrate. It must, however, be converted to carbohydrate before its energy content can be usefully released. Protein, which is used by pea and bean seeds as part of their energy reserve, is also converted to carbohydrate during mobilization of this reserve.

Fuel is usually stored in thin-walled parenchyma cells and starch grains or oil globules are often visible in such cells. Storage tissue can occur almost anywhere in the plant, but is commonest in modified stems, roots and leaf bases. Except in seeds, it is usually a soft, succulent tissue which is easily damaged during harvesting and transport and is prone to disease both before harvest and in store. The handling and storage of such commodities as potatoes, root vegetables and soft fruits therefore demand considerable care and expertise, and an awareness of their vulnerability.

Fig. 1.17 Relationship of normal sowing depth to seed size in different crops.

1.8 Extraction of energy from carbohydrate: respiration

We have seen how the energy of sunlight is used in photosynthesis to make carbohydrate, how this carbohydrate is transported around the plant and how it is stored for later use. We must now consider how the plant extracts useful energy from carbohydrate and other storage products.

1.8.1 Respiratory metabolism

The process of respiration, by which a living cell breaks down carbohydrate to release its stored energy in a usable form, is essentially the same in plants and animals. Like the Calvin cycle of photosynthesis, in which carbohydrate is built up from carbon dioxide, respiration involves long metabolic pathways, requiring many enzymes and forming many intermediate products.

The overall equation for respiration is often written as the reverse of that for photosynthesis:

$$C_6H_{12}O_6 + 6O_2 \rightarrow 6CO_2 + 6H_2O$$

and indeed the most obvious outward sign that the process is going on in a living organism is the familiar exchange of gases – oxygen taken in, carbon dioxide given out. But this equation ignores the whole point of respiration – the liberation of energy in a useful form. As in photosynthesis, the energy is generated in the form of ATP, which can be used to drive any energy-requiring metabolic process or carry out any of the other energy-requiring activities necessary for the maintenance of life and health.

If glucose is burned completely to carbon dioxide and water in the laboratory, it releases 2818 kilojoules of energy per mole ($2818 \, kJ \, mole^{-1}$) as heat. This tells us that every mole of glucose (180 g) contains 2818 kJ of potential energy.

ATP carries about $32 \, kJ \, mole^{-1}$ of useful energy. The complete oxidation of one mole of glucose in respiration generates 38 moles of ATP, representing about 1214 kJ of useful energy. Thus 1214/2818 or about 43% of the total energy content of glucose is used to make ATP; the remaining 57% is dissipated as heat.

The first stage in respiration is known as *glycolysis*. In this process a molecule of glucose is split into two molecules of the three-carbon intermediate product pyruvic acid. It is an oxidation process involving the removal of four electrons. These are accepted not by NADP but by nicotinamide adenine dinucleotide (*NAD*) which is reduced to NADH.

$$\underset{\text{glucose}}{C_6H_{12}O_6} + 2NAD \rightarrow \underset{\text{pyruvic acid}}{2C_3H_4O_3} + 2NADH + 2H^+$$

Glycolysis is a sequence of ten enzyme reactions. Some of these are dependent on a supply of ATP and some generate sufficient energy to make ATP. In glycolysis there is a net gain of two molecules of ATP per molecule of glucose.

Pyruvic acid is oxidized completely to carbon dioxide in a pathway of nine enzyme reactions known as the *Krebs cycle*. For each molecule of pyruvic acid oxidized, four molecules of NAD and one of another carrier, *FAD* (flavine adenine dinucleotide) are reduced.

$$C_3H_4O_3 + 3H_2O + 4NAD + FAD \rightarrow$$
$$3CO_2 + 4NADH + 4H^+ + FADH_2$$

In addition, a molecule of ATP is made in the Krebs cycle.

Thus the complete oxidation of one molecule of glucose via glycolysis and the Krebs cycle is accompanied by the reduction of ten molecules of NAD and two of FAD. NADH and $FADH_2$ are re-oxidized to NAD and FAD in an electron transport chain similar in many respects to that involved in photosynthesis.

The first step in the chain is a transfer of electrons to FAD from NADH, reducing it to $FADH_2$. The electrons lose sufficient energy in this step to make a molecule of ATP by the process known as *oxidative phosphorylation*. $FADH_2$ from the Krebs cycle is fed into the electron transport chain at this point. Two more molecules of ATP are made by oxidative phosphorylation for each molecule of $FADH_2$ oxidized. Thus the reducing power of NADH is sufficient to make a total of three ATP molecules, while that of $FADH_2$ can make two ATP molecules.

In the final step of the chain, electrons are passed to oxygen, reducing it to water:

$$2e^- + \tfrac{1}{2}O_2 + 2H^+ \rightarrow H_2O$$

The necessary H^+ ions are released earlier in the chain during the oxidation of $FADH_2$.

The whole process of respiration is summarized in Fig. 1.18. In total, for each molecule of glucose oxidized, 34 molecules of ATP are made by oxidative phosphorylation, 2 more in glycolysis and 2 in the Krebs cycle.

Respiration, unlike photosynthesis, goes on in every living cell of the plant. Glycolysis takes place in the cytoplasm, but the Krebs cycle and oxidative phosphorylation are confined to *mitochondria*, small membrane-enclosed bodies immersed in the cytoplasm. Actively growing cells or cells performing work such as the transport of food, are especially well endowed with mitochondria.

1.8.2   Fermentation

The respiratory process outlined above is dependent on the presence of oxygen. Without oxygen as the final essential step in the electron transport chain, no oxidative phosphorylation can take place and no NADH or $FADH_2$ can be oxidized. The Krebs cycle quickly runs out of NAD and FAD as these are no longer being regenerated, and without them the cycle cannot operate.

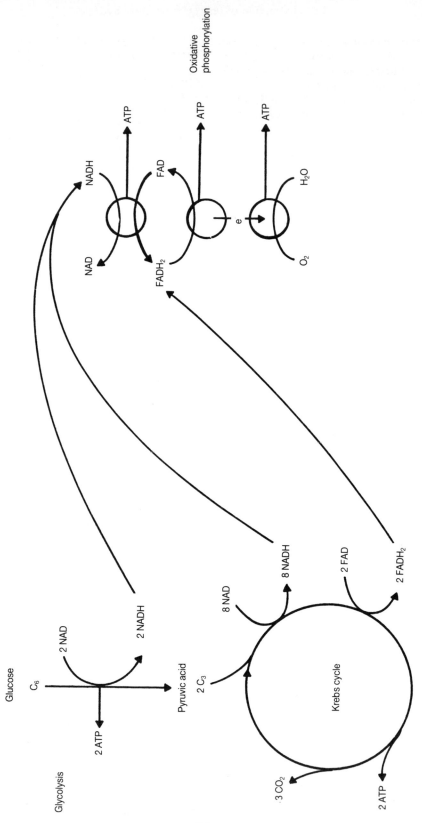

Fig. 1.18 The principal metabolic processes of respiration. Large amounts of ATP are generated, mainly by oxidation of NADH and FADH$_2$ in the process of oxidative phosphorylation.

24

Respiration as described above is properly known as *aerobic respiration*.

In the absence of oxygen, a more limited form of respiration occurs. It is known as anaerobic respiration or *fermentation*. It yields no NADH or $FADH_2$ and does not involve oxidative phosphorylation.

In fermentation, glycolysis proceeds exactly as in aerobic respiration, with the partial oxidation of glucose to pyruvic acid. However, glycolysis results in the accumulation of NADH, the reducing power of which cannot be absorbed in the electron transport chain. The NAD needed for glycolysis is quickly depleted.

NAD is regenerated from NADH in the reduction of pyruvic acid to ethanol:

$$C_3H_4O_3 + NADH + H^+ \rightarrow C_2H_5OH + CO_2 + NAD$$
pyruvic acid   ethanol

The overall equation for fermentation, which is summarized in Fig. 1.19, is therefore:

$$C_6H_{12}O_6 \rightarrow 2C_2H_5OH + 2CO_2$$
glucose   ethanol

The only ATP generated in fermentation is that produced in glycolysis: two molecules per molecule of glucose broken down. This is only one-nineteenth of the ATP generated by aerobic respiration.

The coats surrounding some seeds are rather impervious to oxygen, so that the respiratory metabolism in the earliest stages of germination of such seeds is of the anaerobic or fermentative type. Soon, however, the seed coat is ruptured and oxygen becomes available for aerobic respiration and for the oxidative breakdown of any ethanol which has been produced as a result of fermentation.

Plant roots in waterlogged soil are often so deprived of oxygen that they have to generate their ATP by the fermentation process. Since not nearly enough ATP can be produced by this means for normal growth, such roots are typically stunted. If anaerobic conditions persist for any length of time, ethanol may accumulate to levels which are toxic to the cells producing it. This can kill the entire root system and thereby the whole plant.

Few agricultural crops perform well in poorly drained soil. Plants such as reeds and rushes, which can live in such conditions, have special adaptations which prevent the build-up of ethanol. Some, for example, are able to 'switch off' ethanol production when the concentration becomes excessive.

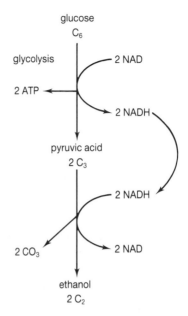

Fig. 1.19 Fermentation. In this form of respiration, which takes place in the absence of oxygen, only a very small quantity of ATP is made. The NADH generated in glycolysis is used up again in the metabolism of pyruvic acid to ethanol.

Others have air channels within the stems and roots to carry oxygen to the underground parts.

Fermentation, as performed by yeasts (a group of fungi) rather than by plants, is of enormous commercial importance, being the basis for all the alcohol produced for industrial and surgical use and for human consumption. Any plant material containing abundant carbohydrates can be fermented by yeasts to produce alcohol; of particular importance are grapes, used to make wine and brandy, and barley grain, used in the brewing of beer and the distilling of Scotch whisky. Barley is normally converted to *malt* before being used by brewers and distillers. This involves allowing the grain to sprout so that the starch reserves are mobilized as sugars.

1.9   The balance of photosynthesis and respiration

In terms of the starting materials and end-products, respiration may be considered to be the reverse of photosynthesis.

$$CO_2 + H_2O \underset{\text{respiration}}{\overset{\text{photosynthesis}}{\rightleftharpoons}} [CH_2O] + O_2$$

If photosynthesis (after accounting for losses in photorespiration) is proceeding faster than true respiration, as will usually be the case in plants in the light, carbohydrate is produced faster than it is broken down in respiration and the net surplus builds up. This is the basis of dry matter production in crops, and is, of course, intimately related to harvestable yield.

Gain in dry weight due to photosynthesis minus loss due to respiration is termed *net assimilation*. The rate of net assimilation (usually expressed as dry weight increase per unit leaf area) is a major physiological characteristic of a crop plant determining its yield, and is therefore of great agricultural significance. In the remainder of this chapter we will consider some of the important factors which influence net assimilation rate.

1.9.1 *Effect of light on net assimilation rate*

Since light provides all the energy needed for photosynthesis, it is not surprising that this should be one of the most important factors influencing net assimilation. Clearly the longer a plant is illuminated the more dry matter it will be able to accumulate, but at any one time the *intensity* of light falling on a plant affects the rate of net assimilation.

In absolute darkness, the rate of photosynthesis is zero. Respiration continues, however, so that net assimilation rate is negative. In light, photosynthesis can proceed, and as light intensity increases, the rate of photosynthesis also increases. At some value of light intensity net assimilation rate changes from negative to positive. This light intensity at which photosynthesis and respiration exactly balance one another, and at which net assimilation rate is therefore zero, is called the *light compensation point* (Fig. 1.20). The value of this depends on temperature, as will be seen later, and also varies from plant to plant, but in general it is relatively high for most agricultural crops. This means that crops are as a rule not tolerant of shading. Shade-tolerant plants, such as those found commonly in woodlands, have low light compensation points.

As light intensity continues to rise, the rate of photosynthesis rises less slowly, until a plateau is reached. Any further increase in light intensity has no effect on the rate of photosynthesis. For most crops in temperate climates the plateau is reached at about a quarter to a half of full summer sunlight. The light intensity at which the plateau is reached

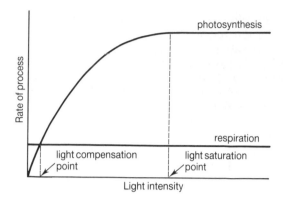

Fig. 1.20 The effect of light intensity on photosynthesis and net assimilation rate.

is called the *light saturation point*, although, as Fig. 1.20 makes clear, this is a less precise parameter than the light compensation point, as light saturation is attained gradually over a range of light intensities.

Even when other environmental factors are ideal for photosynthesis, the phenomenon of light saturation is observed. It means that much of the light energy absorbed by leaves at high light intensity is being wasted, since just as much photosynthesis would go on at lower light intensities. Light saturation comes about for two reasons.

First, in a chlorophyll light trap only a few molecules are of the right configuration to release high-energy electrons. For every special chlorophyll molecule that ejects an electron there are 200–300 whose function is to become 'excited' by light and pass their excitation energy to the special electron emitter. There is thus a bottleneck at the electron emission point in the light trap. This does not matter at low light intensities, when many chlorophyll molecules are needed to gather the energy. At high light intensities, however, there are not enough electron emitters to handle all the available energy.

Leaves which develop in conditions of high illumination adapt to these conditions to some extent by having fewer non-specialized chlorophyll molecules for each specialized electron emitter. They tend therefore to have higher light saturation points. They are, however, less efficient users of low light levels and for this reason also have higher light compensation points than leaves which develop in shaded conditions. There may also be other distinctions between 'sun leaves' and 'shade leaves' giving rise to differences in photosynthetic capacity. Crop leaves which develop in the shade of weeds often do not photosynthesize efficiently even after removal of the weeds, because of low light saturation points.

The second cause of light saturation is photorespiration. Rapid photosynthesis at high light intensities depletes carbon dioxide in the internal air space of the leaf and at the same time produces an abundance of oxygen. As we have seen, these are just the conditions favouring a high rate of photorespiration. Any increase in carbon dioxide assimilation with increasing light intensity is cancelled out by increased carbon dioxide release in photorespiration.

In C_4 plants, the carbon dioxide liberated in photorespiration is reassimilated before it can escape and these plants therefore show responses of photosynthesis to very high light intensities. Their light saturation points are characteristically much higher than those of C_3 plants (Fig. 1.21). This makes C_4 crops such as maize more effective users of high light intensities, as occur in sunny tropical and subtropical climates, than C_3 crops such as wheat. At lower light intensities, however, the greater energy efficiency of C_3 plants gives them the advantage.

At levels of illumination below the light saturation point, light intensity is said to be *limiting* the rate of photosynthesis; that is to say, photosynthesis is going as fast as the supply of light energy will allow.

Fig. 1.21 The effect of light intensity on the rate of photosynthesis in C_3 and C_4 plants. C_4 plants have higher light saturation points than C_3 plants.

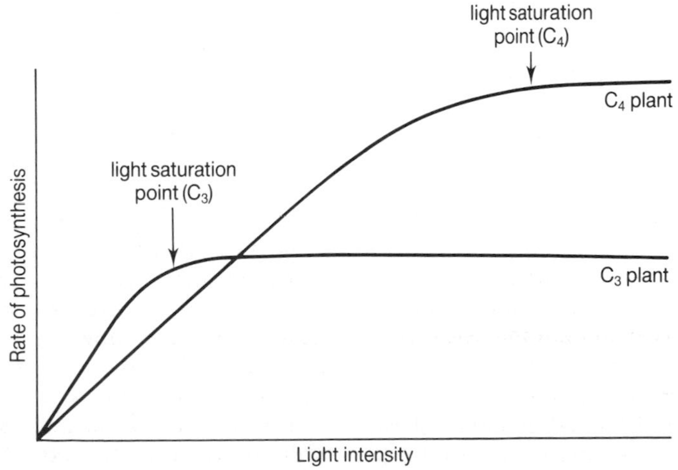

1.9.2   *Effect of carbon dioxide on net assimilation rate*

Just as the rate of photosynthesis can be limited by an inadequate supply of energy in the form of light, so it can be limited by an inadequate supply of the raw material carbon dioxide.

Only a small proportion of the air is carbon dioxide – about 0.03% by volume. However, it is not the carbon dioxide concentration of the atmosphere but that of the internal air space of the leaf that affects the rate of photosynthesis. In still conditions, inward diffusion of carbon dioxide through the stomata may be too slow to prevent the concentration inside the leaf from falling to a level at which photosynthesis is restricted. If the stomata are closed, carbon dioxide in the internal air space is quickly depleted and photosynthesis stops. Plants with the C_4 pathway can photosynthesize at much lower internal carbon dioxide levels than can C_3 plants because of their ability to recirculate photorespiratory carbon dioxide.

Higher concentrations of carbon dioxide in the atmosphere increase the concentration in the internal air space and thereby increase the rate of photosynthesis, provided light or some other factor is not limiting. Higher yields of glasshouse crops are achieved by carbon dioxide enrichment of the atmosphere, demonstrating that in the glasshouse carbon dioxide commonly limits net assimilation rate. Unfortunately it is not practicable to enrich the air above field crops in carbon dioxide. It may, however, be possible to increase crop yields by breeding improved varieties in which the resistance of the stomata to the entry of carbon dioxide has been reduced.

1.9.3 Effect of water supply on net assimilation rate

Water, like carbon dioxide, is a raw material in photosynthesis and one might therefore expect that a shortage of water would reduce net assimilation rate. In practice, however, such a small proportion of the plant's intake of water is used in photosynthesis that the straightforward shortage of water as a raw material never occurs. A leaf will die by wilting long before its water content falls to levels that might directly limit photosynthesis.

It will be shown in Section 2.4.1 that an inadequate water supply, or weather conditions promoting rapid evaporation, induce the stomata to close to conserve water. When this happens the carbon dioxide concentration in the internal air space of the leaf quickly falls to a level that inhibits further photosynthesis. The effect of a water shortage on net assimilation rate is thus an indirect one – it is really a carbon dioxide shortage that slows the process down.

Since photosynthesis in C_4 plants tolerates much lower internal carbon dioxide levels than that in C_3 plants, it also tolerates a greater degree of and more prolonged periods of stomatal closure. It is believed that C_4 metabolism is primarily an adaptation to warm, dry climates rather than to high light intensities as such.

Water stress may also affect net assimilation rate by partially dehydrating the protoplasm of the mesophyll cells. Under these conditions enzymes cannot function with their customary efficiency and all metabolic processes, including photosynthesis and respiration, are slowed down.

1.9.4 Effect of temperature on net assimilation rate

Almost all chemical reactions proceed more rapidly at higher temperatures. Between $0\,°C$ and about $40\,°C$, this is also true of enzyme reactions. Thus an enzyme reaction takes place more rapidly at $30\,°C$ that at $5\,°C$. Above about $40\,°C$, the three-dimensional structure of most enzymes begins to be destroyed, reducing their activity and causing the reactions they control to be slowed down.

Respiration and photosynthesis, as we have seen, both involve long series of enzyme reactions. As the temperature rises from $0\,°C$ to $40\,°C$, the rate of respiration increases progressively, as shown in Fig. 1.22. However, the rate of photosynthesis increases with increase in temperature only up to a certain point, for beyond this point photosynthesis is limited by inadequate levels of light or carbon dioxide supply. This point is typically reached at around 10–$15\,°C$ in most temperate agricultural crops. Above this temperature, respiration increases but photosynthesis remains at the same level. Net assimilation rate is therefore slowed down, and the light compensation point is raised.

In practice, higher temperatures are usually associated with higher light intensities. Any inhibitory effect of increased temperature on net assimilation rate therefore tends to be masked by the stimulatory effect of increased light. The evaporation of water from leaves has a cooling effect, maintaining the internal temperature a few degrees below the external air temperature. For these reasons, the reduction in net assimilation rate due to high temperature is seldom significant, except in dry conditions when stomatal closure inhibits photosynthesis.

1.9.5 Concept of limiting factors

We saw that increasing the light intensity beyond a certain point has little or no further effect on the rate of photosynthesis. No matter how much light is available, photosynthesis is held in check by other factors. It is only at low light intensity that photosynthesis becomes light-intensity dependent; thus at low light intensity, this is the *limiting factor* to photosynthesis. At high light intensity, something else becomes the limiting factor – most likely the carbon dioxide concentration in the internal air space of the leaf.

Figure 1.23 illustrates the fact that when there is an abundant supply of carbon dioxide, light is a limiting factor over a much greater range of intensities than when carbon dioxide is in short supply, as might result from closure of the stomata in dry weather. If the mesophyll cells cannot get enough carbon dioxide, no amount of light will increase the rate of photosynthesis.

In the winter, temperature is the main factor limiting photosynthesis, but during the growing season light and carbon dioxide are more important. It was formerly believed that only one factor could be limiting at a time, but photosynthesis is a very complex process and it is quite possible to have conditions such that increasing any one of the three main factors will increase the rate of photosynthesis.

The concept of limiting factors is applicable not only to photosynthesis but to any process the rate

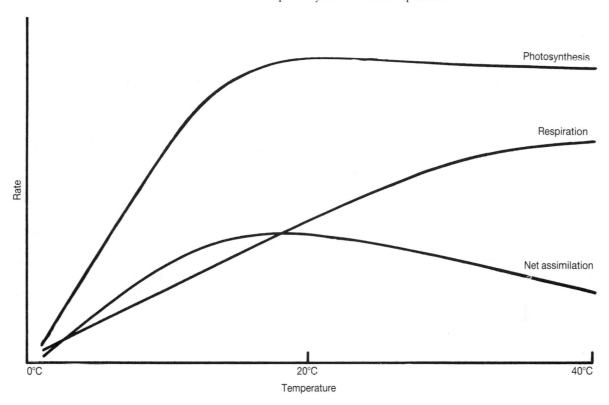

Fig. 1.22 (Above) The effect of temperature on net assimilation rate in C_3 plants. The optimum temperature for net assimilation rate is normally around 10–15°C in agricultural crops of cool temperate regions.

Fig. 1.23 (Below) The response of photosynthesis to light intensity at two levels of CO_2 supply. When CO_2 is abundant, light is the limiting factor over a greater range of light intensities than when CO_2 is in short supply.

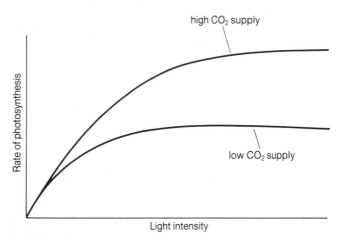

of which is governed by a number of independent factors. In later chapters we shall see how plant growth, an even more complex process than photosynthesis, may be limited by various factors, and the concept can be extended to the most important

of all the agricultural characteristics of a crop – its yield. Both growth and yield are, of course, intimately related to photosynthesis, and the factors such as light, carbon dioxide, water and temperature which we have mentioned in this chapter, will therefore appear again. Photosynthesis itself is a process so fundamental to the life of plants and to the livelihood of farmers that it will be referred to repeatedly throughout this book.

Summary

1. Plants differ from other forms of life in their capacity for photosynthesis, the manufacture of carbohydrates from carbon dioxide and water using the energy of sunlight. They are also characterized by the possession of cellulose cell walls and an indeterminate growth habit.

2. Five levels of evolutionary advancement can be recognized in the plant kingdom: algae, bryophytes, pteridophytes, gymnosperms and angiosperms. All agricultural, horticultural and forestry crops, with the exception of coniferous trees and a few ornamentals, are angiosperms, as are all but a handful of weeds.

3. The angiosperm plant, in the vegetative phase of growth, is constructed of only three types of organ: leaves (of determinate growth), stems and roots (of indeterminate growth). Later, flowers are produced, each an assemblage of organs concerned with sexual reproduction. Following the transfer of pollen, the ovules of the flower develop into seeds enclosed in a fruit. Each seed contains the embryo of a new plant.

Various functional systems can be identified in the plant, the organization of all of them dictated by the genetic blueprint.

4. Photosynthesis is a two-stage process. The excitation of chlorophyll molecules by light leads to the emission of high-energy electrons which are used to make ATP by photophosphorylation and to reduce NADP to NADPH. NADPH and ATP provide the reducing power and supplementary energy for the operation of the Calvin cycle, a metabolic pathway in which carbon dioxide is assimilated into carbohydrate. Chlorophyll replenishes its stock of electrons by withdrawing electrons from hydroxyl ions in water; this results in formation of oxygen which is released as a waste product.

Photosynthesis takes place in specialized parenchyma tissue which has large intercellular air spaces to assist exchange of gases (carbon dioxide in, oxygen out) and which is well provided with chloroplasts, small green bodies in the cytoplasm of the cells. The chloroplasts contain the chlorophyll, the mechanism for ATP and NADPH formation and the enzymes of the Calvin cycle.

The leaf is the principal organ of photosynthesis. The photosynthetic tissue (mesophyll) is sandwiched between an upper and lower epidermis and is permeated by an extensive internal air space which is continuous with the external atmosphere through stomata, pores in the epidermis. Guard cells surrounding the stomata can close the pores to reduce evaporative water loss; this cuts off carbon dioxide supply to the mesophyll and photosynthesis ceases. Veins embedded in the mesophyll contain fibres for structural support, phloem for export of sucrose and other manufactured materials and xylem to bring water to the leaf to replace that lost by evaporation through the stomata.

Carbohydrate, mainly in the form of sucrose, is translocated in the sieve tubes of the phloem from sources, primarily leaves, to sinks, which are actively growing regions of the plant or storage organs such as fruits or tubers. Storage tissues are composed of thin-walled, living parenchyma cells and, except in seeds where they lose most of their water, are easily damaged during handling of agricultural produce. Energy is stored as starch, as sucrose or other carbohydrates, as oil, or occasionally as protein.

5. In respiration, the energy of carbohydrate is used to reduce NAD and some FAD to NADH and $FADH_2$ respectively. These are re-oxidized in the mitochondria (colourless bodies in the cytoplasm of all living cells), making large quantities of ATP. This requires oxygen; in the absence of oxygen a more limited form of respiration known as fermentation occurs in which the end product is ethanol and only a small quantity of ATP is generated.

6. Increase in organic matter due to photosynthesis minus decrease due to respiration gives net assimilation. Net assimilation rate is therefore negative in darkness. The light intensity at which photosynthesis exactly balances respiration is the light compensation point. Photosynthesis shows light saturation at high light intensities, when further increase in light energy input gives no further increase in photosynthesis, even if all other environmental factors are optimal. Leaves which develop under high illumination have higher light saturation points and higher light compensation points than leaves which develop in shade. They therefore utilize high light intensities more efficiently, but low light intensities less efficiently, than shade leaves. C_4 plants such as maize have much higher light saturation points than temperate crops, which are C_3 plants, largely because they show no detectable photorespiration, a process which in C_3 plants prevents the complete assimilation of all the carbon dioxide available in the internal air space of the leaf.

Inadequate carbon dioxide concentration in the internal atmosphere of the leaf can, like low light intensity, limit net assimilation rate. This situation may arise in a rapidly photosynthesizing crop in still weather or in a glasshouse, where carbon dioxide enrichment of the atmosphere increases net assimilation rate. Water stress limits net assimilation by inducing closure of the stomata and preventing gas exchange. If light or carbon dioxide supply are limiting photosynthesis, an increase in temperature will have little effect on photosynthesis but will increase respiration, thereby lowering net assimilation rate. This effect is probably seldom significant in the field.

2

Plants and Water

2.1 Importance of water to the plant

Over 80% of the fresh weight of many plants, and over 90% of some plant parts, is water. Even woody stems and roots contain over 40% water. Of plant parts, only seeds contain as little as 20% or less water. More remarkable still is the appetite of plants for water. The roots of a growing maize plant in hot summer conditions, for example, take up 2–4 litres per day from the soil, about twice the volume of water in the plant at any one time. What happens to this moisture?

Plants give off water vapour continuously from all their above-ground parts but particularly their leaves, and through this process, known as *transpiration*, plants typically lose over 98% of all the water their roots absorb. Only 2% or less is used in building up the plant and maintaining its metabolism. A crop of maize may transpire 80% of the rain falling on it during the growing season, and a deciduous forest around 30% of the entire annual rainfall. Crucial to the understanding of plants, therefore, is an understanding of the functions of water in plants and how and why this profligate flow is maintained.

Land plants, like all other forms of terrestrial life, evolved from marine organisms, and many still show partial dependence on an aquatic environment. For example, in the more primitive land plants such as mosses and ferns, sexual reproduction remains dependent on a film of water through which the male sex cells must swim to reach the female parts. By contrast, in the more highly evolved groups, the conifers and flowering plants, pollen grains carrying the male sex cells journey by air to the receptive female parts, blown by the wind or carried, usually by insects. Water films play no part; these plants have evolved beyond dependence on an aquatic environment.

This evolutionary trend is paralleled by other important developments in the structure and function of land plants, leading to greater independence of water. The most significant of these is probably the development of a water conducting system, namely the *xylem*. Movement of water through unspecialized plant tissues is rather slow, and thus if a plant is to gain any height and size some means of rapidly transporting water from the source in the soil to the points of dissipation in the leaves and stems is required. Xylem is essentially a plumbing system, a set of pipes ramifying throughout the plant, formed by dead tubular cells joined end to end, and providing little resistance to the upward flow of water.

Although with advancing evolution plants have largely escaped from their aquatic origins, internally they have retained a watery environment in and around their cells, which are full of and bathed in water. In that sense, like all living things, they still live in water and directly or indirectly water affects every plant process.

2.1.1 Water and cell structure

The protoplasm within living cells is about 95% water. Its every constituent is suspended in this liquid which not only forms the medium wherein all its innumerable chemical reactions take place, but also hydrates the protein, carbohydrates, nucleic acids and other organic molecules in the cell and modifies their chemical properties. A property of water which makes it particularly suited to housing a vast range of reactions is its ability to act as a solvent for a very wide variety of substances. In addition, water participates as a reagent in many chemical processes in the cell, notably photosynthesis.

Much of a plant's water is contained in the *vacuoles* of its cells. These are sacs separated from the surrounding granular cytoplasm by a membrane, and they often occupy over 90% of the volume of a cell.

The cell holds its water under pressure. This pressure is commonly as much as ten times atmosphere pressure – 10 bars, or around 150 pounds per square inch – pressing against and distending the flexible cellulose walls and giving the cell a firm though not rigid shape, rather like an inflated car tyre. The distension of cells with water is known as *turgidity*. It is this turgidity of packed masses of cells bonded together in the soft green tissues of plants that gives green tissues their strength and thus a leaf its shape and lettuce its crispness. Without turgidity these organs wilt and lose their shape. The plant therefore uses water, a liquid, and the most readily available substance, as a major building material.

2.1.2   Water and translocation

As we saw in Section 1.6.8, the products of photosynthesis are translocated in the phloem from their site of manufacture in the leaves to their sites of storage or utilization. They are carried in solution, mainly in the form of sucrose. Water is the vehicle for translocation not only of sucrose but also of a wide range of organic and inorganic substances, all moving in solution in the phloem.

Similarly, the uptake of mineral nutrients can take place only when these are dissolved in water. As we shall see in Section 3.4.1, roots select and accumulate ions to far higher concentrations than occur in the soil solution, but water uptake is necessary for this to happen.

The xylem is located in the centre of the young, actively absorbing root. Water enters the root and crosses the intervening parenchymatous tissues to the xylem by a route which we shall consider later (Section 2.3.1). In the xylem it is transported upwards through the root and stem to the leaves, perhaps as much as 100 m in tall trees, where it is distributed via the xylem of the fine leaf veins to the leaf tissues. From the veins it migrates in cells and cell walls to the surfaces of the cell walls of the mesophyll and epidermis. Here it evaporates and is transpired out through the stomata and, to a lesser extent, the cuticle which covers the outer surface of the epidermis (Fig. 2.1).

Fig. 2.1 The pathways of water in the leaf. Water is drawn out of the xylem into the living cells of the bundle sheath (a) whence it flows from cell to cell in the mesophyll (b). Thereafter it crosses the mesophyll cell walls and evaporates from the cell wall surfaces (c), crossing the substomatal cavity to exit (d) into the boundary layer of air closely surrounding the leaf. Some water vapour passes out via the cuticle (e). Turbulent eddying (f) in the air flow around the leaf tends to remove the boundary layer and accelerate water loss.

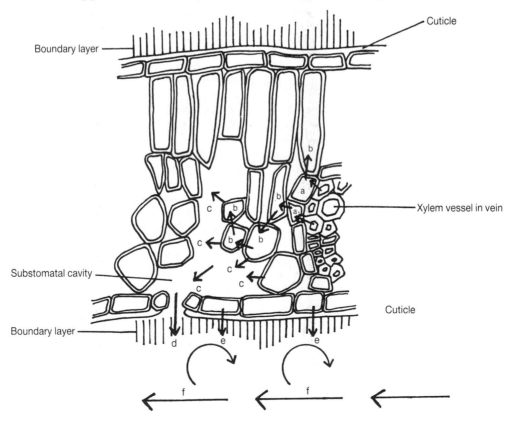

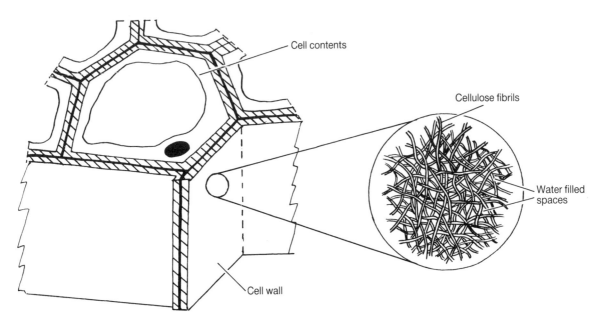

Fig. 2.2 The structure of cell walls showing the cellulose fibrils and the water-filled spaces between them.

Cell walls are composed not of a solid mass of cellulose but of slender fibres known as *fibrils*. Between the fibrils are spaces which are permeated with water, forming a continuous intercellular sea that bathes the outer membranes of the living cells (Fig. 2.2) and provides a medium for the movement of dissolved substances such as oxygen, carbon dioxide and mineral ions from cell to cell. With the exception of one impermeable barrier in the roots (see Section 2.3.1), this moving continuous mass of water extends in an unbroken network from the surfaces of mesophyll and epidermal cell walls in the leaf back into the xylem and thence into the finest roots, the whole flow being the method by which absorbed soil minerals and possibly some soluble organic products of the root are distributed to the leaves.

2.2 Forces causing water to move in plants

Obviously, powerful driving forces are required to maintain this flow and, since the plant is not an open conduit, to overcome what must be considerable resistance to the passage of water at each stage. What is the nature of these driving forces and how do they operate? To understand them, let us first see what causes water to enter or leave plant cells. We can then consider the movement of water in the plant as a whole.

2.2.1 Water movement into and out of cells

In water, as in any fluid, the molecules are in a state of constant random motion. They travel in all directions in straight lines until they are deflected by collision with other molecules or are attracted or repelled by electric charges on these molecules. This spontaneous movement and mixing of molecules, or *diffusion*, is a form of energy, and energy can do work.

To illustrate what is meant by this, consider a body of water in a tubular vessel which is open at both ends (Fig. 2.3). The water is separated into two regions, A and B, by a membrane which allows free passage of water molecules through it in either direction. If for some reason, the water molecules in region A have more energy of movement than those in region B, then it will be more common for a high-energy molecule to cross the membrane from A into B than for a low-energy molecule to cross from B into A. The result will be a net movement of water from A to B. This movement by diffusion will continue until the energy is equal on both sides. Work will have been done in moving the body of water.

Note that such movement can take place only if there is a difference in energy levels, that is an *energy gradient*, between the two regions. The

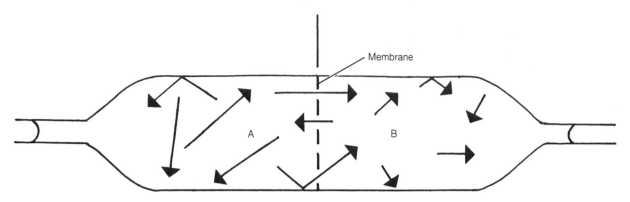

Fig. 2.3 Net diffusion of water molecules across a membrane. When the energy of movement of water molecules is greater on side A than on side B, there is a net movement of water from A to B by diffusion across the membrane.

movement is always *down* the gradient. What might produce an energy gradient between A and B?

It might be gravity. If we turn our vessel so that A is above B, clearly we can expect movement of water from A to B in response to the difference in potential energy. Gravity is not, however, a significant cause of water movement in plants, though it is important in the distribution of water in soils.

It might be pressure. If pressure is applied by a piston to region A (Fig. 2.4), the energy of movement of the water molecules will be increased and more will move from A to B than from B to A. Thus differential pressure on two sides of a membrane causes net movement of water across the membrane by diffusion. Clearly there are no pistons in

Fig. 2.4 Net diffusion of water molecules across a membrane, under the impetus of pressure. If pressure is applied to side A, there is a net diffusion of water from A to B across the membrane.

plants but, as we have seen, the water in cells is held under pressure. The pressure arises from the elasticity of the cell walls, exerting an inward pressure on the cell contents which will increase the more the wall is stretched.

Why, then, is the water not simply squeezed out of the cells? The answer lies in the presence of dissolved materials, or *solutes*, in the water inside the cells. These are another cause of energy differentials on opposite sides of a membrane which can lead to water movement.

To see how this occurs, let us return to our tubular vessel and suppose that region A contains a sucrose solution and region B pure water. Suppose also that the membrane, though not impeding the passage of water molecules, is impermeable to sucrose molecules (Fig. 2.5). Water molecules in A bind loosely to the sucrose molecules, thereby lowering the energy of movement of the water molecules. Meanwhile there is no reduction of energy of movement of the water molecules in B, and the energy difference promotes net diffusion of water from B into A (Fig. 2.6). Any dissolved solute causes similar effects. Movement of water across a membrane in response to the presence of solutes is known as *osmosis*.

The degree to which the energy of movement of

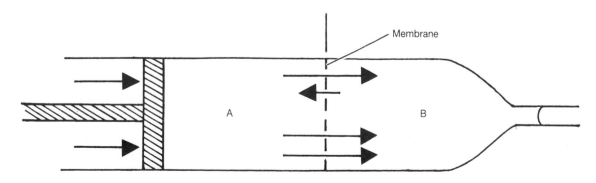

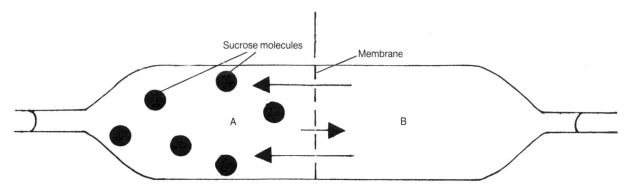

Fig. 2.5 A differentially permeable membrane which allows passage of water molecules but not of solute molecules.

Fig. 2.6 The effect of solutes on the net diffusion of water molecules – osmosis. The presence of a solute such as sucrose lowers the energy of movement of water molecules on side A of a differentially permeable membrane and causes net diffusion of water from B to A.

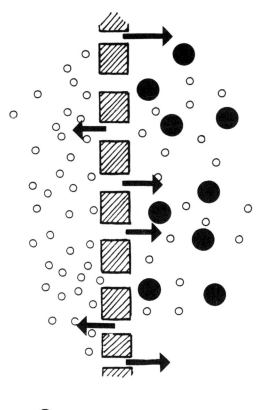

● = solute molecules

○ = water molecules

water molecules in a solution is lowered, and therefore the tendency for water to move into it by osmosis, depends on the concentration of the solute. Osmosis therefore occurs from a region of low solute concentration to one of high solute concentration.

Plant cells contain many solutes, including mineral ions such as potassium and magnesium as well as organic molecules like sucrose and amino acids. The cells are bounded by membranes which, like the one in the tubular vessel, are permeable to water but much less so to most solutes. Such membranes are said to be *differentially permeable*. Unlike the theoretical membrane in Fig. 2.5, the cell membrane does allow some leakage of solutes, but this does not seriously hinder water uptake into the cell by osmosis. The presence of solutes is thus another important driving force for the movement of water into and out of plant cells. Whereas the pressure of the cell wall tends to squeeze water out, the presence of solutes tends to draw water in.

The cell vacuole, which contains most of the cell's water, is separated from the water outside the cell and in the cell wall by a vacuolar membrane, a layer of cytoplasm and a cell membrane. These act together like a single differentially permeable membrane, through which water is given out or taken in depending on the balance of osmotic forces and wall pressure in the cell (Fig. 2.7).

The final factor which can cause an energy gradient leading to water movement from one region to another, is the binding of water molecules on to surfaces, particularly the surfaces of large molecules which are suspended in the water, forming a colloid. The forces of attraction between water molecules and colloidal particles are known as *matric forces* and can cause very large reductions in the energy of movement of the molecules. As we shall see in Section 2.6.3, much of the water in the soil is held there by matric forces and cannot

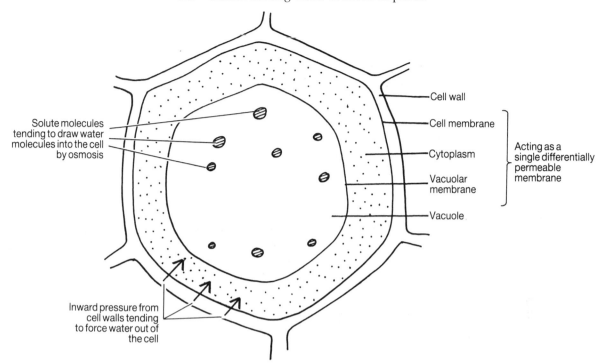

Fig. 2.7 The opposing effects of pressure, exerted by the cell wall, and of osmotic forces, produced by solutes in the cell, on water movement into and out of the cell. Inward pressure from the wall tends to force water out; osmotic forces tend to draw water in. The net movement of water in or out depends on the balance of these opposing influences.

readily be taken up by plant roots. Matric forces are also involved in the uptake of water by dry seeds, a process called *imbibition* (Section 5.1.3). They probably play a part in the movement of water into and out of cells in the growing plant, but it is impossible to distinguish their effect from that of osmotic forces and they are generally not considered separately.

2.2.2 Water potential of plant cells

The tendency of water to move in any system is best expressed in terms of *water potential*, which is a measure of the energy available for movement. The water potential of pure free water subject only to atmospheric pressure is called zero, and all other water potentials are determined with reference to this. Water potential, for which the symbol ψ (the Greek letter psi) is used, is expressed in units of pressure such as kilopascals (kPa) or bars. One bar is equal to 100 kPa and is a useful unit because

it is approximately equal to normal atmospheric pressure at sea level.

Pressure applied to water raises ψ by an amount exactly equal to the pressure applied. The extent to which ψ in a plant cell is raised by the inward pressure of the walls is the *pressure potential* (ψ_p). Thus, for example, if the walls exert a pressure of 5 bars on the cell contents, then the pressure potential of the water in that cell is 5 bars.

A solute present in water depresses ψ by an amount proportional to its concentration. A molar solution of any non-ionized solute such as sucrose depresses ψ by around 22 bars, and a 0.5 M solution by around 11 bars. The influence of solutes on ψ is the *solute potential* (ψ_s) and is always negative. Thus the solute potential of a 0.5 M sucrose solution is -11 bars.

The water potential of the contents of a cell or its vacuole is the sum of pressure potential and solute potential:

$$\psi = \psi_p + \psi_s$$

For example, if ψ_p is 5 bars and ψ_s is -11 bars, the water potential is $5 + (-11) = -6$ bars. If such a cell is immersed in pure water, whose $\psi = 0$, it will take up water, since water always moves from a region of high to one of low water potential.

A cell whose wall is not at all stretched is said to be *flaccid*. In such a cell there is zero pressure potential, that is $\psi_p = 0$, and water potential is therefore equal to solute potential, that is $\psi = \psi_s$. If the cell is immersed in pure water it will take up water by osmosis. How does this affect ψ?

The cell contents are slightly diluted by the incoming water, causing a slight increase in ψ_s towards zero (Fig. 2.8). A far more marked effect, however, is seen in ψ_p. As the cell wall stretches to accommodate the incoming water it begins to exert pressure on the cell contents, this pressure increasing rapidly the more the wall is stretched. Even a small change in cell volume thus has a significant influence on ψ_p. The effect of increasing volume is therefore to increase ψ towards zero, mainly by increasing ψ_p but to some extent also by increasing ψ_s. Once a certain volume is reached, the water potential inside the cell equals that of the external medium (in this case $\psi = 0$) and no more water is taken up. The cell is then said to be *fully turgid*. The effect of wall pressure, tending to push water out, exactly balances the effect of solutes tending to draw water in.

Consider now a line of cells all with the same water potential, as in Fig. 2.9. Suppose that water is drawn out of cell A, perhaps into the xylem. The loss of water causes an immediate drop in ψ, mainly through a fall in ψ_p. Cell A now has a lower ψ than cell B and takes up water from cell B. Cell B now experiences a drop in ψ which it transmits to cell C and thence to cell D in the same way. In this manner, water moves from cell to cell across plant tissues. It should be noted that cell-to-cell water movement in plants is driven primarily by differences in pressure potential resulting from the degree of distension of the cell walls, rather than by differences in solute potential.

2.3 Ascent of water in the xylem

The concept of water potential, which we have used in Section 2.2.2 to explain how water moves

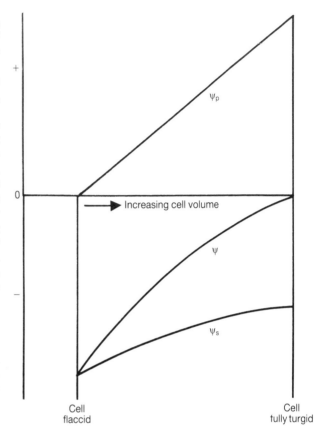

Fig. 2.8 The effects of water uptake on cell water potential. As water is taken up into the cell pressure potential (ψ_p) increases rapidly, causing an increase in water potential (ψ) of the cell towards zero.

Fig. 2.9 Movement of water across tissues in response to changes in pressure potentials (ψ_p) of cells. If water is drawn out of cell A (1), a cascade effect is initiated. ψ_p in cell A falls, causing water to move in from cell B (2). This causes ψ_p in cell B to fall, and hence a movement of water from cell C (3). In turn the resulting fall in ψ_p in cell C draws water from cell D (4), and so on.

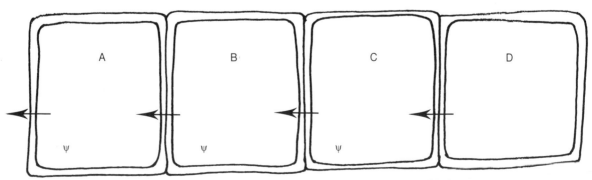

into, out of and between cells, can also help us to understand the forces causing water to rise up the xylem. Xylem cells have no living contents and no membranes. Water cannot therefore move within the xylem by osmosis. It moves instead by *mass flow*, just as in a domestic water pipe. What drives this upward flow?

2.3.1   Root pressure

Part of the answer is revealed when certain plants have the shoot cut off near the base and a glass tube attached to the stump. If the root system is well aerated and healthy, considerable quantities of sap exude from the xylem at the cut end and a column of sap builds up in the tube. The force pushing the water up must clearly be generated in the roots and is known as *root pressure*.

Root pressure is responsible for the drops of water exuded on to the surfaces of leaves of some plants, sometimes through special glands on the leaf. This phenomenon is called *guttation* and can be observed in grasses and cereals on humid mornings, when it is often mistaken for dew. The origin of root pressure lies in the structure of the absorbing root and the way in which water moves across the root from the soil to the xylem.

Although the distance between the root surface and the nearest xylem vessel is seldom greater than 1 mm, this is in fact an important barrier to water movement. In most crop plants, water is absorbed mainly in the region of root just behind the growing tip. Let us look at the nature of the barrier to water flow in this zone of the root (Fig. 2.10).

The young root is enclosed in an outer layer of cells, the *epidermis*, which in some plants has long, single-celled outgrowths called *root hairs* that extend into the surrounding soil. These do not play a major part in water uptake but may be important for the absorption of certain minerals, as we shall see in Section 3.4.1. Beneath the epidermis, most of the volume of the root is occupied by the *cortex*, made up of large parenchyma cells with intercellular spaces. The inner boundary of the cortex is marked by a cylinder, one cell thick, of smaller cells without intercellular spaces. This is the *endodermis*, which, as will become clear, is the key to understanding not only root pressure but also the selection and accumulation of minerals. The part of the root within the endodermis is known as the *stele*. It consists of a narrow parenchyma zone, the *pericycle*, surrounding the vascular tissues which provide the transport pathways between root and

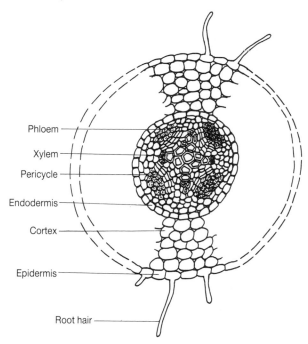

Fig. 2.10 The tissues of a young root as seen in cross-section.

Phloem
Xylem
Pericycle
Endodermis
Cortex
Epidermis
Root hair

shoot. The xylem is massed in a star shape when seen in cross-section, and bundles of phloem lie between the arms of the xylem star.

There are two roads by which moisture and minerals may cross the root. Firstly, all the cells of the root, apart from those of the xylem, contain living cytoplasm which is continuous from cell to cell throughout all the living tissues of the root. This is achieved by innumerable narrow connecting strands of cytoplasm, *plasmodesmata*, crossing the cell walls. The single, uninterrupted network of cytoplasm thus formed is the *symplast*. Like all cytoplasm it is bounded at its surface by a differentially permeable membrane. Within the living, moving mass of the symplast, water and solutes may travel by diffusion and mass flow.

Secondly, outside the symplast, the porous cell walls and the smaller intercellular spaces form another network, the *apoplast*, in which water and solutes may move. However, the cells of the endodermis (Fig. 2.11) have a zone of every radial cell wall impregnated with a waterproof substance called *suberin*. This zone, the *Casparian strip*, renders the endodermis impervious, or nearly so, to water in the apoplast. As the young root continues to develop, this waterproofing is supplemented by further deposits of suberin. Neither the

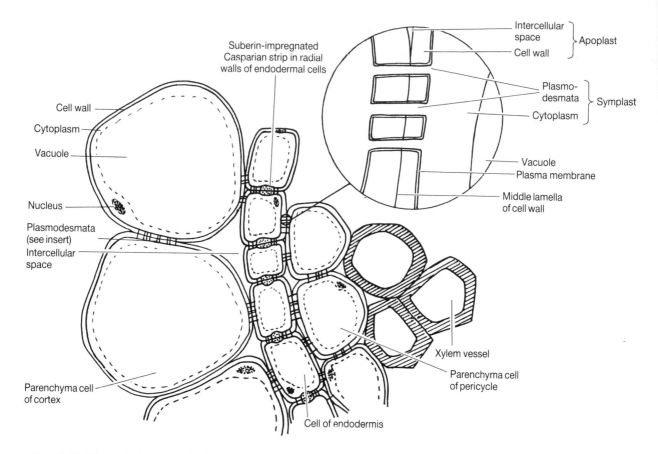

Fig. 2.11 The endodermis and adjacent cells in a young root.

Casparian strip nor the later suberin deposits sever the plasmodesmata linking the cytoplasm of the endodermis with that of the cortex and the pericycle.

Thus the symplast is a single cytoplasmic continuum stretching from the root surface to the pericycle adjacent to the xylem vessels, and bounded throughout by a differentially permeable membrane, but the apoplast is divided by the endodermis into an outer region continuous with the soil solution, and an inner region continuous with the xylem sap. Although water can travel by mass flow within either the outer or the inner apoplast, to cross from the outer to the inner it must enter the symplast somewhere in the cortex and leave it again somewhere in the pericycle. This it can do only by diffusion across the membrane.

Now the concentration of ions in the xylem sap and inner apoplast is much higher than that in the soil solution and outer apoplast. That such a differ-ence in concentration can be maintained is due to the action of the symplast 'pumping' ions in and restricting back-leakage. This difference in solute concentration causes solute potential within the endodermis to be lower than that outside it. Water molecules are drawn across the symplast into the xylem by osmosis in response to this difference.

The water potential of xylem sap under conditions favouring guttation is commonly around −2 to −3 bars, but recent measurements indicate that, exceptionally, values as low as −10 bars may occur. This is the explanation of root pressure. It is entirely dependent on the endodermis, which acts as a valve allowing the entry of water and solutes but preventing the solutes from diffusing back out of the stele and dissipating the pressure.

Osmotic forces, then, draw water into the xylem of the root and push the water up towards the stem and leaves. There are, however, good reasons for regarding this root pressure as being of minor importance in the general water economy of plants.

First of all, the rates of flow it accounts for are only a fraction of those observed in transpiring plants. The stump of a maize plant, for instance, has been observed to yield 500 ml of sap in three days under the influence of root pressure after being de-topped at the peak of its growth, but such a plant would normally transpire 5 litres or more in the same period. Secondly, intact plants can absorb water from more concentrated solutions and from drier soils than de-topped plants, showing that other, stronger forces must be aiding the osmotic forces in drawing water in. Thirdly, some plants such as coniferous trees exhibit no root pressure if cut down to a stump.

Lastly, and perhaps most significantly, if xylem is filled with water by root pressure, that water should be under positive pressure, so that sap should exude from a wound if the xylem is cut. This is certainly the case in many seedlings and in slowly transpiring plants in moist warm soil. In most freely transpiring plants, however, water is absorbed by both faces of freshly cut stems, not exuded by them. Clearly the water in the xylem is not under positive pressure, but under negative pressure, that is, *tension*. In short, it is being pulled from above, not pushed from below, and the source of that pull appears to be transpiration itself.

2.3.2   Role of transpiration in the ascent of water in the xylem

The pulling power of transpiration can be easily demonstrated by cutting across the base of a stem under water, to prevent air entering the xylem, and standing the shoot with its cut end in water. No wilting occurs, and transpiration is not checked, despite the complete absence of root pressure. Indeed in a classic experiment Strasburger cut through the base of a rapidly transpiring oak tree and placed the cut end in a dye solution. In a short time the dye could be observed in the topmost leaves. Even if the living tissues of the stem were killed by poison the transpiration flow continued unabated.

The driving force of transpiration is simply the tendency of water in films on the walls of the mesophyll cells to evaporate and escape into the atmosphere through the stomata. The water evaporates because the water potential of the air in the leaf spaces is lower than that of the water in the mesophyll. Just how much lower depends on the relative humidity of that air, that is the amount of water vapour it contains relative to the maximum amount it can carry at that temperature. During active transpiration, the relative humidity of the air in the leaf space is strongly influenced by the relative humidity of the atmosphere (see Section 2.4.2). If the atmospheric relative humidity is 50%, a common midday value in warm, dry weather, atmospheric water potential is approximately − 1000 bars, about a hundred times more powerful than the lowest water potentials resulting from the osmotic forces in the roots. Even at a relative humidity of 98%, atmospheric water potential is − 27 bars, more than adequate to raise water to the height of the tallest tree.

The picture envisaged, then (Fig. 2.12), is that transpiration from mesophyll to atmosphere exerts a pull which draws water across the mesophyll of the leaf and the pull is transmitted down unbroken columns of water in the xylem to the root, where it draws water into the apoplast of the stele from the apoplast of the cortex via the symplast. The pull is indeed transmitted to the soil, for water flows in the pores of the soil towards the roots of transpiring plants.

There is, however, a difficulty with this explanation of water movement in the plant. Although the forces generated by evaporation from the leaves are unquestionably great enough to haul water from the soil to the tops of the tallest trees, the question arises, why do the water columns in the xylem not snap under the enormous tensions

Fig. 2.12 The pathway of water movement from soil to atmosphere under the influence of the transpiration pull.

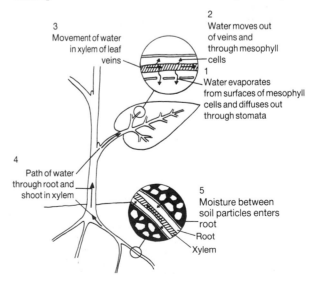

involved? It is impossible to draw water more than about 10 m up a pipe with a suction pump, since atmospheric pressure can only support a column of water of this height. If we try to draw it higher, the water column cavitates (that is, breaks) to leave a partial vacuum (like that at the top of a mercury barometer) containing water vapour. What prevents cavitation occurring in the xylem?

Part of the answer appears to lie in the molecular structure of water. Bonds form between oxygen and hydrogen atoms of adjacent molecules, giving rise to chains and networks of molecules which are surprisingly strong (Fig. 2.13). This strength is, however, evident only in narrow columns of water. Columns of water in tubes of similar diameter to xylem vessels can be shown in the laboratory to cohere so strongly that they can withstand tensions created by transpiration without breaking.

Fig. 2.13 The formation of chains of water molecules by hydrogen bonding. A water molecule carries a negative charge at the oxygen end and a positive charge at the hydrogen end, and there is attraction between the positive and negative ends of different molecules. This attraction causes weak bonds called hydrogen bonds to form between molecules. Numerous molecules thus linked form chains or networks, which account for the high tensile strength of water columns. Note that these hydrogen bonds are not permanent, and are constantly breaking and reforming in different patterns with the random motion of the water molecules.

There is evidence that breakage of water columns does occur in rapidly transpiring trees, at least in the broadest xylem vessels. Probably the narrower vessels carry sufficient water to supply the needs of the tree under these conditions. At night, when relative humidity of the atmosphere increases and transpiration slows down or stops, the broken water columns may be repaired by root pressure forcing water up the vessels.

2.4 Transpiration – the escape of water from the plant

Of the pathway of water through the plant mapped out in Fig. 2.12, only transpiration, the evaporation of water from the cell surfaces of the mesophyll and its migration through and out of the leaf, is left to be considered. In sequence, the steps are: the evaporation of water from the walls of the mesophyll cells; its movement across the spaces within the mesophyll; movement through the stomata; and escape from the outer surface of the leaf.

This section will consider how the rate of transpiration is affected firstly by the structure and functioning of the plant, and secondly by environmental factors, indicating how these two types of influence interact. Some possible benefits of transpiration to the plant will then be discussed.

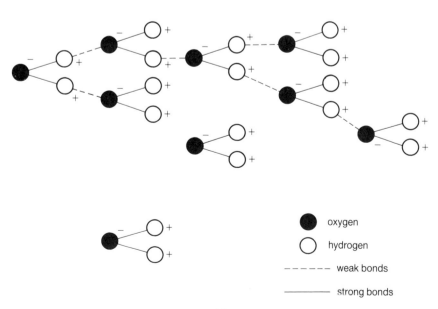

oxygen
hydrogen
- - - - - weak bonds
————— strong bonds

2.4.1   Control of transpiration by the plant

The rapid loss of moisture from the leaves by transpiration, were it left unchecked, would rapidly exhaust the water reserves of the soil around the roots and would then desiccate the plant. Transpiration therefore has to be controlled, but in this the plant faces a dilemma. To carry on photosynthesis, the mesophyll cells must absorb carbon dioxide and release oxygen. This exchange of gases requires the exposure of a large surface area to the air, and that means the simultaneous rapid loss of moisture.

To solve this problem the plant has evolved a number of compromise solutions. Woody tissues are covered in bark, the outer layers of which are impregnated with suberin (see Section 4.5.3), and through this bark little moisture escapes except through special unsuberized areas called *lenticels*. Transpiration through lenticels is, however, slight and probably not significant except during cold spells in winter when it accounts for water loss by deciduous trees.

All young green parts of plants, particularly leaves, are enveloped in a waxy waterproof layer called the *cuticle*, deposited on the outer surface of the epidermis. The amount of water lost through the cuticle depends on its thickness and other properties and varies greatly between species. Shade plants often have thin cuticles and may lose up to 30% of the total transpired water in this fashion, whereas many plants of dry habitats such as deserts lose practically no water through the cuticle. Most crop plants of temperate climates lose approximately 10% of their water by cuticular transpiration.

Most transpiration, however, occurs through the stomata. As was seen in Section 1.6.9, a stoma consists of a small pore in the epidermis surrounded by two special cells, the guard cells. Below each stoma is a space called the *substomatal cavity*, which water vapour must cross to escape through the stomatal pore. Movement across the spaces within the mesophyll is not regarded as a major impediment, but ease of movement through the substomatal cavity varies between species, depending on the design of the stoma. The most important point about stomata is that they are adjustable, and because they are the main route of water loss from the plant their opening and closing effectively controls the rate of transpiration.

The size of the stomatal pore, and therefore the rate of water loss through it, is adjusted by changes in the shape of the guard cells resulting from changes in their turgidity. In some plants other, less specialized, epidermal cells adjacent to the guard cells also play a part in adjusting guard cell turgidity and therefore pore size. In the common type of stoma found in dicotyledons (Fig. 2.14), the ventral walls, that is those adjacent to the pore, are thicker and less flexible than the dorsal walls on the opposite side. When water supply in the plant is adequate, the guard cells take in water and become more turgid, the thinner dorsal walls distending more than the thicker ventral walls. This causes the guard cells to curve, opening up the central pore and allowing free exchange of gases and escape of water vapour. At times of water shortage in the plant, loss of water from the guard cells causes the ventral walls to collapse on the pore, closing it and stopping gas exchange and stomatal transpiration. The stomata of grasses and cereals have dumb-bell-shaped guard cells (Fig. 2.14). The narrow 'bars' of the guard cells have thickened walls, but the inflated ends have thin walls. An increase in turgidity causes the ends to dilate, forcing the middle sections apart, opening the pore. Photographs of these two main types of stomata can be seen in Fig. 2.15.

The numbers and patterns of distribution of stomata on leaves vary greatly with the species. In most dicotyledons there are more on the lower than on the upper surface of the leaf. The shaded lower surface is generally cooler and less exposed to drying winds than the upper surface, hence the rate of water loss is less than would take place from stomata on the upper surface. Many monocotyledons, with more vertical leaves, have stomata equally distributed on both leaf surfaces. In certain grasses stomata are more numerous on the upper surface, where they can be protected by inward folding or rolling of the leaf in dry conditions. Plants native to arid or semi-arid areas have relatively few stomata.

Most plant species have 50–300 stomata mm^{-2} of leaf surface, and when wide open the pores are 3–12 μm across and 10–30 μm long. The pores are thus very small and in total, when fully open, account for around 1.5% of the leaf surface, although in some species this may reach 3%. Stomata, therefore, would not appear at first sight to be a very efficient compromise device. Closed they inhibit water loss, but at the expense of the gas exchange necessary for photosynthesis; open, they still greatly reduce the surface area available for gas exchange between atmosphere and internal air

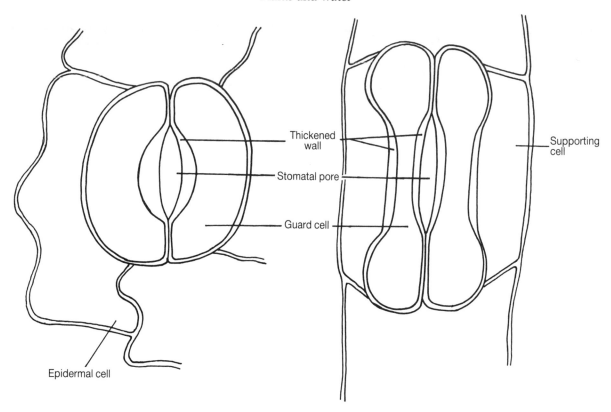

space, while permitting rapid loss of water.

Surprisingly, however, gas exchange through even partially closed stomata has been shown to be not markedly less than would take place if the whole leaf surface were permeable to oxygen and carbon dioxide. Only when stomata are almost or completely closed does photosynthesis slow down or stop for want of carbon dioxide. How effective are stomata in controlling the rate of water loss? It can be demostrated that the progressive closure of the stomata reduces water loss faster than it restricts carbon dioxide entry to the leaf. Thus, the stomata are more efficient than they would appear at first sight, offering a means of regulating water loss while not seriously hindering carbon dioxide uptake, except when they are almost or completely closed.

Obviously the control of the opening and closing of stomata is crucial in the water economy of the plant. Considering the functions of stomata, it should come as no surprise that they tend to close when the plant is under water stress, that is, when it is losing moisture faster than it can take it up from the soil. Water stress results in the development of large negative water potentials in living cells

Fig. 2.14 The structure of stomata in (a) dicotyledons and (b) monocotyledons. The stomata of dicotyledons typically have kidney-shaped guard cells with thickened walls adjacent to the pore. The stomata of monocotyledons, illustrated here by a grass stoma, typically have dumb-bell-shaped guard cells with thickened walls in the narrow portion of the cells.

which, as we have seen, are transmitted from cell to cell largely as a result of reduced wall pressure. Water stress thus causes a reduction in turgidity of the guard cells, leading to closure of the stomata.

Stomata also close when light intensity is low and photosynthesis is reduced, so that there is much less need for gas exchange. Again it is a fall in turgidity of the guard cells that closes the stomata, but how this happens in response to reduced light intensity is a complex procedure which is not yet fully understood. In addition, many plants show daily rhythms of opening and closing of stomata which continue for a while if the normal regular alternation of day and night is replaced by constant darkness.

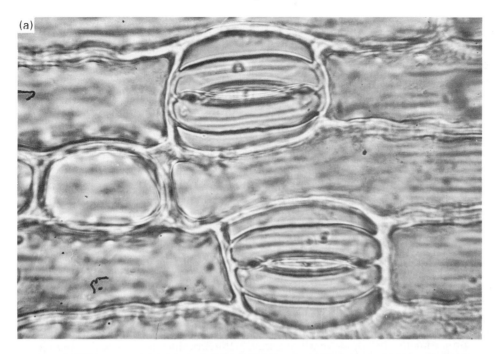

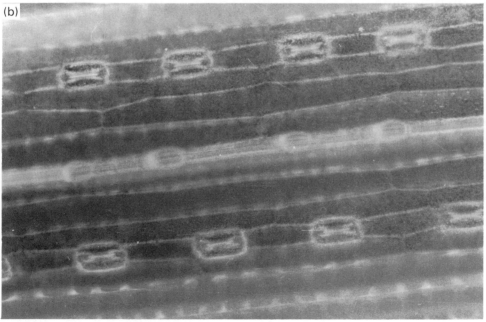

Fig. 2.15 Photographs of (a) typical dicotyledon stomata (× 150 approx.), and (b) monocotyledon stomata (× 300 approx.)

2.4.2 Effects of environmental factors on transpiration

The rate of transpiration depends on a balance among three processes. These are the supply of water to the plant, the supply of energy to evaporate water within the leaf, and the ease with which water vapour can escape from the leaves. The rate of water supply is dealt with in Section 2.5. Here, we deal with the supply of energy and the escape of water vapour. These are dominated by two environmental factors, namely solar radiation and the water potential of the atmosphere (water potential is explained in Section 2.2.2).

Now the initial stage of transpiration, that is, the evaporation of water from the cell surfaces within the leaf, requires energy. The only significant source of this energy is the sun. Solar radiation tends to raise leaf temperature, which in turn raises the energy of movement of water molecules in the mesophyll and therefore the rate of evaporation from the cell surfaces.

The water potential of the air depends on its capacity to accept more water vapour. This in turn depends on the relative humidity and temperature of the air. Air at 100% relative humidity is saturated with water vapour and can accept no more. Air at 60% relative humidity can accept an amount of water vapour equivalent to 40% of that contained in saturated air.

However, the absolute amount of water vapour that air can carry before it is saturated increases with its temperature. Hence, if warmer air and cooler air, both at 60% relative humidity, are compared, the 40% unsaturation in the warmer air represents 40% of a larger total amount of water vapour than in the cooler air. The warmer air can therefore accept more water vapour. The greater the capacity of the air to accept water vapour, the lower is its water potential. Thus the water potential of the air rises towards zero with increasing relative humidity but falls with rising temperature.

The gradient of water potential between the air within the leaf spaces and the atmosphere has a strong influence on the rate of transpiration. The steeper the gradient the higher the transpiration rate tends to be. Where leaf temperature and atmospheric temperature are equal, the gradient is decided solely by the difference between the relative humidities of the air within the leaf and of the atmosphere. Transpiration almost stops in rainy or foggy weather, as the air is almost saturated with water vapour, but increases markedly with even slight falls in relative humidity.

However, leaf temperatures are usually several degrees above or below atmospheric temperatures, and these small differences have a significant effect on the water potential gradient, and thus transpiration rate. The control of leaf temperature is discussed in Section 4.9.1.

It should be noted that the influences of solar radiation and the water potential of the atmosphere on transpiration are strongly modified by the structure and response of the plant. For example, the linkage between solar radiation and transpiration rate is greatly affected by the response of the stomata to light and temperature. With most plants, the size of the stomatal pore increases with temperature up to about 30°C. As mentioned earlier, stomata generally open in the light, at least in well-watered plants, and close in the dark, although plants show a wide range of diurnal patterns of stomatal activity. Thus solar radiation increases transpiration rate through its effects on the stomata as well as by increasing leaf temperature and hence evaporation from the mesophyll cells.

Relative humidity also has its effect on transpiration modified by the response of the stomata. The stomata of most species tend to close when relative humidity is low and open when it is high. This means that the effect of relative humidity on transpiration may be quite the reverse of that mentioned earlier. The two effects of relative humidity, one on the water potential of the atmosphere and one on the stomata, work in opposite directions and the resulting rate of transpiration reflects a balance between them. In most practical cases investigated, transpiration rate increases with falling relative humidity down to a certain point after which, as the stomata gradually close, either no further change occurs, or the rate even declines.

Another feature of environmental influences on transpiration is the interaction of physical factors. Such an interaction occurs between solar radiation and relative humidity. Solar radiation tends to warm the atmosphere. As can be understood from the previous explanation, if the atmosphere is warmed, its level of saturation with water vapour, and thus its relative humidity, falls, even though the concentration of water vapour it carries remains unchanged. It can thus accept more water vapour and its water potential falls, encouraging transpiration to increase.

A rise in solar radiation therefore tends to

increase transpiration in several ways: by stimulating stomata to open, by raising leaf temperature, and by lowering atmospheric relative humidity. Consequently transpiration rate tends to reach a daily peak around midday. In practice, this response is heavily modified by developing water stress within the plant, or the depletion of soil moisture leading to the same, with consequent closure of the stomata (see Section 2.5.1). The result is the commonly observed noon depression in transpiration. This is illustrated in Fig. 2.16. In some species, the stomata reopen as declining intensity of solar radiation permits restoration of the plant's water balance, but more commonly they stay closed for the rest of the day.

Plant species differ widely in how they manage their water balance throughout the day in response to intensifying solar radiation. Some, by means of a combination of stomatal closure, a sustained water supply from an extensive and efficient root system, and use of water reserves within plant organs, steadily maintain a favourable water content. Others, including many herbaceous species that grow in sunny situations, have evolved

a protoplasm that is able to tolerate rapid and extensive fluctuations in water potential, and simply accept water losses resulting from unabated transpiration. Every gradation between these two extremes can be found over the range of species.

The effect of wind on transpiration deserves some special mention. The relationship is not as direct as might be expected. Escape of water vapour from the immediate surface of the leaf is inhibited by a thin layer of stiller air that tends to accumulate close to all objects, simply due to friction between the surface and air molecules. These layers of stiller air, called *boundary layers*, form on leaf surfaces. Escape of transpiring water from such a boundary layer, which may be several millimetres thick, is solely by the slow process of diffusion. Due to the lack of mixing between these layers and the atmosphere, water transpiring from the leaf tends to accumulate in them, and they thereby develop a higher relative humidity than the surrounding air. Naturally this inhibits transpiration by reducing the relative humidity gradient.

However, due to the effects of quite small irregularities on the leaf surface and around its edges, such as prominent veins or serrated margins, air flows, even of quite low speeds, across the leaf become irregular and develop what is called turbulent eddying. One effect of turbulent eddying is to thin down greatly the boundary layers of the leaf and thus increase the rate of transpiration. It is thought that air around crops is seldom so still that turbulent eddying is not a factor in promoting transpiration.

Until recently, increases in windspeed even

Fig. 2.16 The influence of diurnal variation in solar radiation on the transpiration rate of a crop. (a) Solar radiation; (b,c,d) transpiration rates of different crops. (b) Crop growing in soil at field capacity. (c) Crop growing in conditions of water deficiency; a species showing midday closure of stomata, reopening in the afternoon. (d) Crop growing in similar conditions and showing midday closure of stomata but with no afternoon reopening.

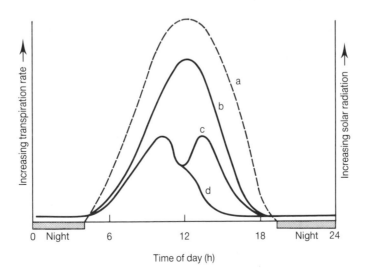

above those that removed boundary layers, were thought to increase transpiration rate, by removing water escaping from the stomata more swiftly and thus steepening the relative humidity gradient. However, wind also has a cooling effect on leaves (see Section 4.9.1) and this tends to reduce transpiration.

The balance between these two conflicting influences, and the resistance to the flow of water vapour through the stomata of the species concerned, decides the effect of wind on transpiration rate. At low levels of solar radiation leaves are often already close to ambient temperature, so that wind has little cooling effect and hence little effect on transpiration. In bright sunlight, leaf temperatures are often well above ambient temperature, and wind has a marked cooling effect, thereby lowering transpiration.

Leaf size has a major modifying effect on the influences of solar radiation, atmospheric water potential and wind. Plants with large flat leaves are much more prone to heating by the sun's rays (see Section 4.9.1), with consequently greater increases in transpiration, while on the whole, small-leaved species respond strongly to changes in relative humidity, but less to changes in wind and solar radiation than do large-leaved ones.

At higher wind speeds, leaf abrasion caused by plants rubbing against one another or by wind-borne particles leads to damage to the cuticle. This leads to rising transpiration rates through loss of control of evaporation by the cuticle and stomata and can cause major loss of water.

In general, then, considering environmental influences on transpiration, what are observed are effects on a physical component, that is the production and dispersion of water vapour from the leaf, effects on a biological component, that is the structure and functioning of the plant, and interactions between these two components. Hence, a simple relationship between an environmental factor such as solar radiation and the rate of transpiration is seldom observed.

It should be borne in mind also that plants exist usually as dense vegetation stands rather than single individuals, especially in agriculture and forestry, and the characteristics of these stands, particularly their height, alter the effects of environmental factors on the transpiration of plants within them.

2.4.3 Benefits of transpiration

Since such a high proportion of the water taken up

by plants is dissipated through transpiration, the question is often asked, what is the function of this voluminous flow from roots to leaves and thence to the atmosphere? When water, or any liquid, evaporates from a surface, the latent heat of evaporation is absorbed from the surface, cooling it down. Keeping the leaves cool is indeed an important function of transpiration in many broadleaved species. Certain features of leaf shape in such species, such as division into leaflets and serration of the margins, may perform the function of helping to cool leaves by promoting turbulent eddying and thereby increasing transpiration, as we saw in Section 2.4.2. However, the plant has other mechanisms for temperature control in addition to transpiration (see Section 4.9.1) and it is doubtful if all of the flow of water is necessary to fulfil this function.

The flow of water up the xylem delivers minerals and certain organic products of root metabolism to the leaves, but water moves in the transpiration stream much faster than is necessary to meet the demand for solutes. Root pressure would probably be sufficient in most plants to meet this demand.

It therefore appears that in many species, at least, much of transpiration is a necessary evil, the unavoidable result, despite adaptations like cuticles and controllable stomata, of the plant's need to expose a large leaf area to the atmosphere for gas exchange and the interception of light. At present little is known that would change that view.

2.5 Water stress in plants

2.5.1 Development of water stress in plants

An increase in transpiration lowers the turgidity of the leaf cells, causing a fall in their water potential which is made up by increased flow of water from the soil. Thus high solar radiation or low relative humidity, by increasing transpiration, may cause large increases in water absorption by roots. At high rates of transpiration, however, this increased uptake is often inadequate to meet the increased demand, and even plants growing in well-watered soil tend to lose water faster than they can take it up. This is thought to account for the water stress that commonly appears about midday in many crops, causing temporary wilting.

Under these conditions the stomata often close to reduce transpiration. Even with stomata fully closed, however, appreciable water loss continues through the cuticle. If the available supplies of moisture in the soil become depleted, the rate of loss through the cuticle will decline, but will persist to some extent even when moisture uptake from the soil has ceased. Thus, even where water supplies are extremely limited, most plants are unable to prevent a lethal leakage to the atmosphere of water from their tissues. Hence, plants are congenitally prone to develop water stress and, when it occurs, almost every plant process is affected directly or indirectly. Among the most sensitive processes is photosynthesis. This is because the stomata are closed, preventing the entry of carbon dioxide, not because of the shortage of water as a reagent.

Pronounced water stress, if sufficiently prolonged, causes permanent wilting and death of the plant, unless it is a species well adapted to an arid climate. It is important to realize, however, that even mild water stress, while producing no outward symptoms, reduces photosynthesis and thereby decreases growth rate and may have other effects on the development of seeds and fruits (see Section 11.6.2). Since more crop plants are damaged by excess of transpiration over water absorption than probably by any other single cause, the subject of water stress and its causes and effects is of profound interest to agriculture and merits closer attention. To begin with, part of the cause of this imbalance lies in the resistance water meets when flowing through plants, and this is considered below.

2.5.2   Resistance of plant tissues to water flow

Water does not move unimpeded in plants, either in parenchymatous tissues or in the xylem. Throughout its passage it encounters resistances. These resistances in themselves, by retarding water flow, induce water stress in vigorously transpiring plants. If a plant growing with its roots in water culture is suddenly placed in conditions favouring rapid transpiration, such as a warm draught in light, transpiration accelerates but uptake lags behind, causing a temporary water deficit in the plant. Indeed if transpiration is maintained at a maximum level, uptake may lag permanently behind, causing wilting and death of the plant. The lag results from resistances within the plant.

Now if a plant is cut at ground level and its cut end placed in water, and if the shoot system is then subjected to the same conditions as above, no lag develops and it becomes very difficult to induce a permanent deficit between water loss and uptake. Apparently, therefore, the main resistance lies not in the shoot system but in the roots.

Resistances of tissues vary considerably, increasing with the rate of water flow, and decreasing with rising temperature . There are also considerable differences between species. Broad beans, for instance, have a much greater resistance per unit of root area than maize which in turn has a much greater root resistance than tomato.

The resistances of stems and leaves to the passage of water through them are appreciable but are much less than root resistance and are not thought to be a significant cause of water deficits. The resistance of roots, however, has important effects in practice and is thought to be the main cause of the temporary water stress that commonly occurs in crops around midday, as described in Section 2.5.1, with the resultant reductions in photosynthesis and other processes.

One of the sources of this resistance is undoubtedly the suberization of the roots as they grow and mature (Fig. 2.17). Not unexpectedly, the zone of maximum uptake of water and also of minerals occurs where the xylem has developed sufficiently to transport these materials away but where suberization of the endodermis has not progressed to the point where root permeability is greatly reduced. In most species this coincides with the region of root hair production. The exact region of maximum uptake is not fixed, however, and if transpiration is increased this region tends to extend further back into the more heavily suberized zones which cannot therefore be totally impermeable.

With age, further deposits of suberin in the cortex and the development of a corky bark on the older roots cause further decreases in permeability. If root growth stops, the zone of suberization gradually encroaches on the tip. Possession of an easily permeable root surface therefore depends on continuous growth by the root system. Any check to root growth temporarily reduces the ability of the plant to take up moisture until after full root growth is resumed.

Suberization is thus a significant source of root resistance, though it is doubtful if roots ever become totally impermeable to water. Trees provide the most convincing evidence of this, for in summer only a small proportion of their root systems consist of white, growing, unsuberized root tips. Most of the mineral and moisture uptake must be through suberized root since no more than a fraction of the plant's water needs could be taken up through such a small proportion of its root

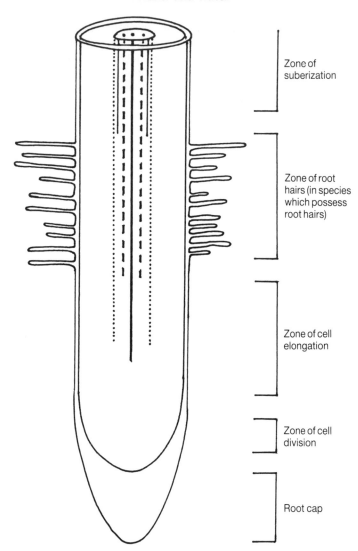

Zone of
suberization

Zone of root
hairs (in species
which possess
root hairs)

Zone of cell
elongation

Zone of cell
division

Root cap

Fig. 2.17 The zones of growth, root hair production and suberization near the root tip.

surface. Further, in cold or dry soil conditions little or no root growth occurs and the area of suberized root tip is almost nil, yet pines and other evergreen trees take up enough moisture to balance transpiration losses during severe winters or droughts.

Experiments, however, have revealed another important source of root resistance which is also of great practical significance. It can be shown that if a root is killed, its resistance to the passage of water is greatly reduced. Rises in temperature increase the permeability of a living root; decreases in temperature reduce it. Furthermore, low oxygen or high carbon dioxide levels around the root reduce

permeability, as do chemicals which inhibit metabolism. Lastly, some root systems show diurnal rhythms in permeability. It would seem, therefore, that root resistance rests primarily in some living part of the root, the condition of which is closely linked to metabolism, particularly respiration. It can be deduced that the seat of this resistance is in the symplast, which, as we have seen, all water must enter in order to cross the endodermis, that gateway to the inner plant.

2.5.3 Root resistance and low temperature drought

The existence and environmental sensitivity of root resistance have profound implications. For example, at low soil temperatures, the ability of plants to take up water is severely reduced. At 5°C the water uptake of many species may be only a fifth of the uptake at 25°C. Even in moist soils, therefore, the plant may be under water stress not because water is absent but because it is, although present, unavailable. This condition is known as *physiological drought*.

Various factors are involved in physiological drought, including decreased root growth, increased viscosity of water and decreased metabolic activity, resulting in lowered mineral uptake and therefore less osmotic uptake of water as the concentration of mineral solutes in the plant roots falls. Undoubtedly, however, increased resistance to the passage of water across cell membranes in the roots is a major contributory factor.

Thus low soil temperatures, through their effects on water uptake, are an important ecological factor and often limit plant growth in cold areas or periods. In prolonged cold spells large numbers of plants may be killed by physiological drought through continued transpiration when low soil temperatures inhibit water uptake. The deciduous habit of many temperate trees and shrubs is, to a considerable extent, an adaptation for avoiding physiological drought by reducing transpiration in winter. Evergreen species of cool temperate climates have waxy leaves, in some cases reduced to needles as in Scots pine, to minimize transpiration in cold periods, in the same way as evergreens of warm, semi-arid climates such as Corsican pine.

Wind, where it accelerates transpiration, can exacerbate the effect of low temperatures, especially in highly exposed places. Thus heather, though adapted to withstand physiological drought, can be killed in the high moors of Scotland even in moist soil conditions by the severe water stress resulting from a combination of wind and cold.

In agriculture, similar effects occur in exposed grasslands and heathlands, resulting in winter kill. Other examples of water stress induced by low soil temperature are known. Orange trees in California are often observed to wilt in cold weather. Cereal seedlings in temperate areas can have their growth restricted by water stress during cold springs. The effects of the temperature of the water supply on soil temperature and therefore on moisture uptake can be considerable due to the very high specific heat of water, and this has implications in the practice of irrigation. For example, the growth of glasshouse cucumbers can be impaired by watering with very cold water. Similarly, cold irrigation water from high dams in the Sierras of California is thought to limit the growth of rice in the Sacramento valley and it is often pre-warmed in warming basins before use. On the other hand, in high soil temperatures the cooling effect of irrigation water may increase the yield of crops such as potatoes, which are more adapted to lower soil temperatures.

There is evidence that where the soil temperature falls gradually, plants develop some tolerance and increase the root permeability they can achieve at low temperatures. This adjustment is not observed where the fall in temperature is sudden. Generally speaking, tropical and subtropical crops display much less tolerance of low soil temperature than temperate crops. Thus effects on root permeability have been demonstrated below 5°C for cabbage, 10°C for citrus and 22°C for cotton and watermelon.

2.5.4   Root resistance and waterlogging

Waterlogging of soils results in poor aeration and therefore depletion of oxygen and accumulation of carbon dioxide in such soils from the respiration of roots and soil organisms. Many plants rapidly wilt if the soil becomes waterlogged, unless they are species especially adapted to marshy situations. Undoubtedly the wilting is caused by increases in root resistance induced by lack of oxygen (Section 2.5.2). Other symptoms such as the distortion and yellowing of leaves may appear in time. The causes of these symptoms are uncertain but the main condition induced is physiological drought in the midst of plenty. In practice it is difficult to distinguish between the effects of oxygen depletion and carbon dioxide accumulation on root resistance but it is generally thought that the former is more important.

In Section 1.8.2, it was described how waterlogging, with the resultant lack of oxygen, induces anaerobic respiration in roots and leads to death or stunting of them. Where waterlogging inhibits root growth, after a few days this too will increase root resistance. Partial waterlogging alters the form of the root system, giving rise to shorter, shallower, less branched roots often lacking root hairs. New roots may arise at the base of the stem. Within the root the air spaces between cells in the cortex are larger than normal and the switch to anaerobic respiration results in accumulation of ethanol and other toxic by-products. These may be responsible for some of the symptoms such as growth distortion and yellowing described above in the shoot system.

The susceptibility of plants to waterlogging injury varies greatly with species and environmental conditions. High soil temperatures increase the damage, probably because of the greater oxygen demands of growing and rapidly metabolizing roots. Dormant plants such as trees and herbaceous perennials are much less susceptible than when actively growing. Species normally found in moist habitats such as marshes are little affected by waterlogging. This group includes few crop plants; rice, which thrives in flooded soil, is the obvious exception. By contrast, tobacco is permanently injured by waterlogging after heavy rains for one or two days even in well-drained soil. Such variations between species reflect differences in such characteristics as the efficiency of internal air circulation from the shoot through the intercellular spaces of the root cortex, and the tolerance by roots of anaerobic respiration.

Deficient aeration may be induced by soil texture rather than by excessive moisture. Heavy clay soils tend to be poorly aerated because their fine pores stay filled with capillary water after rain, preventing air circulation. In these conditions, or where the water table is very high, the root system is confined to a shallow layer near the soil surface. This not only limits the volume of soil that the plant can exploit for mineral nutrition but, through the absence of roots deeper in the soil, may make the plant more susceptible to true drought at a later date.

Drainage is of cardinal importance in agriculture because by lowering soil moisture content it not only promotes aeration but also decreases the specific heat of the soil, thereby increasing its average temperature. Through these effects it encourages the healthy growth and respiration of roots. The resulting improvement in root permeability leads to increased water uptake by crops, and the increased volume of soil explored by the roots improves mineral nutrition.

2.6 Water stress and soil moisture supply

Despite the role of root resistance, water stress occurs most frequently in crops where neither soil temperature nor aeration are important contributory factors. The most frequent causes of water stress lie in the supply of moisture to the soil and in the nature of the forces tending to hold that moisture in the soil. Of these forces the most important are the matric forces (Section 2.2.1) holding water on the surface of soil particles. Less important is the solute potential of the soil solution, which we shall consider first.

2.6.1 Osmotic drought

In most soils the concentration of salts in the soil solution and therefore its solute potential are too weak to inhibit water uptake by roots and to make a significant contribution to plant water stress. However, there are a few situations where salt concentration in the soil is so high that this generalization does not apply. In these situations, the solute potential of the soil prevents or decreases water uptake, causing a condition called *osmotic drought*.

The most obvious example is the sea-washed shore where sodium chloride is the chief solute. Here species of plants known as *halophytes* are found, specially adapted to resisting osmotic drought. Beet, as a crop, is derived from such a species, and still shows a yield response to applications of sodium salts. Some halophytes have glands on their leaves through which excess salt is excreted, while others simply have devices to cut down transpiration. Where grass turf is sea-washed, halophytic grasses flourish. Many of these are highly nutritious and form the useful shoreline grazings known as saltings.

In most other situations where high salt concentrations occur in soils, this is due to water evaporating from the soil, leaving behind its load of dissolved salts instead of their being flushed out with the drainage water. For example, rivers flowing into arid areas lacking precipitation to replace water losses by evaporation and seepage may eventually dry up leaving their salt load in the soil. Over time the salts accumulate giving a saline soil. Osmotic drought is induced, preventing uptake of water by plants and thereby creating salt of alkali deserts. The problem occurs especially where water flows into internally draining land basins, which are areas with no drainage outlet to lower areas, and water is left to evaporate in the bottom of the basin (Fig. 2.18). Sizeable salt basins of this type are found in the arid centres of large continental land masses such as in the southwestern United States. The Dead Sea in the Middle East is also the result of just this effect.

Osmotic drought also occurs where accumulation of salts near the soil surface, left behind by eva-

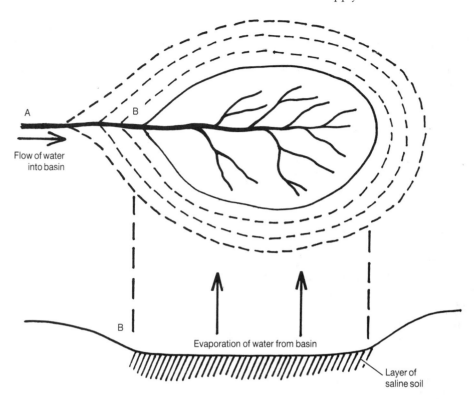

Flow of water
into basin

Evaporation of water from basin

Layer of
saline soil

Fig. 2.18 Salt accumulation in an internally draining land basin. Water carrying the normal load of dissolved salts enters the river at A, and fans out in a delta as it reaches the flat bottom of the basin at B. Thereafter, it soaks into the soil but gradually evaporates leaving behind its salt load to accumulate in the soil.

porated or transpired water, is not balanced by the downward leaching effect of drainage water. This has arisen very extensively in irrigation schemes over much of history, when they have been mishandled, especially if the drainage of them has been inadequate. It is also possible for salinity of soils to arise through mishandling of groundwater supplies where fresh water overlies salt water, and the salt water is brought up to the surface.

2.6.2   True drought

The widespread occurrence of water stress in plants is, however, due more to true drought, resulting from inadequate rainfall and the strong matric forces holding water in soils. Over the land masses of the earth the distribution of vegetation is governed more by water supply than by any other single factor.

Many plant species are adapted to survival in drought-prone areas; such plants are known as *xerophytes*. In some of the more primitive plants such as mosses and terrestrial algae the protoplasm can survive desiccation unscathed, but this is a rarer characteristic among higher plants, except for their seeds. Nonetheless, higher plants growing in arid or semi-arid regions show a wide variety of adaptations for surviving drought (Fig. 2.19).

Some species evade drought by germinating rapidly in response to rainfall and completing their life cycle in a very short space of time before the water supply dries up. The drought-resistant seed survives in the dry soil until the next fall of rain. Other species have very deep roots, tapping zones of the soil which remain moist even during prolonged drought. Still others shed their leaves during the dry season to reduce transpiration.

Cacti and other succulent plants store large quantities of water in their swollen stems or leaves. They gather moisture through extensive root systems and retain it by cutting down on moisture loss through devices such as thick waxy cuticles and sunken stomata. Such species grow only when the soil is moist, and transpire almost as freely as non-xerophytes under these conditions, but are

Fig. 2.19 A range of xerophytes, in this case part of a collection growing in a greenhouse.

inactive during periods of drought.

Many succulents, including pineapple, use the crassulacean acid metabolism (CAM) mechanism of photosynthesis. As we saw in Section 1.6.6, this enables them to assimilate carbon dioxide at night, when the stomata can be open with less danger of excessive transpiration. CAM requires a large pool of malic or aspartic acid to assimilate carbon dioxide at night and release it during the day for reassimilation into carbohydrate. A considerable volume of water is required to hold sufficient of these acids in solution. The water reserves of succulent plants are thus necessary to CAM photosynthesis as well as for water storage, and in both ways help such plants survive and grow in arid climates.

However, few crops exhibit, to any marked degree, the characteristics of xerophytes. Nearly all crops have their growth and yield restricted by

water stress at some stage during their development, unless they are continuously irrigated. In many crops, inadequate water supply from precipitation is the main factor limiting growth and yield.

The water supply problem is further aggravated by the fact that not all the water a soil contains can be taken up by plant roots. For this reason we shall now look briefly at the nature of soil moisture and its availability to plants.

2.6.3 The soil moisture reservoir

Except in highly organic soils such as peats, most of the solid material consists of mineral particles. Many of these particles adhere loosely together in groups called 'crumbs', aided by the organic matter and its breakdown products acting as a sort of glue between the particles. This *crumb structure* is an important feature of soil structure in the upper layers, and soils in good tilth have it in good measure. Between and within the crumbs, many small spaces or pores occur. The finer pores within the crumbs act as capillaries and retain their moisture, forming an important soil moisture reservoir, while the large pores between the crumbs drain freely under gravity after inundation of the soil with water and provide the main route of soil aeration. Thus after a few days of free drainage following rain or irrigation the moisture content of a soil in which no plants are growing reaches a fairly stable level at which its capillary pore space is replete with water but the larger pores are air-filled. This moisture content is known as the *field capacity* of the soil.

Deeper in the soil is a zone in which all pores are permanently filled with water; the upper surface of this zone is called the *water table*. Some upward capillary movement of water occurs from the water table but it makes little contribution to the moisture content of the soil layers inhabited by plant roots.

At moisture contents above field capacity, gravitational water in the larger pores is available to plants, but at lower moisture contents the plant draws exclusively on the reserves of capillary water, not all of which is available to the plant due to the matric forces binding it in the soil. As transpiration continues, these reserves are depleted until, at a certain level of soil moisture content called the *permanent wilting point*, the plant wilts irreversibly, though it may continue to extract significant quantities of water even beyond this point. Before the permanent wilting point is reached, temporary or reversible wilting may

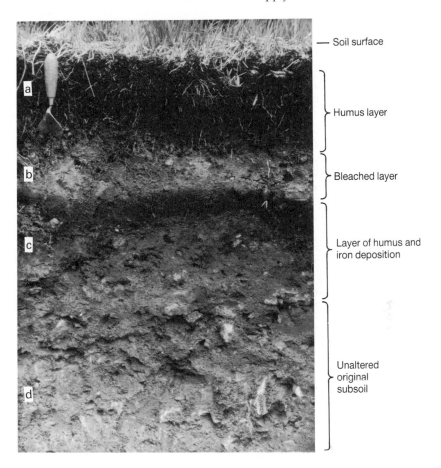

Soil surface

Humus layer

Bleached layer

Layer of humus and
iron deposition

Unaltered
original
subsoil

Fig. 2.20 The soil horizons developed in a soil type known as an iron humus podzol. The top layer (a) consists mainly of roots and the organic matter produced by death of plants. The decay of this produces humus and organic acids. The acids and humus are washed down by rainwater to the upper mineral layers where the acid dissolves out the iron in the soil giving a light-coloured, bleached layer (b). Further down, soil conditions encourage deposition of the humus and precipitation of the iron, giving a dark-coloured layer (c). Below that, the subsoil (d) remains largely unchanged by these processes (courtesy of E A Fitzpatrick, University of Aberdeen).

occur under conditions of rapid transpiration.

The picture we have given is very simplified. Soils are not uniform throughout their depth but form distinct layers called *horizons* which behave in different ways (Fig. 2.20). Also, the permanent wilting point of any one soil is not a fixed value of soil moisture content but differs considerably between plant species and between plants of the same species with different moisture requirements.

Other complications could be enumerated. Nonetheless, the point remains that soils have the capacity to retain a limited store of moisture in a form available to the plant. This reservoir plays a vital role in evening out the disparities between a continuous water demand by the plant and a discontinuous supply to the soil by erratic or seasonal rainfall.

Deep soils have a greater storage capacity than shallow soils, and soils already fully pervaded with extensive root systems are depleted more rapidly than those less thoroughly colonized. In maritime climates such as in Great Britain, rain usually occurs in appreciable if sometimes inadequate amounts in all months. Soil moisture deficits in such climates are therefore often short-term and the buffering effect of the soil moisture reservoir is highly significant. In regions with alternating dry and rainy seasons, as occur in many tropical and subtropical areas, the ability of the soil moisture reservoir to sustain growth through much or all of the dry season is vital. Tropical soils are, however, often much deeper than those in temperate climates, and their large moisture reservoir can often sustain crops for a considerable period without rain.

At permanent wilting point in many soils there is still a considerable amount of moisture present, and there are various reasons for this. As they dry both roots and soil (especially soils rich in clay) tend to shrink, opening up gaps across which water cannot migrate except as vapour. Also, as soil moisture content declines, root growth slows down and suberization approaches the tip, greatly increasing root resistance. The most important cause, however, is the matric forces holding water in the soil.

As capillary reserves are depleted and the finer pores empty, curved films of water develop between adjacent soil particles (Fig. 2.21) and the matric forces holding these films in position resist the withdrawal of further moisture despite the

intimate contact between the fine roots and soil particles (Fig. 2.22) In addition, the parting of columns of water in the pores leads to a loss of continuity. When this happens the ability of water to move through the soil to replenish moisture removed by roots, that is the *hydraulic conductivity* of the soil, declines with increasing rapidity. Indeed, at certain stages in soil moisture depletion, a fall of 2–3% in soil moisture content can enormously increase the impact of the water-retaining forces.

In soils much below field capacity, then, since the moisture will not go to the root, the root must go to the moisture. In drying soils, failing replenishment by rainfall or irrigation, the root system must expand into new untapped areas of soil to obtain adequate water. This is generally possible for annual crops as they often have a large volume of uncolonized soil available for root growth. Under perennial crops such as grass the root system

Fig. 2.21 In soils at field capacity (a) the smaller pores between soil particles are full of water, but as moisture reserves are depleted (b) these pores empty and the remaining water is held as curved films of water between particles, and much more strongly. Fine-textured soils such as clays (c) have much finer mineral particles and pores than coarse-textured ones such as sands (a). Within these finer pores the water is held more strongly and a smaller proportion of it is available to plants. However, this is more than offset by the fact that there are many more pores per unit volume of soil in fine-textured soils. Thus they have a greater field capacity and a greater total amount of water available to plants.

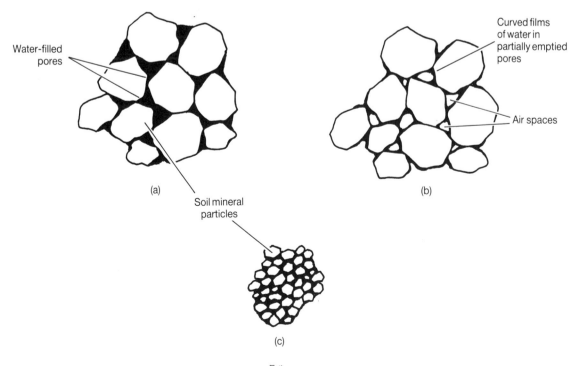

Water-filled pores

Curved films of water in partially emptied pores

Air spaces

(a)

(b)

Soil mineral particles

(c)

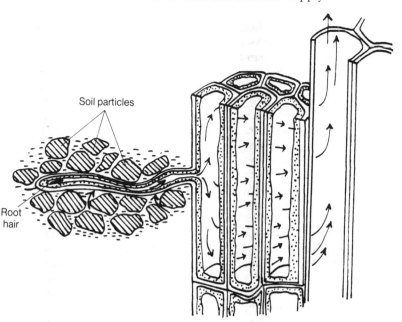

Fig. 2.22 Diagram of a root of perennial ryegrass in soil showing the close contact between soil particles and root hairs.

will already have explored fully the accessible soil and this expansion is not possible. In deep soils under deep-rooted crops like tropical grasses soil moisture is depleted by roots at progressively deeper levels.

The lowering of hydraulic conductivity with decreasing moisture content also has a beneficial effect. Once the surface layers of soil have dried out by evaporation, little water moves up from below and the result is a powdery surface layer with fully moist soil below. This effect is increased where layers of organic detritus occur on the surface, as in forests, and is exploited in the practice of *mulching* of crops where layers of such material are applied to the soil surface to conserve moisture.

When plants grow in soils their roots extract moisture from considerable depth and dissipate it through transpiration. As a result the rate of water loss by the combined effects of transpiration and evaporation, commonly called evapotranspiration, greatly exceeds the rate of loss by evaporation from bare soil. In semi-arid climates it is normal practice to leave land uncropped for a year to avoid loss by transpiration so that soil moisture levels can be replenished adequately to support a crop in the following year. It is important in these *fallow* systems to control weeds, either by cultivation or with a herbicide, to further limit transpiration losses.

Undoubtedly the most important factor influencing the capacity of soils to store available moisture is soil texture. Sandy soils with large pore spaces hold little capillary water, but nearly all of it is available to plants. Clay soils have more total pore space than sandy soils, most of it being in the form of fine capillaries. They thus have a high field capacity but much of their water is held too strongly by matric forces to be available to plants. Even so, the total amount available for uptake is greater than in sandy soils. Organic matter absorbs water in an available form and, as we have seen, aids the formation of soil crumb structure and thus of pore spaces. It thereby increases the capacity of soils to store available moisture. The available water capacities of various soil types are given in Table 2.1 and it can be seen that the highest capacities are found neither in very coarse nor in very fine soils but in intermediate loams.

Not all the moisture between field capacity and permanent wilting point is equally freely available to plants. It seems that about the first half of this water is freely available, but that the plant extracts the second half with increasing difficulty and with developing water stress. This last situation is thought to be extremely common in crops, limiting their growth rate before any sign of wilting occurs.

Table 2.1 Field capacities of soils of different textures, expressed as the volume of water retained in the rooting zone after 48 h free drainage of a saturated soil, as a percentage of the total soil volume. Note that there is no single value for field capacity that might be considered typical, even within a broad textural group.

	Field capacity (v/v %)	
	Topsoils	Subsoils
Fine-textured soils (> 20% clay)		
> 35% clay	40–60	35–50
20–35% clay	30–50	25–40
Medium-textured soils (< 20% clay)		
< 10% sand + coarse sand	25–40	20–35
10–30% sand + coarse sand	20–35	15–30
Coarse-textured soils (< 20% clay)		
30–50% sand + coarse sand	15–25	10–20
> 50% sand + coarse sand	10–20	5–15

Clay defined as particles smaller than 0.002 mm diameter, sand defined as particles from 0.1 to 0.2 mm diameter, coarse sand defined as particles from 0.2 to 2 mm diameter.

From *Irrigation*. MAFF Reference Book 138. Ministry of Agriculture, Fisheries and Food.

2.6.4 Frequency of water stress due to true drought

Because of the constant appetite of the growing plant for water and the limited nature of the soil reservoir, plants tend to create conditions of water stress for themselves. Since water stress is such a major limiting factor to photosynthesis and therefore to growth and to yield it is important to know how frequently it occurs.

The only significant natural source of water is precipitation, either as rain or as snow. Water is lost from the soil by drainage, evaporation and transpiration. Significant losses by drainage occur only when the soil is above field capacity; at other times the water balance of soils is determined by the difference between precipitation and evapotranspiration. So long as precipitation exceeds evapotranspiration, the soil reservoir will remain full and growth will not be limited by water stress, whereas if evapotranspiration exceeds precipitation a moisture deficit will develop, leading to plant water stress.

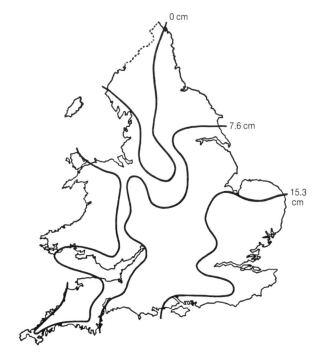

Fig. 2.23 Average soil moisture deficit (mm) in England and Wales in late summer. Reproduced from S. Laverton (1964) *Irrigation – its profitable use for agricultural and horticultural crops.* OUP, Oxford, with permission of the publisher.

Worldwide, soil moisture deficits limiting crop growth occur in most agricultural areas, sometimes due to low overall rainfall but often due to its seasonal distribution. A few coastal temperate, equatorial and island regions receive adequate rainfall to balance evapotranspiration all year round but most areas receive adequate rain only in part of the year. This is especially true in large continental land masses such as North America, where the season of maximum evapotranspiration often has little rainfall. In such areas irrigation is essential to realize good yields.

Deficits are usually expressed in millimetres of shortfall of precipitation. Fig. 2.23 shows a typical pattern of soil moisture deficits occurring in England and Wales in the summer months.

2.6.5 Resistance to drought in crop plants

Since water stress is so common, the ability to withstand it is an important characteristic in crop plants. Considerable differences exist between crop species in their ability to maintain turgidity and to yield acceptably under conditions of drought. Sorghum yields, for instance, are far less reduced by continued water stress than those of maize. Studies show that a wide range of characteristics of root and shoot are responsible for such differences.

Plants usually develop a much larger root system than is necessary for survival under normal soil conditions. Up to half the root system of many plants such as trees and grasses may be removed without any apparent serious injury to them, though there may be a temporary reduction in growth rate. This excess of roots is an important aid to the survival of plants during periods of severe water stress.

The ability to regenerate quickly after severe drought may be equally important. Many herbage grasses possess this ability. Not only do their roots restart growth quickly but even after most of the foliage is desiccated and dead, buds protected deep in the crown of the plant survive and regenerate.

However, most crops do not have this demand for regeneration placed upon them. In the prolonged or short-term partial drought conditions experienced by most of the world's crops, the ability to give reasonable yield under water stress is more important. We may draw a distinction between those characteristics which allow the plant to make the most efficient use of moisture once it has been absorbed and those which make the plant more efficient at foraging for water in the soil.

The efficiency of water use is effectively measured by comparing the amount of water taken up with the amount of dry matter produced, the ratio of these two quantities being known as the *transpiration ratio*. Representative values of transpiration ratios for some crops are shown in Table 2.2.

Table 2.2 Transpiration ratios for eight crops, grown under good conditions of soil fertility in Colorado, USA

Crop	No. of varieties tested	Transpiration ratio
Millet	5	276
Sorghum	5	324
Maize	8	329
Wheat	7	510
Barley	4	514
Cabbage	1	518
Horse bean	2	746
Flax	7	783

From H L Shantz and L N Piemeisel (1927) *J. Agric. Res.* **34**, 1093.

However, growing conditions, including such factors as moisture supply and application of nitrogen fertilizer, can greatly influence the transpiration ratio. Thus although the differences between crop species are relatively fixed, there is no fixed value of transpiration ratio for any crop. Nor can the ratio be taken as the only guide to the efficiency with which a species utilizes soil moisture on a site, as other factors also strongly influence this.

Shoot characteristics are more important than root characteristics in affecting the transpiration ratio. In particular, the ability to reduce water loss, by partial closure of stomata or other methods, while maintaining a high rate of photosynthesis leads to a low transpiration ratio. This ability is possessed by C_4 plants such as maize and sorghum, which, as was seen earlier, continue to photosynthesize with stomata fully closed, until all the carbon dioxide in the internal air space of the leaf is used up.

Root characteristics are more important in influencing the ability of the plant to forage for water. The roots of annual plants, starting from germination, expand continuously into new soil and therefore new reservoirs of water. Species which are efficient at this process may show little response to irrigation except in severe drought. In addition, since the season is far advanced before an annual crop establishes full leaf cover, maximum water demands on the soil are delayed. Perennial crops, on the other hand, begin the season with an already well-developed root system and, in many cases, including trees and grasses, a well-developed shoot system as well. Their capacity for further root expansion is very limited and they suffer large transpiration losses from an early stage in the season.

Rooting depth is very important in influencing the water foraging ability of crops. For example, in one experiment comparing two grasses, the deeper-rooted meadow fescue not only absorbed more water from below 35 cm in the soil than the shallower-rooted timothy, but it absorbed a greater total quantity of water. One of the best known examples of a crop species able to withstand drought is lucerne (alfalfa), the roots of which may extend 3 m or more down into the soil, reaching permanently moist subsoil.

Root branching increases the surface area available for water absorption since several small roots have a greater surface area in total than one large root of the same weight. In addition, the low hydraulic conductivity of a soil where the reserves of available moisture have been partially depleted means that water shortages may be extremely local around roots. Numerous fine rootlets are therefore more efficient at tapping a bulk of soil than fewer larger ones.

An important characteristic of roots which affects crop tolerance of water stress is their flexibility in functioning and development. A plant growing where an adequate water supply is continuously available will develop a much shallower and less extensive root system than one of the same species in a moisture-depleted soil. Roots tend to grow into and develop in areas of soil where moisture is more readily obtainable. In the practice of trickle irrigation, for instance, water is released in small quantities at only a few points, through nozzles on water pipes lying along the ground. As the root systems develop they soon cluster round these points, demonstrating their ability to take in most of the water the plants require from a few such points.

In general, soil conditions favourable to root development, such as good crumb structure to a considerable depth, moderate but not excessively high fertility and an even distribution of mineral nutrients, tend to produce a more drought-resistant plant. On the other hand, soil conditions which restrict root development, such as compaction, deficient aeration through poor drainage, slow water penetration and low temperature produce a more drought-susceptible plant.

Within an established root system the plant shows considerable flexibility in its response to changing soil conditions. Plants tend to extract moisture first from the upper layers of soil, where roots are most prolific. As the upper layers dry out, water absorption shifts more to the roots at deeper levels, and so therefore does mineral absorption. Since, in field crops, fertilizer applications are usually concentrated in the upper layers, this has serious implications for the growth and nutrition of crops.

Water stress, however, is the result neither of water loss nor of water uptake alone but of the balance between the two, and the ratio of root to shoot is therefore important. It is noteworthy that the more drought-resistant crop species such as sorghum have high root/shoot ratios.

Summary

1. Of all the water absorbed by a plant, only about 2% is used in building up the plant, acting as the universal medium for metabolic activity, participating as a reagent, providing support through cell turgidity and being the vehicle for transport of organic and inorganic materials within the plant. The remaining 98% is lost from the leaves as water vapour in the process of transpiration.

2. Water moves across cell membranes by diffusion in response to differences in pressure on opposite sides, caused by the resistance of cell walls to stretching, and also to differences in solute concentration which give rise to osmosis. These differences may be expressed in terms of water potential. Wall pressure increases water potential in a cell; the presence of solutes lowers water potential. The effect of solutes is usually greater than that of wall pressure so that most cells have a water potential below that of pure water. Small changes in cell volume resulting from intake or loss of water considerably affect wall pressure but have little effect on solute concentration. It is through these changes in pressure that water movement from cell to cell is largely regulated.

3. Water ascends from root to shoot in the xylem, a system of dead cylindrical cells forming pipes in which water is moved by mass flow. In order to enter the xylem of the root, however, all water must enter and leave living cells, the protoplasm of which forms a continuous cell-to-cell pathway, the symplast, across the root. This is because the other, non-living, pathway of water movement in the cell walls, that is the apoplast, is blocked by deposits of waterproof suberin in the endodermis of the root. The symplast 'pumps' mineral ions into the xylem and prevents back-leakage. The resulting high concentration of ions in the xylem draws water in by osmosis. The effect may be to push water up the xylem by positive pressure, known as root pressure.

4. For various reasons, it is known that root pressure cannot account for more than a small proportion of the water rising in the xylem. The main driving force is transpiration, drawing water up the xylem under tension. As water evaporates from the surfaces of leaf cells, more water is drawn out of the xylem in the leaf to replace it, and this pull is transmitted all the way down to the root. It appears that the narrow columns of water in the xylem have sufficient tensile strength to withstand the transpiration pull without breaking.

Transpiration rate depends basically on the steepness of the water potential gradient between the inside and the outside of the leaf, on the effect of any barriers to the free movement of water vapour, and on the effectiveness of the plant's mechanisms for controlling water loss. All influences on transpiration act finally through one or more of these.

Most of the green shoot surface is covered with waxy cuticle, and woody tissues with a suberized bark, to minimize water loss by transpiration. The main route for water loss is therefore through the stomata which can be closed during periods of water stress and also in darkness when they are not needed for photosynthetic gas exchange. The rate of transpiration is thus to a large extent controlled by the plant.

Solar radiation is the driving force of transpiration, chiefly through its tendency to raise leaf temperature and thus the rate of evaporation of water molecules from the surfaces of the mesophyll cells. Changes in atmospheric relative humidity, and wind, themselves also caused by solar radiation, are the other two chief environmental influences on transpiration. Falls in atmospheric relative humidity encourage transpiration by steepening the humidity gradient. The effect of wind is a balance of its stimulatory effect through assisting removal of water vapour from the leaf surface and its inhibitory effect through cooling of the leaf. Solar radiation and relative humidity also influence transpiration through their effects on the plant. Stomata tend to open in the light and with rising temperature. Plant structure, especially leaf size and shape, also affects the plant's response to these influences. Hence, there is seldom a simple relationship between an environmental factor and the rate of transpiration.

Transpiration has a useful cooling effect on the leaves, and the flow of water through the plant helps to distribute minerals from roots to leaves. In part, however, transpiration appears to be no more than the unavoidable side-effect of the large leaf area necessary for efficient photosynthesis.

5. When transpiration exceeds water uptake from the soil, water stress develops in the plant. This situation is favoured by low atmospheric relative humidity, high solar radiation, high air temperature and in some circumstances wind, which increase transpiration. As water stress develops the stomata close but water loss cannot be stopped completely because some transpiration continues through the cuticle. An increase in transpiration leads to increased water absorption by the roots but there is a lag before this happens, during which temporary wilting may occur. The lag is caused by resistances of water movement in the plant, mainly in the root. Root resistance increases with suberization of the root and is therefore greater where the growth of new non-suberized roots is restricted.

The main seat of root resistance, however, is in the symplast. Water crosses the symplast from the outer apoplast to the root xylem more readily when metabolism is proceeding actively. Root permeability is thus adversely affected by low soil temperature and by deficient aeration as in waterlogged soil, both of which therefore result in water stress in conditions favouring rapid transpiration. Plants in such situations are said to suffer from physiological drought.

6. Water stress may arise from osmotic drought when high salt concentrations in the soil solution prevent water uptake by roots. More commonly, however, it is the result of true drought, when there is an inadequate supply of water available in the soil for plant uptake. Some plant species, known as xerophytes, have adaptations of various kinds to reduce transpiration in order to survive in drought-prone areas, but few crops have xerophytic characteristics.

Soils have, between the solid particles, pore spaces of a range of sizes, all of which may contain water. The larger pores drain under the influence of gravity and are seldom a significant reservoir of water for plant uptake. The smaller pores hold water by means of the matric forces binding the water molecules on to solid surfaces. The amount of such water that a soil can hold is known as the field capacity of the soil. Not all of it is available to plants; as it becomes depleted the matric forces become stronger and roots are therefore less able to extract water. Eventually the resulting water stress is such that the plant wilts irreversibly, and the moisture content of the soil when this stage is reached is called the permanent wilting point.

As soil moisture content falls, the ability of water to move through the soil to replenish that extracted by roots declines rapidly. Thus continued water absorption depends on continued root growth to tap new areas of soil. In perennial plants the root system may already have explored the soil fully and the scope for further expansion is very limited. In annuals, resistance to water stress is often correlated with root growth, branching and depth of penetration in the soil.

3

Plants and Minerals

3.1 Importance of minerals to the plant

In Chapters 1 and 2 we examined the relationships between plants and two of their basic requirements, energy and water. In this chapter we consider a third essential ingredient for plant life – a supply of *mineral elements*. Here, we include nitrogen which is not, strictly speaking, a mineral nutrient, since it is derived directly or indirectly from the air.

Plants obtain nearly all the minerals they require from the soil, from which their roots absorb them as ions. A small proportion may be absorbed by the leaves from atmospheric dust and rainwater.

The character of the mineral supply in the soil is a major factor in deciding which plants grow where. Under natural conditions plant growth is commonly limited by mineral supply, and from this derives the importance in crop husbandry of the almost universal practices of liming and fertilizing to boost yields.

Only about 6% of the dry weight of plants is minerals, but analysis of that 6% reveals a very wide range of elements to be present, and the question that naturally arises is what, if anything, they are all doing there. About thirteen are known to be essential for the growth of the plant. Another seven, including silicon, selenium and nickel are not essential but perform some useful function when present. Silicon, for example, improves the resistance of cereals to lodging, and in some cases their resistance to certain diseases. The remainder, such as silver or lead, have no known plant function, but are there simply because of the ability of water to dissolve a wide range of materials, and because the plant cannot exercise complete control over what minerals enter it in solution from the soil. Most of the essential and functional elements are also essential or functional in animals which live off the plants, as indeed are some of the chance inhabitants of the plant, such as cobalt and iodine. Fig. 3.1 shows approximate ranges of concen-

tration in which the essential and some of the other elements occur in plants.

Plants require minerals for three main types of function. Firstly, some mineral elements are essential components of molecules vital to plant life. Nitrogen is present in all proteins (including enzymes), nucleic acids, energy carriers, chlorophyll and other pigments, and a wide range of other substances required in minute amounts. Phosphorus is an essential ingredient of nucleic acids and energy carriers, and is a component of phospholipids, of which cell membranes are partially composed. Sulphur occurs in proteins and other substances. Calcium, as calcium pectate, forms the cement that binds adjacent cell walls together. Magnesium is a component of chlorophyll.

Secondly, mineral elements may be directly involved in plant cell metabolism. Phosphorus is required for the synthesis of ATP, important both in photosynthesis and in respiration, as we saw in Chapter 1. Iron is involved in the oxidation-reduction reactions by which electrons give up their energy to make ATP. Certain other elements, including copper and molybdenum, function as *coenzymes*, complementing the protein component in certain enzyme systems.

Thirdly, some minerals, in the form of ions, preserve various equilibria in the plant cell. By moving into and out of the vacuoles they regulate the solute potential of the cell, which, as we have seen in Section 2.2.2, influences the movement of water in and out. Positively charged ions (cations) help to preserve electrostatic equilibrium by balancing negative charges on organic molecules. Ions may also have a buffering action, regulating the pH balance of the cell contents, which is important because many enzyme reactions are extremely sensitive to pH. Potassium, calcium and magnesium are the most important mineral ions involved in preserving equilibria, and in this respect are to some degree interchangeable.

The mineral content of plants is not a good

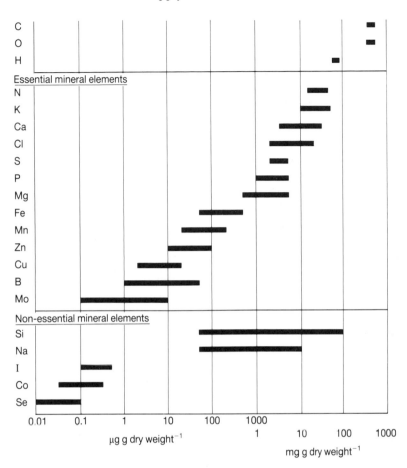

Fig. 3.1 Normal ranges of concentration of elements in healthy plant tissue. Note the logarithmic scale. The position of a bar on the horizontal scale indicates the concentration within which the element occurs in plants. Thus there is more nitrogen than magnesium, and more iron than boron, in healthy plants. The length of a bar indicates the range of concentration which is normal for an element in plants in general. Thus there is great variation between plants in their molybdenum content, but much less variation in their sulphur content.

indication of the proportions in which different minerals are required for healthy growth. Certain minerals may be accumulated greatly in excess of the plant's requirement. For example, where high levels of fertilizers containing potassium are applied, crops often take up much more potassium than they can utilize, a phenomenon known as luxury uptake. It is not unusual for silicon, a non-essential element, to be the most abundant mineral element in plant tissue.

There are enormous differences between min-erals in the amount required for healthy plant growth. The nitrogen requirement is about a million times greater than the molybdenum requirement. The other elements fall between these two extremes. The six elements required in greatest quantity – nitrogen, phosphorus, sulphur, potassium, calcium and magnesium – are often termed the *macronutrients*, and the others are known as the *micronutrients*. The term 'trace elements' is often used to refer to the micronutrients, but more properly refers to all minerals present in small amounts in plants and soils.

3.2 Supply of minerals in the soil

Undoubtedly, the main source of mineral nutrition for plants is the soil, but at least one other source of supply is significant, and that is the atmosphere. We will consider this source briefly in Section 3.3, but in the meantime we will examine the supply of minerals from the soil in some detail because of its major importance.

3.2.1 Sources of minerals in the soil

With the single exception of nitrogen, the ultimate source of all soil minerals is the parent rock from which the soil is itself derived. The continuing decay of rock particles by physical and chemical weathering releases a steady trickle of minerals to the soil.

However, there is hardly any nitrogen in rocks, and virtually all the nitrogen in the soil derives ultimately from that present in the air. Although nitrogen gas is present in the air spaces between soil particles, it cannot be used directly by plants. Certain microbes, mainly bacteria, convert this nitrogen gas to nitrate ions, which can then be used by plants as a nitrogen source. Some of these microbes live freely in the soil, while others form a close relationship with certain plants. A considerable range of plant species have such a relationship with a microbe which fixes nitrogen, but undoubtedly the most important example is the association between plants of the family Fabaceae (the legume family) and bacteria of the genus *Rhizobium*. Peas, beans and clovers, for example, form swellings called *nodules* on their roots, and within these nodules live nitrogen-fixing *Rhizobium* bacteria. These plants obtain their nitrogen supply from that fixed by the bacteria in the nodules and can be entirely independent of nitrates in the soil. In return the bacteria obtain carbohydrates from the plant. Thus the relationship is of mutual benefit to both partners, and is known as a *symbiotic* association. A further example is the growth of nitrogen-fixing bacteria present on the leaves of some tropical grasses.

Minerals incorporated in plants, animals and soil organisms are eventually returned to the soil through death and subsequent decay. Excretion by animals is also part of this cycle. Thus decaying organic matter in soil is an important secondary source of minerals for plants and the agricultural practice of adding materials such as farmyard manure and slurry to soils simply augments this supply. Because of the lack of nitrogen in rocks, decaying organic matter is particularly important as a source of nitrogen.

Artificial fertilizers form an important additional source of minerals for agricultural crops and are considered in Section 3.2.3. Nitrogen fertilizers, usually ammonium or nitrate salts, are produced chemically from atmospheric nitrogen using the energy of fossil fuels. Other fertilizers are derived from geological deposits rich in the sought-after mineral – potassium and phosphorus are by far the most important of these.

3.2.2. Availability of soil minerals to plants

The level of mineral nutrients naturally present in soils depends greatly on the mineral richness of its parent rock, and on its history during its evolution as a soil. With very few exceptions, soils contain insufficient amounts, in forms available for uptake by plants, of the macronutrients nitrogen, phosphorus and potassium to sustain maximum growth rate, hence the practice of applying fertilizers containing these nutrients to crops.

The position regarding micronutrients is quite different. Most rock types contain in themselves and release enough trace elements for plants' needs, but much depends on the extent to which these stores are depleted over the ages by weathering and by downward washing of nutrients out of the rooting zone, a process known as *leaching*. Thus, old soils tend to have lower reserves than relatively young ones, and heavily leached ones have lower reserves than those evolving in dry climates. Where soils derive from relatively mineral-poor rocks such as granites, these effects are exacerbated. To demonstrate this, in temperate areas, most soils are young, often being derived from recent glaciations, and their micronutrient contents are high enough to supply the needs of many successive crops. As a result these nutrients are not so frequently required in fertilizers. In tropical areas, most soils are relatively ancient and have weathered over a long time, under regimes of high rainfall and consequent leaching. In these situations, micronutrient deficiency is a much more prevalent problem in crops, often aggravated by local conditions of drainage and soil pH (see Section 3.5.2).

Sulphur and magnesium occupy an intermediate position between nitrogen, phosphorus and potassium on the one hand and micronutrients on the other hand, as regards their supply in soils, and the requirement for application of fertilizers containing them.

Whatever the mineral content of a soil, the great bulk of it is not immediately accessible to plants. The soil contains a great volume of water, and a small fraction of the soil mineral content is dissolved in this to give a very dilute solution called the *soil solution*. Almost all the mineral ions in this solution can be taken up by plants and such minerals are said to be *available*.

Minerals in the soil solution are the only ones to

which plants have direct access, but there are various means by which this supply is sustained. The weathering of rock particles and the decay of organic matter steadily release the minerals they contain. The flow of mineral nutrients from these two sources is strictly one-way; that is to say they are not reversible processes. The bulk of minerals from these sources do not immediately dissolve but change into forms still unavailable to plants. Some are precipitated as insoluble salts which nonetheless continuously exchange ions with the soil solution. If the solution becomes depleted in a certain ion there is an overall release of that ion from the insoluble to the soluble form. Similarly, if the concentration of an ion dissolved in the soil solution is increased, for example by the addition of fertilizer, there is a trend for more to be precipitated as an insoluble salt. Thus a chemical equilibrium is maintained between dissolved ions and insoluble salts and such reversible reactions are important in regulating the balance of available and unavailable minerals in soil.

Other chemical equilibria contribute to the supply of available cations in soil. A proportion of the soil cations become loosely bound within the structure of clay particles and, from there, these too interchange ions with the soil solution, as described for insoluble salts.

Notwithstanding, a depleted soil solution is only slowly replenished from the above two sources owing to their slow rate of release of ions. However, a further proportion of cations become loosely attached to the surface of soil particles or decayed organic matter (*humus*). Such *adsorbed* cations are readily released to the soil solution and thus establish a sensitive equilibrium by which a depleted soil solution is rapidly replenished. The sites vacated by the release of adsorbed cations are then occupied by other cations, including hydrogen, from the soil solution. The process is one of exchange between the soil solution and the adsorption sites and for this reason the cations involved are known as *exchangeable cations*.

Exchangeable cations are so readily released to the soil solution from this source that to all intents and purposes they are part of the available fraction of minerals in the soil. They form the main reservoir of cations on which a crop draws during its growth, and the capacity of soils to store exchangeable cations is an important factor in fertility.

The flows of mineral ions between the various reservoirs in the soil are illustrated in Fig. 3.2.

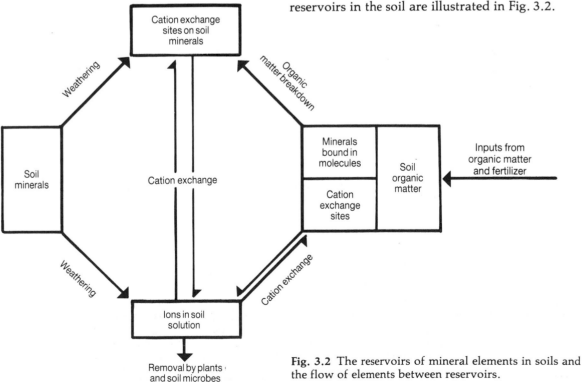

Fig. 3.2 The reservoirs of mineral elements in soils and the flow of elements between reservoirs.

One of the main effects of adding organic manures to a soil is to increase its humus content, and hence its capacity to store exchangeable cations. This is particularly important in sandy soils where there are few clay particles to adsorb cations. In acid soils, a large proportion of the exchange sites are occupied by hydrogen ions which are of no nutritional value to plants. Application of lime removes them and frees the sites for occupation by useful cations, thereby increasing soil fertility. Other effects of liming on the mineral nutrition of plants are mentioned in Section 3.5.2.

The capacity of soils to adsorb and store exchangeable anions is so limited as to be insignificant, and this poses special problems of crop nutrition, in fact of plant nutrition in general. Nitrate supply is the chief example of these problems, which are considered in Section 3.2.3.

Thus in soils the great bulk of the mineral content is present in forms unavailable to plants and only the very small proportion dissolved in the soil solution is immediately available to them. This proportion would be inadequate for the growth of a crop were it not continuously replenished from other sources. The various forms in which minerals occur in soils, their availability for plant uptake, and the forms in which they are supplied as fertilizers are summarized in Table 3.1.

Table 3.1 Minerals in soils and fertilizers

	Forms in soil	Available	Fertilizers
Nitrogen	amino group (NH_2) bound in organic molecules; ions in soil solution: ammonium (NH_4^+) (scarce), nitrite (NO_2^-) (scarce), nitrate (NO_3^-) (abundant)	ions in soil solution	organic manures slowly releasing NH_4^+; simple or compound fertilizers in the form of NH_4^+ or NO_3^- salts; occasionally urea ($CO(NH_2)_2$) for foliar application
Phosphorus	phosphates in order of increasing solubility: PO_4^{3-}, HPO_4^{2-}, $H_2PO_4^-$; very small proportion in soil solution	ions in soil solution	basic slag, ground rock phosphate for long-term slow release; superphosphate ($Ca(H_2PO_4)_2$); quicker-acting $NH_4H_2PO_4$ and $(NH_4)_2HPO_4$
Sulphur	bound in organic molecules; insoluble and soluble sulphates (SO_4^{2-})	ions in soil solution	SO_4^{2-} salts when necessary
Potassium	bound in clay particles; K^+ ions adsorbed on clay particles (exchangeable K^+) and in soil solution	exchangeable ions; ions in soil solution	K^+ salts (mainly KCl)
Calcium	bound in clay particles; exchangeable Ca^{2+} ions adsorbed on clay particles (abundant); Ca^{2+} ions in soil solution (less abundant)	exchangeable ions, ions in soil solution	ground limestone, chalk or shell sand rich in $CaCO_3$ for slow release; pure lime ($CaCO_3$); quicker-acting CaO or $Ca(OH)_2$
Magnesium	bound in clay particles; exchangeable and soluble Mg^{2+} ions	exchangeable ions, ions in soil solution	dolomitic limestone rich in Mg^{2+}; common impurity in K fertilizers and lime
Iron	soluble and insoluble Fe^{2+}, Fe^{3+} salts; organic complexes (chelates)	Fe^{2+} ions; certain chelates	occasionally $FeSO_4$ or chelated Fe
Other micronutrients	bound, exchangeable, soluble cations (Cu^{2+}, Zn^{2+}, Mn^{2+}, Mn^{3+}); bound and soluble anions (molybdate (MoO_4^{2-}), borate (BO_2^-), chloride (Cl^-))	exchangeable and soluble cations (except Mn^{3+}); soluble anions	soil application or sprays to correct deficiencies, e.g. $CuSO_4$, $ZnSO_4$, $MnSO_4$, Na_2MoO_4, borax ($Na_2B_4O_2 . 10H_2O$); Cl never deficient in soil

3.2.3 Supply of mineral nutrients by fertilizers

We have touched on the topic of fertilizers where appropriate throughout this chapter, but must give it a little more attention at this point. Application of fertilizers, usually containing the macro-nutrients nitrogen, potassium and phosphorus, aims to prime soil supplies of minerals to levels that will support maximum growth rates.

In some fertilizers, the minerals applied are in forms not immediately available to plants as, for example, the phosphorus in ground rock phosphate or basic slag. But the bulk of minerals applied in intensive agriculture is in forms readily available for uptake, as with the nitrogen in ammonium nitrate or the potassium in potassium chloride. Organic manures, the chief of which is animal manure, contain a mixture of available nutrients and those still bound in the structure of the organic matter until released by its decay.

Inherent in the practice of fertilizer application are the questions of when to apply and how much to apply, since too little results in reduced yield, and too much is wasteful and may damage the crop. The degree of success in answering these questions depends very much on the mineral in question, and its behaviour in the soil.

Potassium has presented the least difficulty in answering these two questions. It is usually applied in one of two available forms, potassium chloride or potassium sulphate, and is taken up as potassium ions from the soil solution, which tends to be depleted by this and by leaching. However, considerable potassium is stored in the cation exchange system of the soil, and when it is applied in any of the immediately available forms, much is quickly absorbed into this system. Good assessments of the available potassium supply from these two sources can be made.

Thus, given that reasonable estimates of the potassium (and phosphorus and nitrogen) requirements of most crop species are available, a reasonable estimate of any crop's fertilizer requirements for potassium in any soil can also be attempted. Since the potassium is readily released to the soil solution from cation exchange sources, the crop's supplies in the soil solution are replenished on demand, and no problem of timing occurs.

Phosphorus, which is taken up as phosphate from the soil solution, presents problems because of the strong tendency of available phosphate to be rapidly precipitated as unavailable forms in the soil. This process is known as phosphate fixation and soils differ considerably in their tendency to fix phosphate. Phosphorus is usually applied in a readily available form as salts derived directly or indirectly from phosphoric acid. Other forms that become available only over a period of years are derived by grinding up phosphate-rich rocks, or are a by-product of the steel industry, called basic slag. Even when sizeable amounts of readily available phosphate are applied as superphosphate or other fertilizer, the bulk of the application generally becomes unavailable before the growing crop can utilize it. Some phosphate becomes available through the breakdown of organic matter in the soil, but is strongly competed for by soil animals and microbes. Further, being an anion, phosphate is not stored in the cation exchange system, which can thus have no buffering effect between supply and demand as it does for potassium.

The concentration of phosphate in the soil solution is thus very low; this reduces losses due to leaching to a minimum but creates special problems of phosphate nutrition for the plant. In practice, each annual application of phosphate fertilizer is largely precipitated into unavailable forms in the soil. The crop's intake for that year includes only a very small proportion of the current year's application, and is largely derived from the cumulative total of equally small proportions from all previous years' applications, plus some of the natural reserves of the soil, all being released into available forms to a very limited extent.

Estimates of a soil's total phosphate content give a very general guide to the crop's fertilizer requirement, but it is not yet clear whether with successive applications we are priming a deep well, or adding unendingly to a bottomless pit. Since on average 80% of any application rapidly becomes unavailable, both timing and placement of phosphate fertilizer are important to give the crop maximum benefit before this happens.

Nitrogen is the mineral applied in the greatest quantity as artificial fertilizer, almost always as readily available salts containing nitrate or ammonium ions. Indeed, seedlings may show retarded germination, injury to young roots, and leaf scorch resulting from the elevated concentration of salts around them, particularly if the fertilizer is combine-drilled with the seed, that is placed below the seed at the time of sowing by machine.

The business of regulating the supply of nitrogen to the plant is beset with difficulties that have never been adequately resolved. The main store of nitrogen in the soil is that bound up in the organic matter. This is considerable in quantity and can be estimated, but the proportion of it which is likely to become available at any one time, or even in any one season, cannot be estimated as it depends heavily on variable factors such as soil moisture and temperature. When released from organic matter or applied as fertilizer, nitrogen appears in the soil mainly as nitrate ions, in which form plants take up most of their nitrogen. Nitrate is readily soluble and is rapidly leached out of soils in drainage water. Certain species of soil bacteria also break down nitrate to release gaseous nitrogen (*denitrification*), and appreciable losses can occur by this route also. Since nitrate is an anion, the soil particles cannot store it in an exchangeable form. There is thus no reserve of this ion which can be released to buffer these destabilizing influences. Consequently, nitrate levels in the soil fluctuate widely, even from day to day. Some ammonium ions occur, and can be taken up by the plant, or stored in the soil in an exchangeable form, but their concentration is seldom enough to influence matters greatly. It follows from all the above that neither by examination of available dissolved nitrate nor of bound nitrogen can even good approximations of a soil's nitrogen status, and thus the nitrogen requirements of a crop, be made. At best, recommendations for the rate of application of nitrogen fertilizers are derived from rough estimates of need based on previous cropping history of the site, known crop requirements and other factors.

This behaviour of nitrogen in the soil also leads to problems in the timing of applications. The rate of decay of soil organic matter, and thus of release of nitrogen in a form available to plants, is regulated by soil temperature and moisture, being favoured by warm moist conditions. In temperate regions, in the spring, young crops have a high nitrogen demand to build up their foliage, but low soil temperatures inhibit the decay of soil organic matter and the supply of nitrogen from this source. Later in the summer, with high soil temperatures, organic matter decays more rapidly, and nitrogen from this source is more readily available. The matter of estimating required inputs of nitrogen from fertilizer is thus complicated by fluctuations in the supply of organic matter, as well as short-term fluctuations in nitrate concentration in the soil solution.

For these reasons, chronic problems arise in synchronizing the supply of available nitrogen with the crop's demand. Correct timing of applications, so that nitrate is not leached out before the crop can utilize it, is important, hence the practice of splitting the needs of some crops over several applications, with some going on at sowing or planting, and more at later dates as 'top dressings'. It is possible to obtain fertilizers from which the nitrogen is only slowly released, but their rate of supply is too low to meet the needs of a rapidly growing crop. Such fertilizers are nonetheless useful where vegetation is being re-established on difficult, disturbed sites and a steady supply of nitrogen is required over several seasons.

Under continuous vegetation cover, as in natural vegetation, or under perennial crops such as grass, nitrogen is taken up by plants as it is released from organic matter, and losses by leaching are minimal. Where harvests remove the vegetation cover, then the decay of organic matter, often aided by subsequent cultivations such as ploughing, releases much nitrogen that may be quickly lost by leaching or denitrification. Under warm soil conditions and heavy rainfall, both of which encourage rapid decay of organic matter and consequent release of nitrogen, this can be a particularly acute problem, especially in many tropical and subtropical areas where both of these conditions occur.

In recent years, it has become apparent that in certain soils at least, especially with crops such as brassicas that have a relatively heavy sulphur demand, sulphur too may have to be treated as a major nutrient and applied as fertilizer. Previously, this demand may have been hidden by incidental inputs that have now been reduced. Chiefly, these sulphur inputs were in atmospheric pollution (see Section 3.3) and as constituents or impurities in fertilizers such as potassium sulphate and ammonium phosphate.

3.2.4 How soil minerals reach plant roots

The rooting system of a plant, although profusely branching, is in close contact with only a small proportion of the total soil volume. It is necessary, therefore, to consider how mineral ions reach the root surface from the surrounding soil even over very short distances such as a fraction of a millimetre.

Most of the ions available for absorption by roots are in the soil solution. It follows that the methods by which the root is supplied with water must also supply minerals. As we have seen in Chapter 2, water is supplied to the root surface by three main processes: root growth, enabling the plant to explore ever greater volumes of soil for water extraction; mass flow through the capillary spaces of the soil in response to the pull of transpiration; and gravitational percolation of water following rain. In the case of minerals these processes are augmented by one other process – *diffusion*. Ions in solution are in a state of constant motion, drifting randomly about. As a root depletes the soil solution around it, ions tend to diffuse in this manner towards the root from further away.

But are these processes, together with the slow release of ions from bound and adsorbed forms, sufficient to provide the root with an adequate mineral supply? This depends on how great is the demand of the plant for a mineral, in relation to the rate at which these methods supply it. If the demand for a mineral exceeds the supply in the soil solution reaching the root, then the soil solution becomes depleted in that mineral in the immediate vicinity of the root.

The relative importance of these different processes – root growth, mass flow in response to transpiration pull, gravitational percolation, diffusion and release from bound and adsorbed forms – depends on soil, weather and plant characteristics, but it is especially dependent on the mineral in question. For example, mass flow probably supplies enough calcium and magnesium for the needs of most crops in a fertile soil, but this is not true of phosphorus and potassium. Although potassium is fairly plentiful in the soil solution, the plant has a high demand for this element and mass flow cannot supply enough. Diffusion is thought to be ten times more important than mass flow in bringing potassium to the root surface.

The scarcity of phosphate ions in the soil solution in relation to the requirements of the plant has already been mentioned, and phosphorus nutrition poses special problems for the plant. Mass flow can supply only 1–2% of the phosphate needed by plants, and diffusion is therefore a more important process. Because of the tendency of phosphate to be precipitated as insoluble salts, however, these ions probably do not diffuse far. Root growth, enabling the plant continually to reach new supplies of the mineral, is thought to be particularly important in phosphorus nutrition. Plants with root hairs can exploit a very large proportion of the soil volume; thus root hairs are considered to be of especial value in the supply to the plant of phosphorus, and also of potassium, since by making intimate contact with clay particles they can tap the exchangeable potassium content of the soil very effectively.

The roots of many plants form symbiotic associations with fungi called *mycorrhizas*. It is not fully understood how mycorrhizas function, but if a normally mycorrhizal plant is grown in sterile soil so that the association cannot be formed, its mineral uptake, and in particular its phosphate uptake, is seriously impaired. It appears that the fungus draws minerals from a large zone of soil around the plant root, or perhaps converts them to more available forms, and releases them to the root in return for a supply of carbohydrates from the plant. This is discussed further in Section 6.1.6.

3.3 Supply of minerals from the atmosphere

The supply of minerals reaching plants from the atmosphere, excluding nitrogen fixed in leguminous root nodules, is often overlooked, but although the total weight thus arriving is small, especially relative to the amounts of nitrogen, phosphorus and potassium supplied in fertilizer, it can be significant. Minerals can arrive from the atmosphere in three different forms: as solid dust particles, either dry or trapped in rain drops or snow flakes; as ions dissolved in rain; or as gases.

Forests are efficient dust traps, and this source of supply of minerals plays an important part in their nutrition. In coastal areas, certain minerals, especially sodium, derived from wind-borne sea spray, reach plants in significant amounts. Ironically, atmospheric pollution by industrial emissions brings appreciable amounts of sulphur and nitrogen dissolved in rain to crops. The sulphur often arrives in sufficient quantities to make an appreciable contribution to the nutrition of crops such as brassicas with high sulphur demands. There is some evidence that where plants give off some minerals such as sulphur in a volatile form, other plants absorb some of this in the gaseous phase through their leaves. The full extent of this route of arrival is uninvestigated, but it is unlikely to be very great. Certain specialized plants that live on the branches of trees, and are called epiphytes, must draw a large part of their mineral supply from atmospheric sources.

3.4 Mineral uptake, circulation and loss

3.4.1 Uptake of minerals by plant roots

If we compare the mineral content of a plant with the available mineral content of the soil in which it is growing, two interesting observations can be made. Firstly, there is a far higher concentration of minerals in the plant than in the soil; indeed the concentration of salts in the xylem of a plant root is often around forty times greater than that in the soil solution. Plants can therefore *accumulate* and hold minerals against a massive concentration gradient.

Secondly, some minerals which are plentiful in the soil are scarce or absent from plants, while others which are relatively scarce in the soil are abundant in plants. Roots can therefore show some degree of *selectivity* in the ions they absorb from the soil. Potassium, for instance, is almost always far more plentiful in the xylem than in the soil solution, while the reverse is true of sodium.

The region of root just behind the growing tip is where most of the water enters the root (Section 2.5.2), and this is true also of minerals. Similarly, the two pathways (non-living apoplast and living symplast) by which water crosses the root are also pathways for mineral uptake. Although ions in the soil solution can penetrate freely in the apoplast as far as the endodermis, to get beyond this and reach the xylem they must at some point en route cross into the symplast. This means crossing a differentially permeable cell membrane.

The soil solution percolates into the root, totally permeating the outer apoplast and bathing all the membrane surfaces it encounters in a supply of mineral ions. The total surface area available for absorption into the cells of the cortex is therefore enormous in proportion to the volume of the root (Fig. 3.3). Water can move passively across these membranes in response to solute or pressure potential but mineral ions cannot. By a process which is

Fig. 3.3 Cross-section of a young dicotyledon root. The cortex consists of a large number of cells, all in direct contact via the apoplast with the soil solution outside the epidermis. This gives an enormous surface area for mineral absorption.

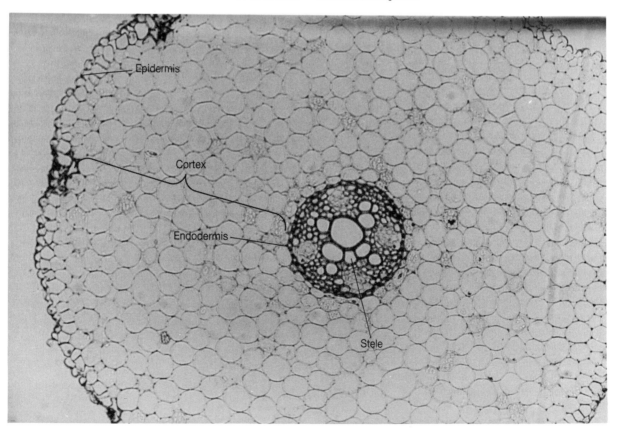

little understood, but which requires energy from respiration in the root, the membranes actively absorb ions, both selecting particular minerals and accumulating them to concentrations many times greater than those in the soil solution. Once within the symplast, they are carried along with the water flow across the endodermis and into the stele. Here they are actively secreted, by a process again requiring metabolic energy, either into the inner apoplast or directly into the xylem vessels. Since ions are prevented from leaking back into the cortex by the waterproof Casparian strip of the endodermis, they can become concentrated in the stele.

Thus it is the membranes of the living root cells, supported by respiratory energy, that are responsible for the selection and accumulation of ions in the plant, but it is the Casparian strip of the endodermis, acting as a cylindrical ion barrier in the apoplast, that allows them to exercise those powers.

In passing, it may be noted that all living plant cells, by virtue of their differentially permeable membranes, have the ability to absorb and secrete ions. Thus leaf cells can secrete salts and can also absorb them, as when micronutrients are sprayed on to crop foliage to correct a deficiency.

The uptake of minerals is, of course, closely interrelated with the uptake of water by plant roots. When the cells of the cortex absorb ions, some are passed to the vacuoles. Here, as we saw in Section 2.2.2, the ions give rise to a negative solute potential and water is taken in by osmosis. Water may pass osmotically from vacuole to vacuole across the root under the influence of differing mineral contents. Similarly, the accumulation of ions in the xylem sap creates a large negative solute potential which draws water into the xylem. This, as we showed in Section 2.3.1, is the cause of root pressure which may, under conditions of low transpiration, push water up the xylem and cause exudation of water, or guttation, from the leaves.

Thus, from soil to xylem, water and dissolved mineral ions follow the same path, continuously interacting but moving under the influence of quite different forces. Ions are 'pumped' in by the expenditure of metabolic energy; water enters passively by osmosis and by the pull of the transpiration stream.

3.4.2 Factors affecting mineral uptake

Assuming an adequate supply of minerals in the soil, a number of important factors can influence the ability of the plant to absorb ions. Much hinges on the fact that the passage of ions across cell membranes requires energy from respiration. Any reduction in the rate of respiration in the root will markedly reduce mineral uptake. The low soil temperatures, for example, which are experienced by young crops in the spring do not allow rapid respiration and therefore restrict mineral uptake. This, together with a restriction on water uptake, is probably an important factor limiting the growth rate of crops at this time of year. As oxygen is necessary for respiration, poor aeration of soil or roots will also restrict the uptake of minerals. Clay soils, which have a very small pore size, and compacted or waterlogged soils tend to be poorly aerated. Cultivation to break up compacted soil and good drainage are thus important for mineral as well as water uptake.

Unless there is a previously stored reserve of carbohydrate in the root, a flow of carbohydrate from the leaves will be necessary to maintain root respiration. Thus climatic factors which slow down the rate of photosynthesis and reduce the flow of carbohydrate to the root will often thereby reduce mineral uptake.

An adequate soil water supply is also necessary for mineral uptake. As we saw in Section 2.5.1, the rate of transpiration is increased when soil moisture is plentiful. This causes a greater mass flow of ions to the root surface and also apparently makes the membranes of root cells more permeable to mineral ions. One of the benefits of irrigation is thus to improve the supply of minerals, in addition to the supply of water. Climatic factors such as high relative humidity or low air temperature which reduce the transpiration rate can be expected to reduce mineral uptake.

Examples such as this demonstrate how important it is to think of the plant as a single functioning unit, not as an assemblage of leaves, stems and roots each performing their separate functions in isolation. What affects the shoot affects the root.

Another important factor affecting mineral uptake is the degree of development of the root system. This depends on the species and age of the plant. Root growth takes place by extension at the root tip, and behind the tip as the root develops there is progressive suberization, first of the endodermis and later of a sheath of cork tissue produced in the cortex of most plants. The capacity of the root for mineral uptake reduces as suberization progresses, but even heavily suberized roots remain to some extent permeable to water and ions.

Up to 95 % of the roots of a tree are in this condition but they must absorb a large proportion of its mineral requirement. Unsuberized areas of the cork (lenticels) and ruptures caused by the development of lateral roots are probably where most minerals enter older roots.

In annual plants, such as most agricultural crops, most of the mineral absorption takes place 20–50 mm behind the root tip where the only deposit of suberin is in the Casparian strip of the endodermis. Since this is a passing phase in root development, continuing growth is necessary to maintain the surface area of the root in this phase. Branching, to increase the surface area of a young root, is necessary to cope with the increasing demands of the growing shoot. If the root growth of a crop is restricted, as by soil compaction (see Section 6.1.8), mineral uptake is reduced and the shoot suffers – what affects the root affects the shoot.

Plant species differ in their ability to select and accumulate ions, even if their roots show a similar degree of development. The ability of certain plants to accumulate high concentrations of copper, nickel and even gold has been used in prospecting for deposits of these minerals in the underlying rocks. Certain range plants in North America are toxic to grazing cattle because they accumulate high levels of selenium. Other plants, given heavy applications of nitrogen fertilizer, accumulate levels of nitrate which are harmful to cattle. Differences between species in their mineral content are an important aspect of herbage quality, and as such are discussed further in Section 12.1.9.

3.4.3 Circulation of minerals in plants

Minerals taken up into the xylem remain for the most part in the form of simple ions. An important exception is nitrogen. Absorbed as nitrate and to a lesser extent as ammonium ions from the soil solution, nitrogen is incorporated as the amino (NH_2) group into small organic molecules, which account for most of the nitrogen in the xylem. This incorporation takes place in the root symplast. Nitrogen in the ammonium form can be incorporated directly, but nitrate ions must be first reduced, a process requiring energy from respiration. It is for this reason that plants respond more quickly to ammonium than to nitrate fertilizers. A small proportion of the phosphorus and sulphur taken up by plants is also incorporated into organic compounds before entering the xylem.

Once in the xylem, minerals ascend rapidly in the transpiration stream. In a typical annual crop plant, transpiring rapidly, ions move from base to top in about an hour. Rates of ascent of 60 m h^{-1} have been measured in rapidly transpiring trees. Root pressure may aid the ascent of ions in the xylem when transpiration is very slow.

During the ascent of minerals in the plant, they are also dispersed throughout it. The xylem is rather like a collection of porous pipes from which minerals may leak into the apoplast of surrounding tissues at any point along its length. From the apoplast, or from the xylem itself, minerals may be absorbed into the symplast of adjacent living tissues. Anywhere in the root, stem or leaf, minerals may cross over in this way from the dead xylem to the living phloem, for these two transporting systems are always close together. Having entered the phloem, minerals are swept along in the flow and are distributed to all actively growing parts of the plant. Some will be returned to the root, and some will travel to the shoot apex, which is not reached by the transpiration stream in the xylem. Minerals which travel all the way to the leaf in the xylem may remain in the leaf apoplast or may be absorbed by the symplast of the leaf cells. Here they may fulfil their vital functions or be stored in the vacuoles; some may be re-exported from the leaf in the phloem. Thus not only has the plant a means of carrying absorbed minerals from the root to all parts of the shoot, it has a transport system which allows continuous redistribution of minerals between its different parts.

The organs of a plant differ in their demand for minerals and the demand of any one organ varies during its life. There is a net transfer of minerals from organs of low demand to organs of high demand. The source-to-sink movement in the phloem which we observed in the translocation of carbohydrates (Section 1.6.8), is thus also characteristic of mineral transport. For example, young expanding leaves have a high demand for, and a considerable capacity to accumulate, minerals. Mature and senescent leaves gradually lose their ability to retain minerals and they export minerals via the phloem to the younger, more demanding leaves. This redistribution is important in the plant's mineral economy.

Most minerals do not remain in tissues where they no longer have an important function, but are moved to tissues where they can be used over again. This reduces the overall requirement of the plant for absorption of minerals from the soil. As we shall see, the transfer of minerals from older to younger leaves affects the expression of mineral deficiency symptoms in a plant.

Another example of source-to-sink movement of minerals is their accumulation in storage organs at the expense of stems and leaves. This accounts for the relatively high mineral content of such organs as the grains of cereals and the roots of carrots.

Plants cannot, however, redistribute all types of mineral with equal ease; there are in fact great differences between minerals in their mobility within the plant. Calcium, for example, is rapidly precipitated in the phloem, and its distribution throughout the plant takes place almost entirely in the xylem. It cannot therefore be exported from older leaves, and it cannot be accumulated, like other minerals, in storage organs. Sulphur is also an immobile element, but for a different reason. Once in the leaf, most sulphur is irreversibly incorporated into molecules which cannot be translocated. By contrast, nitrogen, phosphorus and potassium are highly mobile in the plant, and undergo continuous redistribution.

The supply of minerals to the plant is irregular due to the vagaries of weather and soil conditions, and the plant's demands vary with growing conditions and its stage of development, but these fluctuations in supply and demand do not necessarily coincide. In times of abundant supply, minerals accumulate in the vacuoles of cells of root, stem and leaf, and during periods of restricted supply these minerals are released into circulation. Accumulation and release are to a certain extent controlled by the plant, allowing it to regulate mineral flow to even out the supply to organs requiring nutrition.

Systems for redistribution of minerals within the plant thus allow it to make the most efficient use of a limited and fluctuating mineral supply. It is, perhaps, more than a coincidence that nitrogen, phosphorus and potassium, the commonest limiting factors to plant growth, should be among the most mobile elements in plants.

3.4.4   Loss of minerals from plants

The accumulation of minerals by plants is to some extent offset by various losses. These losses, however, need not be permanent, for the minerals sooner or later return to the soil where they may again become available for uptake. There are three principal means by which plants lose minerals.

Firstly, when dead leaves, roots and other parts are shed, the minerals they contain are eventually returned to the soil by decomposition. The shedding of leaves is a particularly important cause of loss of the less mobile elements such as calcium, which the plant cannot withdraw from aging tissues.

Secondly, the leaf has no mineral barrier like the endodermis of the root, dividing its apoplast into inner and outer regions. The entire leaf apoplast is therefore continuous with the xylem and contains similar concentrations of minerals. The leaf cells, by selection, accumulation and release, modify the mineral content of the apoplast. When the leaf surface is wetted by rain, minerals can be brought on to the surface. In grasses and some other plants, under conditions of low transpiration, root pressure may force water, with its dissolved minerals, out of the leaf apoplast on to the leaf surface, a process known as guttation. In either case, rain can then wash these minerals off the leaf and return them to the soil or possibly to lower leaves which may absorb them. This process is of particular importance for potassium.

Finally, roots may exude minerals into the soil, but it is not clear how significant this process is in the loss of minerals from plants.

3.5   Deficiency, sufficiency and excess of minerals

The supply of a mineral element to a plant may be said to be *sufficient* when the growth or development of the plant is not affected by increasing the supply, that is the plant has as much as it needs. When the plant has less than it needs, so that an increase in the supply of the element improves growth or development, it is said to be *deficient* in that element. Sometimes, at the other extreme, a plant has too much of an element supplied to it, and growth or development is impaired. The supply of the element may then be said to be *excessive*.

A plant's appetite for a mineral varies with growing conditions and its stage of growth. A well-watered and therefore vigorously growing crop has a much greater uptake, especially of the major nutrients, than one in which growth is restricted for lack of water. Also, the plant's requirements for different minerals changes as it develops. Cereals have a particularly high potassium requirement at the stem elongation stage, but during grain development their uptake of it is much reduced. Instead, an additional supply of nitrogen and phosphorus is needed. In contrast, crops with fleshy fruits or storage organs, such as grapes, potatoes or sugar beet, need a high nitrogen supply during their early phase when they are building up leaf tissue, but require a high potassium supply in their later stages when their fruits, tubers or roots are developing.

Generally, a deficiency in mineral nutrition at one stage leads to a loss of potential yield that cannot be retrieved by an increased supply of the nutrient at a later stage. The mineral nutrition of crops, therefore, has to be adjusted to their stage of growth, and it is partly for this reason that supplies of an easily leached nutrient like nitrogen are commonly applied as several dressings of fertilizer during a season, to maintain the supply of the element at successive stages of the crop's development.

The range of available mineral content of the soil over which the supply to a crop is sufficient for optimum growth and development thus depends on the mineral in question, on the crop species or variety, and on other soil and weather factors which may be limiting growth. Below the range of sufficiency, a deficient supply can have two effects. First, the growth rate and hence the final yield may be reduced but no disorder arises. The plant continues to function normally, though at reduced efficiency. Second, the crop may show abnormal development in the form of *mineral deficiency symptoms*. Either or both of these effects may be produced, depending again on the minerals, the crop and other factors. Similarly, above the range of sufficiency, an excessive supply can cause yield reduction, the appearance of *mineral toxicity symptoms*, or both.

Deficiencies of nitrogen, phosphorus and potassium more commonly show themselves as quantitative yield reductions than as deficiency symptoms. Disruption of normal development, which gives rise to symptoms, is normally found only in cases of extreme deficiency of these elements. By contrast, with other minerals, quantitative yield may remain constant even when deficiency or toxicity symptoms appear. Often, for these minerals, there is a wide range of supply over which yield is constant. In the lower part of this range there will be a level of supply at which the crop shows deficiency symptoms but no quantitative yield loss, and at the other extreme a level at which toxicity symptoms appear but again no yield loss occurs. Between these two will be a band within which the crop shows normal development.

The band within which there is no disruption of development due either to deficiency or excess of a mineral, may be termed the *tolerance range* of the crop for that mineral. Within this range there may or may not be a yield response to increasing supply of the mineral.

The width of the tolerance range depends mainly on the mineral in question. For instance, plants characteristically have a wide tolerance range for manganese but a much narrower one for boron. However, for any one mineral, there are important differences between crops. Compare the tolerance ranges of cereals and swedes for boron (Fig. 3.4). Not only is the tolerance range of cereals more than ten times wider than that of swedes, but these ranges are in different positions on the scale. Cereals tolerate low levels of boron that would cause deficiency symptoms in swedes, while swedes tolerate relatively high levels of boron that would result in toxicity symptoms in cereals. Thus boronated fertilizers intended for swedes and turnips can cause serious damage if applied to cereals, despite the wider tolerance range of the latter.

Differences in sensitivity to mineral supply may extend to varieties of the same crop species. In specialized habitats, where deficient or toxic levels of particular minerals occur, species may evolve local strains tolerant of these conditions. For example, certain strains of some pasture grasses colonize mine spoil heaps containing high levels of

Fig. 3.4 (Top) Boron tolerance ranges of cereals and swedes. (Bottom) Zinc tolerance ranges of two varieties of soya bean. Note logarithmic scales. From E B Earley (1943) *J. Am. Soc. Agron.* **35**, 1012–3.

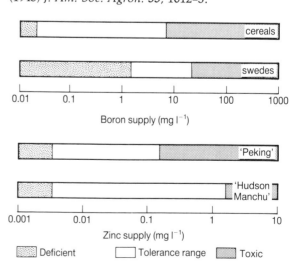

metals such as lead and copper that would be toxic to most strains of these species.

3.5.1 Mineral deficiency symptoms

The growth reduction effect of mineral deficiency is seen clearly in the enormous improvements in yield that are possible by applying fertilizers to crops. In the remainder of this chapter, however, we shall focus on mineral deficiency symptoms rather than growth reduction effects.

Four main categories of deficiency symptom can be recognized in plants.

1. *Death* is a common result of mineral deficiency when the supply is so limited as to disrupt completely the vital functions of plant life. In extreme cases, whole plants may die, but more frequently the premature death of older leaves or of susceptible tissues such as the shoot apex is observed. Commonly there is localized death of groups of cells in the leaf.

2. Mineral deficiency often causes *stunting* in the form of abnormally small leaves or shortened internodes.

3. Deficiency in certain minerals causes a general or localized yellowing of leaves, a condition called *chlorosis*. This results from impairment of the ability of the plant to synthesize chlorophyll.

4. Some plants react to certain mineral deficiencies by increased production of red or purple pigments (anthocyanins) similar to those seen in autumn leaves.

As we have seen in Section 3.4.3, elements differ in their mobility in the plant. Immobile elements tend to remain in older leaves, while mobile elements are exported from older leaves to the young developing leaves where demand for them is greater. When a plant is deficient in an immobile element such as calcium, sulphur or iron, the young leaves show symptoms more strongly than the older leaves, the demand of which may have been met before soil supplies became depleted. Deficiency in a mobile element such as nitrogen, phosphorus, potassium or magnesium produces symptoms in older leaves first because their mineral supply and content have been depleted to augment the supply to younger leaves.

Table 3.2 lists the symptoms most commonly associated with deficiency in particular macronutrients and in iron. It should be appreciated, however, that symptoms of any mineral deficiency

Table 3.2 Typical symptoms of deficiency in macronutrients and iron in crop plants

Nitrogen	stunting ('little leaf' condition) chlorosis purple pigments produced lower leaves affected first (N mobile)
Phosphorus	stunting no chlorosis sometimes necrotic patches on leaves lower leaves affected first (P mobile)
Sulphur	stunting chlorosis sometimes purple pigments produced upper leaves most seriously affected (S immobile)
Potassium	mottled chlorosis of leaves necrosis at tips and margins of leaves lower leaves affected first (K mobile) sometimes flowering inhibited
Calcium	premature death of shoot and root apices malformation of leaves (tips often hooked back) marginal chlorosis, then necrosis, of leaves upper leaves most seriously affected (Ca immobile)
Magnesium	chlorosis, then necrosis, between veins of leaves sometimes purple pigments produced lower leaves affected first (Mg mobile) flowering inhibited
Iron	chlorosis, without necrosis, between veins of leaves upper leaves most seriously affected (Fe immobile)

can differ greatly in different crop species. This is especially true of deficiencies in micronutrients other than iron. Micronutrient deficiencies tend to give rise to specific disorders which differ enormously from crop to crop. A few examples of such disorders are listed in Table 3.3.

3.5.2 Causes of mineral deficiencies

Mineral deficiencies in plants, far more than mineral toxicities, are widespread and important agricultural problems. They arise from various causes. For example, vigorously growing plants sometimes develop temporary deficiencies simply

Table 3.3 Some micronutrient deficiency disorders of crops

Name of disorder	Crop	Symptoms	Deficient nutrient
Grey speck	oats	Irregular grey-brown streaks or specks on leaves	manganese
Speckled yellows	sugar beet	chlorosis between leaf veins, inward curling of leaves	manganese
Marsh spot	pea	brown area in centre of seed	manganese
Little leaf	apple	small, malformed leaves, shortened internodes	zinc
Wither tip	cereals	chlorosis of leaves, withering of tips of leaves and inflorescences	copper
Raan (brown heart)	swede, turnip	rotting of centre of root	boron
Heart rot	beets	death of centre of 'crown', rotting of centre of root	boron
Hollow stem	cauliflower	rotting of centre of stem	boron
Whiptail	cauliflower	reduction or suppression of leaf blades	molybdenum

due to their inability to absorb ions fast enough. In addition, any impairment of the functioning of roots can lead to mineral deficiency, especially of the macronutrients. Thus plants growing in cold, waterlogged or dry soil often exhibit symptoms, mainly of nitrogen deficiency.

Perhaps the most straightforward cause of mineral deficiency is where there is simply a low content of the mineral in the soil. This can be termed an *absolute deficiency*. Very often, however, mineral deficiencies occur in plants growing in soils with an apparently adequate content of the mineral. Such deficiencies are brought about by particular soil conditions and are termed *induced deficiencies*.

One very important kind of induced deficiency can be explained by the ability of many mineral elements in soils to exist in plant-available and unavailable forms. The unavailable forms may be bound or insoluble, or they may be in the soil solution as ions which plants cannot readily absorb. For example, the trivalent ions of iron (Fe^{3+}) and manganese (Mn^{3+}) are not nearly as easily taken up by plants as the divalent ions (Fe^{2+}, Mn^{2+}). Any soil factor which favours the formation of relatively unavailable forms decreases the fraction available to plants. If too low a concentration is left in an available form, deficiency results.

Prime among such soil factors is the influence of pH, the measure of acidity or alkalinity of a solution. Changes in the pH of the soil solution alter the relative proportions of available and unavailable forms of minerals. Iron, for example, is more available at pH 6 than at pH 8. This is because in alkaline conditions the bulk of the ions are in the trivalent form, but with increasing acidity and therefore declining pH an increasing proportion are converted to the divalent form. The effects of soil

pH on the availability of minerals are summarized in Fig. 3.5. All the macronutrients and one micronutrient, molybdenum, become less available as pH falls, but the other micronutrients become less available as pH rises.

So profound is the influence of pH on mineral nutrition that a crop grows only over a pH range within which the supply of minerals in available forms is adequate for that species. Indeed, the pH ranges of most crops reflect not their tolerance of acidity or alkalinity as such, but the points at which the concentrations of various minerals in available form become inadequate or excessive for normal growth. Thus the symptoms of pH damage to crops are usually those of specific mineral deficiencies or toxicities.

Crops differ in their mineral requirements, in the maximum concentrations they will tolerate in soils and in the efficiency with which they can extract the available minerals from soils. Thus, although most major crops grow best at pH values between 5.5 and 7.5, considerable differences exist between them in their pH ranges (Fig. 3.6). Barley, for instance, requires a minimum pH of about 6 for successful growth, whereas oats, which tolerate lower supplies of macronutrients than barley, can be grown on soils with a pH of around 5. Barley growing at the upper limit of its pH range exhibits signs of manganese deficiency, but at its lower limit exhibits manganese toxicity, because the available manganese content of soil is influenced more strongly by pH than by the total manganese content of the soil.

In very acid soils, phosphorus deficiency and aluminium toxicity may be the limiting factors to plant growth. Aluminium, like manganese, becomes much more available to plants at low pH, and it can accumulate in the roots even though

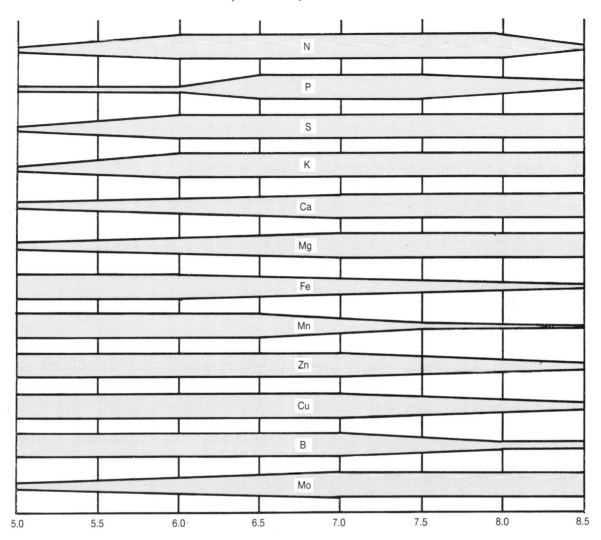

Fig. 3.5 Effect of soil pH on availability of minerals to plants. N = nitrogen; P = phosphates; S = sulphur; K = potassium; Ca = calcium; Mg = magnesium; Fe = iron; Mn = manganese; Zn = zinc; Cu = copper; B = boron; Mo = molybdenum. Redrawn from E. Truog (1959) *Mineral Nutrition of Plants*. University of Wisconsin Press, Madison, WI, with permission of the publisher.

plants have no requirement for it. The acid-loving moor matgrass is unusual in tolerating high levels of available aluminium without damage. Many acid-loving species, including heather, rely on mycorrhizas for adequate phosphorus nutrition. Heather cannot tolerate soils of high pH, probably as a result of manganese or iron deficiency.

The significance of all this for the balanced liming of agricultural soils is obvious. Liming an acid soil raises pH and increases the availability of macronutrients and molybdenum, but excessive liming reduces the availability of all the other micronutrients. With the exception of manganese, however, induced deficiencies of a micronutrient tend to result from liming only where the mineral is already in rather short supply in the soil. For example, lime-induced copper deficiency in cereals is seen mainly on soils low in copper.

Even where the supply of all minerals in available forms is adequate, deficiencies may be induced by other means. The plant cannot exert complete control over the ions that enter it. To some extent, the higher the concentration of an ion in the soil, the more of it will enter the plant. However, it would appear that the total cation uptake that plants can achieve is relatively constant

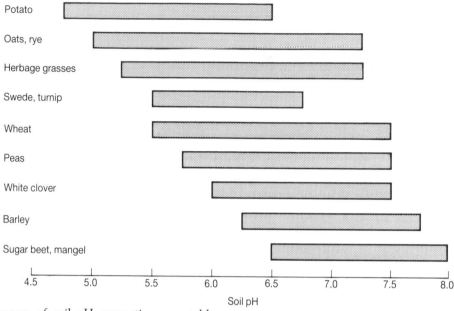

Fig. 3.6 The ranges of soil pH supporting acceptable growth and yield of crops.

for any one plant species. There is therefore competition between cations for entry into the plant. A very high uptake of one cation causes the uptake of one or more of the others to be proportionately low. The uptake of a cation may be so restricted in this way that a mineral deficiency is induced. For example, heavy applications of potassium or ammonium fertilizers can induce calcium deficiency.

Similar competitive effects between ions also occur within the plant. High uptakes of one mineral may exclude another from its appropriate site of action and prevent it carrying out its vital metabolic function in the plant. For instance, high copper or manganese uptake may induce iron deficiency even in plants with a fairly high iron content.

There are, by contrast, situations in which adequate uptake or proper utilization of one mineral can take place only if the plant has a plentiful supply of another. Low levels of phosphorus, potassium or calcium, for example, may induce iron deficiency. Molybdenum deficiency can induce nitrogen deficiency, but only if there is an inadequate supply of nitrogen present as ammonium rather than as nitrate ions. The uptake of nitrate is not inhibited, but molybdenum is necessary for reduction of nitrate to a usable form in the plant, a step unnecessary for ammonium ions.

Soil physical conditions also play a significant part in determining the adequacy of available mineral supplies in soils, not least through their influence on soil chemistry. Low soil temperature, soil compaction, or waterlogging are all examples of factors that can induce or aggravate deficiencies or toxicities. Often, in a cold wet spring, cereal crops will display the undoubted symptoms of magnesium or manganese deficiency, which disappear when warmer weather conditions arrive.

Surveying the various causes of mineral deficiencies in plants, it would appear that for balanced, vigorous growth several conditions must be fulfilled. Not only must certain minerals be present in soils, but they must be present in available forms, in certain minimal concentrations and in certain balanced proportions, and preferably in the correct soil physical conditions. These requirements have obvious implications for the judicious fertilizing and liming of soils.

3.5.3 Diagnosis and correction of mineral deficiencies

A mineral deficiency condition in a crop may be recognized on the appearance of the characteristic symptoms. Often, however, the symptoms are not distinctive enough for the condition to be diagnosed with absolute certainty. It may indeed be too late to try to correct a deficiency by the time symptoms appear. Other means of diagnosis of mineral deficiencies are therefore necessary.

Analysis of plant parts, usually foliage, or of the soil for mineral content can provide useful indications of deficiencies. In general, an abnormally low content of a mineral will indicate that the element is in deficient supply. As explained, however, a crop may be suffering from an induced deficiency even if the leaf or soil content of the mineral in question is fairly high.

Confirmation that a particular disorder is caused by mineral deficiency comes when the crop is sprayed or the soil fertilized with the mineral thought to be deficient. If a treated crop grows normally while an untreated crop alongside shows the disorder it is fairly certain that the diagnosis was right. Corrective measures may then be applied routinely to subsequent crops without waiting for symptoms to appear.

The method adopted to correct a deficiency depends on the cause of the deficiency as well as on the mineral concerned. When a deficiency is induced by a factor under the grower's control, it is probably best corrected by removing or adjusting that factor. For example, copper deficiency in cereals may be relieved by reduced liming. Where it is impossible or impracticable to make such adjustments, or where the deficiency is absolute, the deficient mineral must be applied as a fertilizer.

Absolute deficiencies can often be rectified by soil application of the deficient mineral. Boron, for example, is usually incorporated as borax in compound fertilizers for swedes and turnips. Similarly, conventional fertilizers and lime augment the supply to crops of nitrogen, phosphorus, potassium and calcium.

Induced deficiencies seldom respond satisfactorily to this simple approach, as the applied fertilizer may be converted to unavailable forms under the influence of pH, excluded from the plant by competitive effects, or in some other way remain unavailable. Methods of circumventing the problem have to be sought. Spraying micronutrients on to crops for direct uptake by leaves, a method also used for the correction of absolute deficiencies, may be effective. For example, copper sulphate is often sprayed as a routine measure on cereals growing on low-copper soils.

Alternatively, minerals may be applied to the soil in forms available to the plant but unaffected by soil pH. Many cations form complexes with certain organic molecules. These complexes, called *chelates*, can be taken up by plants even at pH levels where the unattached cation would be converted to an unavailable form. Embedding minerals in small particles of glass, called *frits*, also places them largely beyond the influence of soil conditions. The fritted elements are slowly released in an available form, often partly by the action of plant roots etching the glass.

The timing of corrective action is extremely important. When certain deficiencies are diagnosed and corrected early in the development of a crop, it resumes a normal pattern of growth and develops to maturity. However, even where a very quick-acting corrective is used, such as foliar spraying with urea to correct nitrogen deficiency, the final yield may still be irreversibly reduced.

More serious is the situation where the damage itself is of a type, or reaches a stage, where it is irreversible. Where, for example, the shoot apex is damaged by mineral deficiency, as in acute molybdenum deficiency (whiptail) of cauliflower, quantitative yield loss is usually catastrophic. Even in cases where quantitative yield is not reduced, cash losses due to reduction in quality of the produce may be very serious. Where symptoms are expressed in harvested organs such as seeds, fruits or tubers, the resulting blemishes reduce the market value of the produce out of all proportion to any loss in nutritional value. This is the case with blossom-end rot of tomatoes, which can be caused by calcium deficiency. Particularly insidious are those deficiencies which do not affect the outward appearance of the crop but which produce internal damage. Swedes suffering from boron deficiency (raan) are brown in the centre and become bitter and fibrous. Peas with manganese deficiency (marsh spot) also have brown centres and tend to give rise to seedlings with poor vigour.

In mineral deficiency, as in most things, prevention is better than cure. Timely corrective action needs to be taken before the crop is planted or before the symptoms appear. The prediction of mineral deficiency, whether by soil mineral analysis or on the basis of past experience of the site or soil type, is therefore even more important than its diagnosis.

3.5.4 Mineral toxicity

Toxic effects are not confined to the essential elements. Many non-essential elements such as lead, nickel and aluminium cause toxicity if present in sufficient concentration.

Soils with toxic levels of one or more minerals occur naturally, but increasingly often the situation arises as a result of human activities. Mineral toxicities can be produced by the spreading on the land of certain waste materials such as some forms of sewage sludge, by the accumulation of mine wastes in spoil heaps, or by faulty techniques or systems of agriculture.

Atmospheric pollution is an increasingly important source of toxic elements. Fluorine, for example, can damage crops growing near aluminium smelters, which generate fluorine as a waste product.

Minerals present as soluble salts in the soil can have two types of deleterious effect on plants: a general one due to lowering the solute potential of the solution around the roots leading to osmotic drought (as discussed in Section 2.6.1); and a specific one due to the toxic activity of particular ions.

Specific toxic effects of minerals are of several kinds. Some operate at the root surface, as in aluminium toxicity, which causes damage to root hairs. Others operate within the plant, an example being the inactivation of certain enzyme systems by heavy metals such as lead. Many toxicities manifest themselves as induced deficiencies, similar to the copper-induced iron deficiency mentioned earlier in this chapter. Aluminium, for instance, causes the precipitation of phosphorus as insoluble phosphates in the plant. Another example is iron deficiency induced by nickel. It is impossible to separate entirely the twin topics of mineral deficiency and mineral toxicity.

As we have seen, plants can tolerate very high concentrations of available macronutrients, provided they are not so abundant as to cause osmotic drought. Excessively heavy fertilizer applications may, however, produce injurious effects on crops. For example, a very heavy phosphate application may reduce the uptake of some micronutrients, including zinc, copper and boron. Excessive nitrogen application produces sappy growth, giving plants with weak stems and an increased susceptibility to some diseases. Thus cereals receiving high levels of nitrogen may contract foliar fungal infections more readily.

The commonest symptom of mineral toxicity is some form of chlorosis, but stunting of growth and necrosis of tissues also occur frequently. The symptoms may be sufficient for accurate diagnosis of the toxicity. Where visual diagnosis is doubtful or not possible, crop material or soil may be analysed for their mineral content, as in the diagnosis of mineral deficiency.

Summary

1. Thirteen minerals elements (N, P, S, K, Ca, Mg, Fe, Mn, Zn, Cu, Mo, B, Cl) are essential for plant life. The macronutrients (N, P, S, K, Ca, Mg) are required in fairly large amounts, but only small quantities of the other elements (micronutrients) are necessary. The soil supplies minerals as ions to plant roots.

Water in the soil contains ions in solution, and these are mostly available for uptake by plants. Minerals in the soil solution are in equilibrium with those which are bound in, or adsorbed on, soil particles, or which are precipitated as insoluble salts. Depletion of the soil solution by plant roots results in the slow release of minerals from bound, adsorbed and insoluble forms. Adsorbed cations may also be available for uptake by plants; these exchangeable cations are replaced by hydrogen ions from roots.

2. The soil is the ultimate source of all mineral nutrients for plants, with the exception of nitrogen which must be fixed from gaseous nitrogen in the atmosphere. Small amounts of minerals can be derived as dust or dissolved in rainwater from the atmosphere and this is a significant source of supply of minerals for forests. The ability of soils to supply minerals depends greatly on how mineral-rich were the parent rocks they were derived from, and on the weathering of the soils since then.

For all soils, the macronutrients N, P and K must be augmented by fertilizers to achieve maximum crop growth rates. Micronutrients are usually in adequate supply in most soils, especially young soils, but less often in old, heavily weathered and leached soils. Judging the timing and quantity of fertilizer application presents difficulties with N and P, but little difficulty with K.

3. Ions are supplied to the root surface by root growth, continually exploiting new areas of the soil, by mass flow in water drawn to the root by the pull of transpiration, by gravitational percolation of water, and by diffusion. Diffusion and root growth are especially important for phosphate and potassium ions. Phosphorus nutrition poses special problems of supply, as only a very small proportion of the soil phosphorus is available as soluble phosphates. Root hairs, in those species in which they occur, improve phosphate supply by increasing the volume of soil exploited. Many plants form mycorrhizas (symbiotic associations with fungi) which allow more efficient uptake of phosphorus. Nitrogen nutrition of peas, clovers and related plants is improved by their ability to form a symbiotic association with bacteria living in their roots; these bacteria convert atmospheric nitrogen gas to plant-utilizable nitrate ions.

Plant roots accumulate ions to much higher concentrations than they exist in the soil solution; they are also selective in the ions they absorb. Because of the Casparian strip, a barrier in the endodermis of the root which blocks radial movement of ions in the cell walls, ions must travel at least part of the way across the root in living cytoplasm (the symplast). It is here that selection takes place and that accumulation of ions is accomplished by the expenditure of respiratory energy.

Mineral uptake is reduced when respiration in the root is inhibited, whether by poor aeration due to heavy, compacted or waterlogged soil, or by lack of carbohydrate due to reduced photosynthesis. It is also reduced when the transpiration rate is low. Most mineral absorption takes place in the young, unsuberized region of the root; since this is a passing phase continued root growth is necessary to maintain efficient absorption.

Minerals enter the xylem of the root where they are carried in the transpiration stream to the shoot. At any point they may cross from the xylem into adjacent tissues, including the phloem, which transports them to all parts of the plant. Minerals reaching the leaves may be re-exported, and in general there is net movement out of older leaves and into younger leaves. S, Ca and Fe are fairly immobile in the plant and are not redistributed in this way. Plants lose minerals by shedding organs, washing from leaves, and exudation.

4. Any mineral may be present in the soil in deficient, sufficient or excessive supply for a crop. Also, the appetite of a crop for a particular mineral varies with growing conditions and stage of growth. Deficiency causes either yield reduction, or the expression of mineral deficiency symptoms, or both, while excess may cause mineral toxicity. Crops differ in their tolerance range to the supply of any mineral. Mineral deficiency symptoms include death of plants, organs or small areas of leaf, stunting, yellowing (chlorosis) and production of red or purple pigments. Deficiencies in mobile elements show first in lower leaves, while deficiencies in immobile elements affect upper leaves more seriously.

A mineral deficiency may be absolute, resulting from low soil content, or it may be induced by soil conditions. Low pH can induce deficiencies of macronutrients and Mo; high pH can induce deficiencies of Fe, Mn, Zn, Cu and B. The range of soil pH in which a crop will grow reflects its tolerance of mineral deficiencies and toxicities. A deficiency may also be induced by very high or very low supplies of one or more other minerals or by poor soil physical conditions.

Fertilizers boost yields that are restrained by low supplies of macronutrients; they are also used to correct deficiencies showing as symptoms. Induced deficiencies are corrected by adjusting pH or other inducing factors, or by applying the deficient mineral as a frit, chelate or foliar spray. It is usually essential to correct deficiencies before symptoms appear in fruits, tubers or other harvested organs.

Excessively high concentrations of salts in the soil solution give rise to osmotic drought; this is a general form of mineral toxicity. Specific toxicities, which may be caused by any essential or non-essential element, can damage roots, interfere with metabolism, or cause induced deficiencies of other essential elements.

4

Plant Strength and Integrity

In the foregoing chapters three fundamental requirements for plant life have been reviewed – energy, water and minerals. Another set of requirements may be grouped under the heading of *structural integrity* – the ability of a plant to support itself mechanically, to resist breakage and to maintain its shape. These aspects are often conveniently described as plant 'strength' but, as will be pointed out in Section 4.1.1, this term is used in mechanical engineering in a much more restricted sense to refer to just one of several properties contributing to structural integrity.

The integrity of a plant, however, is not only a question of mechanical design. Plants live in a hostile environment and require defence systems to protect themselves against mechanical injury, desiccation, waterlogging, extremes of temperature, invasion by pests and pathogens and grazing by animals. These defence systems will be considered later in this chapter, after an outline of structural integrity. The topic of structural integrity of plants has been neglected in recent times, though it once received much attention and remains of great practical importance.

4.1 Structural integrity of plant tissues

There are two basic structural materials in plants: cellulose and, surprisingly, water. The distension of cells with water, that is, *turgidity*, gives support to and maintains the shape of leaves, young stems, fruits and fleshy storage organs. As shown in Section 2.2.1, turgidity depends on the differentially permeable cell membrane which allows solutes to accumulate in the cell. The solute potential thus generated draws water into the cell, thereby distending the walls and packing masses of cells tightly together. The plant therefore uses water, a liquid and the most readily available substance, as a major building material.

The second, more important source of structural support is the cell wall system itself. Consisting mainly of cellulose, this system is responsible for the structural integrity of the plant both directly as a building material and indirectly by containing the pressurized cell contents and thereby giving support through turgidity as outlined above. We therefore need to examine the mechanical properties of cellulose on which all of this is founded.

4.1.1 Cellulose as a structural material

The cellulose molecule (Fig. 4.1) consists of a chain of hexose units joined end to end by covalent bonds, which are the strongest kind of chemical bond. These chains are aggregated in bundles of several hundred, linked by lateral bonds which form between chains. The lateral bonds are of the kind known as hydrogen bonds, not nearly as strong as covalent bonds, but there are so many of them that the bundles of cellulose molecules, known as *fibrils*, are comparable in strength to steel.

Fibrils of cellulose are arrayed in sheets or layers, embedded in a matrix of other substances, to make up the cell walls. Cells in turn are organized into tissues, the walls of adjacent cells being cemented together with a gelatinous material called *pectin*. The importance of pectin can be seen in the reduction of tissue strength that occurs when the pectin breaks down, as in overripe fruit or in potato tubers infected with bacterial soft rot.

At each of the four levels of organization – molecular, fibril, cell wall and tissue levels – the cellulose structures are organized to provide the maximum support to the plant for the minimum investment of dry matter. In order to understand how cellulose maintains the structural stability of the plant in the face of the various forces which act on the plant, it is necessary to borrow some concepts and terms from mechanical engineering.

A force per unit area applied to a material is

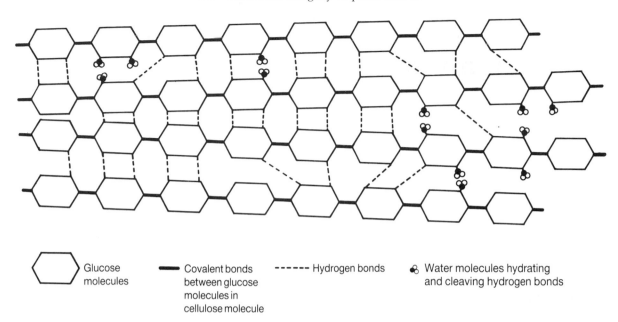

| Glucose molecules | Covalent bonds between glucose molecules in cellulose molecule | Hydrogen bonds | Water molecules hydrating and cleaving hydrogen bonds |

Fig. 4.1 Grouping of cellulose molecules into microfibrils.

called a *stress*. In the case of cellulose, the stress may come from outside the cell, as when a stem is bent by the wind or a potato is knocked about during harvesting. In other situations, the stress may originate within the cell due to the positive pressure of water in a turgid cell or the negative pressure (tension) of water in xylem during transpiration.

When stress is applied to a material the material tends to deform by changing shape. For example, a pulling or *tensile* stress tends to cause the material to stretch, while a squeezing or *compressive* stress tends to cause compression or shrinkage. The degree of deformation induced, for example the increase or decrease in length as a proportion of the original length, is called the *strain* (Fig. 4.2). In a simple material the strain induced is directly proportional to the stress – that is, the greater the stress the greater the degree of deformation.

Many of the stresses that have to be withstood by cellulose are not simple tensile or compressive stresses but tend to deform by slippage or shearing (Fig. 4.3). These are known as *shear* stresses.

The amount of stress required to produce a given strain in a material is a measure of the *stiffness* of the material. Under a given stress, a stiff material such as steel will stretch or bend much less than a pliant or flexible material such as rubber. It will

be seen later how cellulose can provide either flexibility as in young stems and leaf stalks or rigidity as in old, woody stems.

Stiffness should not be confused with *strength* in the mechanical engineering sense. Strength is the maximum stress that can be supported by a material. Stresses in excess of this lead to *failure* of the material. For example, excessive tensile stresses lead to breaking, compressive stresses to buckling or squashing and shear stresses to splitting or cracking. The resistances of a material to these three kinds of stress are often very different. Cellulose, for example, has great tensile strength (similar to steel) but much lower compressive strength.

Another important mechanical property of materials is *toughness*. This is defined as the amount of energy required to fracture the material. It is more difficult to cut or break a tough material such as leather than a brittle material such as glass. Note that the toughness of a material is completely independent of its stiffness: leather, for example, is tough but flexible while glass is brittle but stiff.

The important properties that cellulose confers on plant structures are tensile and shear strength, flexibility or rigidity (depending on the presence of other substances in association with the cellulose, as will be seen in Section 4.1.2), and toughness. Strength can be understood by examination of the structure of cellulose at molecular and fibril levels, while the other properties depend on the cell wall and tissue levels of organization.

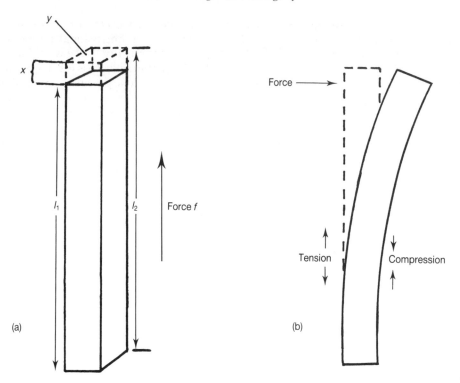

(a)

(b)

Fig. 4.2 (Above) When a material is acted on by forces, the responses of the molecular bonds cause it to deform. In (a) a piece of material of cross-sectional area y and anchored at the base, is placed under tension by a force f. Its length increases by x from l_1 to l_2. The stress is the force per unit area (f/y) and the strain is the amount of stretch per unit length (x/l_1). If the force is applied from the side, as in (b), the side under tension lengthens and that under compression shortens, causing bending of the material.

Fig. 4.3 (Below) Mechanical stresses and the kinds of deformation resulting from them.

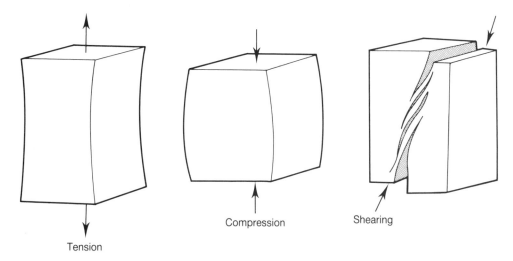

Tension

Compression

Shearing

The strong covalent bonds linking the hexose units together are responsible for the great tensile strength of cellulose. These bonds stretch as tensile stress is applied, and when the stress is removed they rapidly return to their original length. In other words, the deformation is *elastic*. The more the bonds stretch the greater becomes their resistance to further distortion. This resistance exerts a counter-force to the imposed stress. Thus strain in the cellulose molecule generates counter-forces which balance the stress.

The shear strength of cellulose fibrils depends on the hydrogen bonds cross-linking the molecules. They allow strain in the form of slippage of cellulose molecules over one another when shear stress is applied (Fig. 4.1). Again the deformation is elastic, the bonds reverting to their normal orientation when the stress is removed.

Water permeating the fibril weakens the cross-linkages as hydrogen bonds between adjacent cellulose molecules are partially replaced by hydrogen bonds between cellulose and water (Fig. 4.1). Thus shear strength is greatly reduced when cellulose is wetted, but the greater freedom of slippage leads to increased flexibility.

4.1.2   Strength and stiffness of the cell wall

In the cell walls of leaves and young stems and roots the most abundant component is not cellulose but water, which occupies most of the space between fibrils and also permeates the fibrils themselves. This, as has been pointed out, gives improved flexibility and, while it reduces shear strength, does not affect the great tensile strength necessary for the cell wall to withstand the stresses imposed by turgidity.

Plants can vary the properties of cell walls in several ways. Simply by thickening the wall with additional deposition of cellulose, stiffness and strength can be increased. However, the most significant variations are produced by the deposition of other materials in the cell wall, notably hemicelluloses and lignin.

Hemicelluloses have many more cross-bonds in their structure than cellulose and their role appears to be to bind fibrils together, reducing slippage of fibrils over one another under shear stress. They thereby stiffen the cell wall.

Where still greater rigidity or shear strength are required, the walls of certain cells become *lignified*. This means that lignin is deposited around the cellulose fibrils. Lignin molecules, like those of

cellulose, are very large but are much branched and form a three-dimensional network. This in itself gives some rigidity to the wall, but additionally the lignin prevents access of water to the cellulose fibrils and thus protects and stabilises hydrogen bonding between cellulose molecules. This is probably its main effect. Lignin is often described as the main strengthening agent of lignified cells, but the strength really lies in the cellulose. The role of lignin is to restrict slippage and promote cross-bonding in the cellulose fibrils. Thus shear strength is improved but tensile strength is not affected. The three-dimensional rigidity of lignin also gives a considerable improvement in compressive strength by helping to resist buckling. In effect, deposition of lignin permits the plant to dispense with turgidity as a means of support and it is thus most often observed in aging or maturing supportive structures such as stems, as their moisture content is declining.

Lignification is almost invariably accompanied by death of the cell; it is also often associated with increased deposition of cellulose, producing thicker walls. In commercial fibres where flexibility is needed, lignin is undesirable, hence the value of flax which has stem-supporting thick-walled fibres of unusually low lignin content, and cotton, which is made from very long unlignified epidermal hairs. In the making of fine paper from wood pulp, as much as possible of the lignin is removed during pulping.

4.1.3   Types of strengthening cell

The main strengthening cells of many leaves are of a type known as *collenchyma*. These are elongated, unlignified cells with bars of cellulose forming thickened portions of wall running the full length of the cells (Fig. 4.4). The cellulose fibrils are mostly oriented along the length of a collenchyma cell. Tissues composed of collenchyma have enormous flexibility, and are particularly important in leaf stalks which are continually subject to bending stresses owing to the large surface area presented by leaves to the wind. Heavily lignified tissues would probably fail under such stresses.

The water-conducting cells of the xylem are heavily lignified. They need the rigidity of lignin so that they can withstand the strong forces acting on them from the great tension in the xylem water created by the transpiration pull. To help them resist collapse, xylem cells are also internally buttressed with thickenings of the walls which

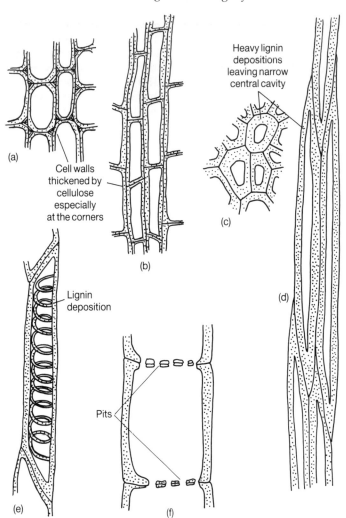

Fig. 4.4 Some types of plant cell with thickened walls. Collenchyma, shown in cross-section (a) and in longitudinal section (b), has additional cellulose deposition mainly at the cell corners, forming bars that run the length of the cells. These cells give great flexibility to young herbaceous stems and leaf stalks. Fibres, shown in cross-section (c) and in longitudinal section (d), have greatly thickened walls, usually heavily lignified, with only a narrow cavity remaining in the middle of each cell. Tracheids (longitudinal section, e) are also lignified, with cell wall thickenings often forming characteristic patterns. They are dual-function cells, providing mechanical support as well as conducting water. Vessels (longitudinal section, f) are also thick-walled and heavily lignified, but having a much larger cavity they are mainly for water transport, giving little structural support. Their walls, however, need to be stiff enough to withstand the great forces imposed by the tension of water being pulled through the vessels in the transpiration stream.

often form characteristic patterns (Fig. 4.4).

Being lignified, xylem is able to combine its water-conducting function with some support for the plant. The wider-bore xylem cells which form the *vessels* (the main channels of water flow) are of less support value than the narrower *tracheids*, which are truly dual-purpose cells.

The lignified cells known to plant anatomists as *fibres* (note that not all commercial fibres derived from plants are fibres in the strict botanical sense) are longer, narrower, and thicker-walled than tracheids, although intermediate types of cell occur as well. With few exceptions, fibres (Fig. 4.4) are the main strengthening cells of stems and roots. They are members of a whole class of lignified non-conducting cell types known collectively as *sclerenchyma*. Other members include *stone cells*, non-elongated or only slightly elongated, which form a very hard, dense tissue as, for example, in the walls of nuts or the 'stones' of fruits such as plums and cherries. Similar cells form the 'grit' in the flesh of pears.

4.2 Distribution of strengthening tissues in plant organs

4.2.1 Leaves

The *veins* of the leaf provide a skeleton on which the outstretched photosynthetic leaf blade is carried. In addition, the leaf edge is bounded by a network of small veins that prevent the edge fraying in the wind. As we saw in Section 1.6.9, the veins contain xylem and phloem for transport of water and the products of photosynthesis respectively. The xylem of the larger veins may contain vessels as well as tracheids, but in the smaller veins the dual-purpose tracheids give the necessary support while delivering all the water needed by the limited area of leaf blade that they serve.

The phloem, it will be remembered, lies on the underside of the vein. Below it there may be further strengthening tissue so that the phloem is effectively sandwiched between this and the xylem. The strengthening tissue below the phloem may consist of collenchyma, or fibres, or both. Collenchyma is often particularly well developed in the larger veins, especially the midrib and the leaf stalk.

Most dicotyledons have veins forming a network covering the entire leaf blade, whereas most monocotyledons have approximately parallel veins. The veins are, however, only the framework of the leaf; as we have seen, it is the turgidity of the parenchyma cells of the mesophyll, the bundle sheaths around the veins, and the epidermis that keeps the leaf flat and outstretched for maximum interception of light. A leaf demonstrates the importance of turgidity for support. The epidermis of the leaf, acting rather like the skin of a balloon, is often particularly important in maintaining structural stability.

4.2.2 Stems

Stems are sometimes classified as *woody* or *herbaceous*. These terms are imprecise, and no clear dividing line can be drawn between them, but they are useful nonetheless. For the purposes of this discussion a herbaceous stem will be defined as one in which no new tissues are formed after extension growth is completed, although existing tissues may or may not become heavily lignified. All the tissues present in a herbaceous stem are thus *primary* tissues, that is they are derived from the apical meristem (see Section 6.2). In a woody stem new tissues are formed after extension growth is completed, some of them becoming heavily lignified. In many species these tissues may add to the girth of the stem and greatly increase its mechanical stability. They are known as *secondary* tissues because they are derived from new meristems which arise in parenchyma tissues of the primary stem. Monocotyledon stems, with few exceptions, show no secondary tissue development, whereas dicotyledon stems may or may not become woody as they get older.

Herbaceous stems. The conducting tissues of herbaceous stems are aggregated in *vascular bundles* arranged in one or more circles around the stem not far beneath the epidermis, as shown in Figs 4.5 and 4.6. The parenchyma tissue lying outside the ring of vascular bundles is the *cortex* and that within the ring is the *pith*. Sometimes the pith is partly or wholly replaced by a large air-filled cavity, giving a hollow stem. In maize the vascular bundles are scattered throughout the stem as seen in cross-section. Within the bundles are xylem, towards the centre of the stem, phloem, lying outside the xylem, and in most cases fibres outside the phloem. There may be additional bundles of fibres, without xylem or phloem, in the cortex. Some plants, such as broad bean, have additional strengthening in the wings of the stem, acting as buttresses (Fig. 4.5).

The strengthening tissues of the herbaceous stem are thus largely peripheral, this being particularly well marked in cereal stems such as wheat (Fig. 4.6). This more or less tubular structure has a high strength-to-weight ratio and, as scaffolders know, is a very efficient load-bearing system. However, the importance of turgidity in the cells of the cortex and pith, especially in very young stems, should not be underestimated.

As stems grow, stress is built into their tissues in such a way as to increase the strength of the whole structure. This is known as *pre-stressing*, and is exactly analogous to the use of pre-stressed concrete in modern building technology. It is too complex a subject to explain fully here, but in principle a tissue which is pre-stressed in compression shows increased resistance to tensile stresses imposed later, while one pre-stressed in tension is

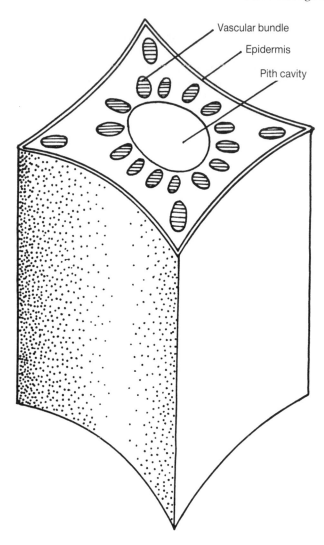

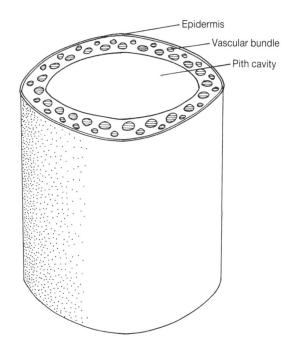

Fig. 4.5 (Left) The distribution of vascular bundles in a broad bean stem.

Fig. 4.6 (Below) The distribution of vascular bundles in a wheat stem.

Woody stems. In general, however, herbaceous stems are not strong enough to withstand the stresses resulting from the greater leverage of a plant taller than about 2 m. This may be especially true where the stem is branched and the plant therefore presents a large surface area to the wind. The development of the woody stem, with increased girth resulting from secondary tissue growth, is one way in which plants have evolved to resist such stresses and thereby grow to great heights.

The main strengthening and stiffening tissue of woody stems is, of course, the *wood*. Wood is secondary xylem. It is produced from a cylinder of meristematic cells, the *vascular cambium*, which is to be found at the periphery of the wood (Fig. 4.7). The most recently formed cells of the wood are therefore those adjacent to the vascular cambium, while the oldest wood cells are in the middle of the stem. At the same time as it produces secondary xylem towards the inside, the cylinder of vascular cambium produces secondary phloem towards the outside. This replaces the primary phloem which is destroyed by the great increase in girth that takes place with the development of the wood. In a tree, the secondary phloem forms the inner bark. The

similarly more resistant to compressive stresses. Peripheral tissues of a herbaceous stem are pre-stressed in tension. The effect of a side force, as applied for example by wind, is to put tensile stress on the upwind side and compressive stress on the downwind side. Cellulose has more than enough tensile strength to resist the increased stress on the upwind side of the stem, but does not have such great compressive strength. Failure of stems subject to strong winds is therefore more likely by buckling on the downwind side than by breaking on the upwind side. However, because the peripheral tissues are pre-stressed in tension, they can withstand much more compressive stress and the overall strength of the stem is greatly enhanced.

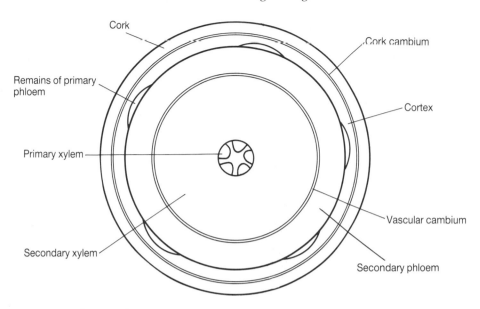

Fig. 4.7 The main tissues of a woody stem as seen in cross-section.

outer bark, or *cork*, has a protective function and arises from a second cambium, the *cork cambium*. The structure and function of cork are dealt with in Section 4.5.3.

Wood consists very largely of lignified cells – vessels, tracheids and fibres. In addition it usually contains a substantial amount of lignified and unlignified parenchyma. The anatomy of wood differs greatly from species to species, hence the different commercial properties and uses of different timbers. In climates where growth is seasonal, wood formed from the vascular cambium early in the growing season is different from that formed at the end of the previous season, so that the boundary between the two types of wood is visible, and each year's growth is marked off as a *growth ring*. In cut timber these growth rings appear as the 'grain' of the wood. Spring wood of temperate trees tends to have more and larger vessels, and thinner-walled fibres, than summer and autumn wood (Fig. 4.8).

In trees, the younger (outer) wood is known as *sapwood*, and is involved in the transport of water as well as giving strength to the stem. The older (inner) wood is known as *heartwood*, and conducts little or no water but assists the sapwood in supporting the great weight of the tree.

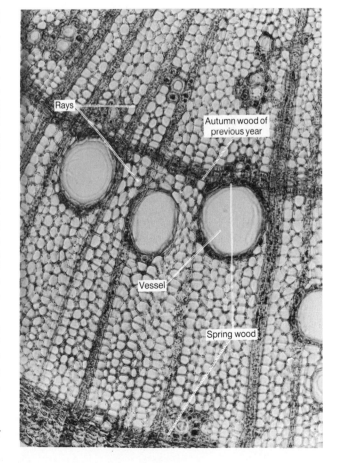

Fig. 4.8 Cells of autumn and spring wood in a woody stem.

Radial series of parenchyma cells in the wood form the *rays* (Fig. 4.8). These provide a living pathway for translocation of organic and inorganic materials across what is otherwise an almost completely dead tissue. In addition to transport, they often store carbohydrates and other materials.

The secondary phloem, though not a strengthening tissue, contains fibres which help to prevent it being squashed between the ever-expanding wood on the one side and the cork on the other.

The development of secondary tissues enables not only great size but also great age to be attained because of the annual production of new functional tissues, especially in the phloem. The specialized living cells of plants, like those of animals, have a limited useful lifespan, but the constant renewal of phloem in a tree means that the tree can live for long periods, up to thousands of years in some species.

The high quality of wood as a construction material derives from a combination of features that enhance the strength, stiffness and toughness of tree trunks. Wood, being composed mainly of cellulose, has a tensile strength considerably greater than the stresses likely to be generated by wind and gravity, but there is no such comfortable margin for compressive stength. Pre-stressing, as outlined for herbaceous stems, helps trees to resist the great compressive stresses caused by high winds, and gives wood the compressive as well as tensile strength that makes it such a valuable construction material.

Another important structural property of wood is its toughness, and in particular the difference in longitudinal and transverse toughness. Longitudinally, wood is rather brittle – for example, it is fairly easily split with an axe along the grain, giving quite a clean fracture. At right angles to the grain, however, it is much tougher, typically requiring about one hundred times as much energy to fracture as is required to split it longitudinally.

This is of great importance in enabling wood to resist transverse cracking. In a more homogeneous material such as glass, cracks have a natural tendency to propagate themselves because stresses become concentrated at the points of the cracks. In wood, however, a transverse crack does not become self-propagating because failure occurs preferentially along rather than across the grain. As Fig. 4.9 shows, this spreads rather than concentrates the stress and only local damage is done. The result is to make wood a very tough material.

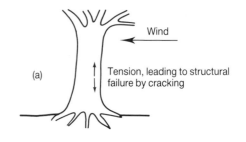

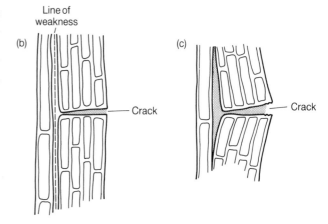

Fig. 4.9 Crack-stopping in a woody stem. A horizontal crack is initiated as a result of tension in the wood caused by wind forces (a). The crack tends to self-propagate across the wood (b) until it reaches a line of weakness, along which the wood ruptures. The crack then expands vertically (c) and is no longer self-propagating. Note that in this diagram tissue structure is simplified.

Given the range of design features built into wood during the course of evolution, it is not surprising that it remains unrivalled as a construction material for its combination of stiffness, strength, toughness and low density.

4.2.3 Roots

In young roots the support and conducting tissues are in the centre rather than at the periphery. This makes roots more flexible than stems. Roots do not have to resist bending in the way that stems do, the effect of wind being to pull or tug at roots rather than to bend them. The distribution of tissues in the young root has already been described (Section 2.3.1).

Dicotyledon roots may undergo secondary tissue development, just like stems. The wood of roots is very similar to stem wood but may contain more storage parenchyma.

4.3 Mechanical stresses on plants

The main causes of mechanical stress and therefore of structural failure in plants are gravity, wind, impact and freezing.

4.3.1 Stress due to gravity

A vertical stem is a load-bearing structure, which suffers compression stress. A stem, however, hardly ever fails under gravitational forces alone, even when water is added to the normal weight of the plant.

The branches of a tree are not vertical, so that gravity produces more than simple compression stress. Compression stress is present, being greatest on the underside of the branch, but it is accompanied by tension stress which is greatest on the top side and shear stress which is greatest in the middle (Fig. 4.10). These stresses cause a bending strain. The forces are greatest at the base of the branch, where the leverage is greatest, being distant from the centre of gravity of the branch.

The base of a branch is usually considerably thickened. This increase in cross-sectional area lowers the force per unit area, that is, the stress. The result is that the maximum stress usually develops some distance along the branch from the base where the cross-sectional area is less. Thus if a branch breaks under a heavy crop of fruit or when excessively loaded with water, snow or ice, the break tends to occur a short distance from the base. Shear stresses may, however, induce cracking or splitting back to the base.

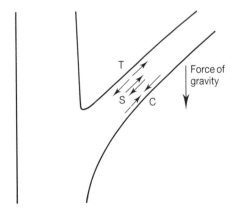

Fig. 4.10 Mechanical stresses imposed on a tree branch by gravity. Tension stress (T) is greatest on the top side, compression stress (C) on the underside and shear stress (S) in the middle.

As an adaptation to gravitational stress, trees often have *reaction wood* in their branches. This is wood of differing tissue composition from that found in the vertical trunk. In conifers it takes the form of 'compression wood', with increased compressive strength, on the underside of the branch. In broadleaved trees 'tension wood', with increased tensile strength, forms on the top side of the branch.

4.3.2 Stress due to wind

The forces on plants due to wind are normally more or less horizontally applied, that is, at right angles to gravitational forces. Turbulent air movement or freak conditions like whirlwinds can, however, sometimes impose considerable upward forces which tend to pull plants up by the roots.

A horizontal wind force is proportional to the area of the plant presented to the wind and to the square of the wind velocity. Thus an 80 km h^{-1} gale will produce stresses sixteen times greater than a 20 km h^{-1} breeze. As with gravitational stress on branches, wind stress on a vertical stem produces a bending strain, the greatest forces being at the base of the stem. Again, however, because of increased girth at the base, the greatest stress may be elsewhere, the exact position depending on the detailed geometry of the stem. In shallow-rooted trees such as spruces, or in maize which has special prop-roots buttressing the stem, structural failure usually takes place not in the plant at all but in the soil. The effect is uprooting, or *windthrow*, a mass of soil being lifted with the roots. Trees present an enormous surface area to the wind. A gale in Scotland in January 1953 felled more trees in one day than would have been felled commercially in five years in the whole of Great Britain. Dry soil can withstand about twice as much stress without shearing as soil at field capacity; flooded or waterlogged soil is even more liable to mechanical failure. Trees are therefore more easily blown down, and maize more readily laid flat, on wet sites or following heavy rain.

To uproot a deep-rooted tree such as birch would involve moving an impossibly large mass of soil. Structural failure, when it occurs, generally takes place some distance up the trunk from soil level. It takes the form of breakage, usually with extensive splitting.

The laying flat of agricultural crops through structural failure resulting from wind stress is known as *lodging* and can be a major cause of yield loss. Lodging in maize, as we have seen, occurs most commonly through structural failure in the soil, but in wheat, barley and oats it is failure of the stem that typically produces lodging.

The stems or straws of wheat, barley and oats have solid nodes but are otherwise hollow. The greatest wind stress on these stems is not at the base because the nodes are closer together there and the hollow portions of straw are broader and thicker-walled (Fig. 4.11). The maximum stress occurs one-fifth to one-third of the way up the straw. Wind causes lodging of these cereals by buckling of the stem on the side opposite the wind direction at the point of maximum compression stress. A second zone of weakness occurs, especially in ripe cereals, at or near the base of the ear. Breakage here is called *brackling* and causes ear loss.

Once a stem has been bent by the wind, the centre of gravity of the above-ground portion of the plant is shifted to one side. Gravitational forces are then added to the wind forces. The greater the bending strain, the greater is the gravitational stress promoting further bending. Thus if a stem is extremely flexible, it is likely to suffer the full combined effects of wind and gravity, and the limit of stress that can be withstood may be exceeded. Stems carrying heavy weights therefore need to be fairly rigid.

A high-yielding cereal or a ripening ear laden with moisture after rain is more likely to lodge than a lower-yielding or dry cereal crop because of the greater gravitational stress added to the wind stress. Rain falling as drizzle is more likely to cause lodging than heavy rain because the smaller droplets are not shed from the ears and foliage as are larger droplets and a greater load of water is retained on the plant. Modern cereal varieties are relatively short-strawed and thus produce less leverage on the point of maximum stress. These varieties are less likely to lodge even under heavy loads. Growth regulators such as chlormequat are frequently applied to wheat crops to reduce straw length and thereby improve standing ability.

Many plants are capable of reducing the surface area they present to the wind, thereby minimizing the danger of structural failure. The leaves of trees behave like flags, turning away from the wind on their highly flexible stalks. The heads of two-row barley present their narrow edges to the wind; when ripe, the heads hang down on the side away from the wind, rather like a wind vane. A similar thing happens with one-sided oat heads.

It is now being increasingly held that plants have

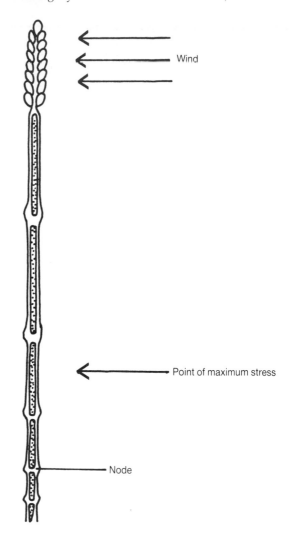

Fig. 4.11 Wind stress on a cereal stem. Though the greatest leverage produced by horizontal wind forces is at the base of the stem, the greatest stress occurs at a point part way up the stem, where the nodes are farther apart and the hollow portions of stem between the nodes are thinner-walled than near the base.

the ability to bring about considerable adaptive changes in response to quite limited exposure to wind. This topic is being more thoroughly researched.

In exposed areas, shelter belts of trees can give great benefits in agricultural crop performance. This benefit is not only through reduced lodging, but also through reduced loss of ripe grains by shaking out of the ear and through reduced water stress because of lower transpiration and other benefits of the less windy conditions in the shelter provided.

4.3.3 Stress due to impact

The main impact forces met by agricultural crops are the effects of trampling (for example by grazing animals) and wheel-tracking (for example during fertilizer or pesticide application). Mechanical handling during storage and marketing is also an important source of impacts. The morphology of grasses and cereals in the vegetative stage is such that they can withstand trampling and wheel-tracking. They have no elongated stem to bend or break, only leaves which are highly resilient and quickly resume their original position after the treading foot or rolling wheel has passed on. Established crops of grass or cereals are frequently subjected to heavy rolling to consolidate the soil; when correctly timed, this has no deleterious effect on crop performance. With the progressive development, however, of a more rigid, lignified flowering stem, damage akin to lodging results from trampling or wheel-tracking.

The impact forces of rain and snow, even when driven by strong winds, are relatively small. It has been estimated that in cereals subjected to wind-blown rain, the stresses due to rain impact are only about one-thousandth of those due to wind. Rain accompanying wind *does* increase lodging risk, but only because of the greater weight of the ear when soaked and because wet straw is less stiff (see Section 4.4).

4.3.4 Stress due to freezing

The freezing of soil water causes an expansion of the soil called *frost heave*, tending to raise the soil surface. This places a great tension strain on roots, particularly those that go deep in the soil. Shallow-rooted seedlings may thereby be uprooted, or roots may be snapped, with subsequent death or checking of the plant. This is particularly liable to happen in temperate areas with late autumn-sown seedlings that have inadequate time to establish well before the winter, or with seedlings showing poor root development or those in waterlogged soils, which expand greatly on freezing because of their high moisture content. The effects of ice formation within plants themselves are considered later in this chapter.

4.4 Factors affecting the strength of plant support systems

Any support system can only be as strong as its weakest part. As has been seen, plants have become adapted in different ways to meet mechanical stresses, but within any one system there are variations in strength due to plant and environmental factors.

Variety and species. In general, wheat straw is stronger than barley straw, which in turn is stronger than oat straw. These differences are partly the result of differences in the diameter of the central cavity, which is smallest in wheat and largest in oat straw. In addition, there are differences in degree of lignification, wheat being most and oat least heavily lignified. For this reason, wheat straw is the poorest and oat straw the best of the three for animal feeding (although none is particularly good, as we shall see in Chapter 12). There are also marked differences in these characters between varieties within each species. Plant breeders have put considerable effort into improving straw strength to reduce the risk of lodging and thereby improve the reliability of cereal crop yield. They have also, as we saw earlier, selected in favour of shorter straw length for reduced leverage.

Growth regulators. Synthetic growth regulating chemicals such as chlormequat and mepiquat chloride can be used to shorten cereal straw, but they may also strengthen the straw, and help prevent ear loss by breakage near the top of the straw (brackling).

Fertilizer. Heavy applications of nitrogen fertilizer tend to decrease straw strength by reducing lignification; they also increase straw length and ear weight, all tending to reduce standing ability.

Sowing density. High plant density causes mutual shading and a common effect of this is that the plants tend to grow taller though without proportionate strengthening of support tissues. In cereals the longer, often narrower, straw resulting from excessively high sowing rates is more prone to lodging; the straw may also be weaker because of increased internode length near the base.

High temperatures. If these occur during stem elongation weakened straw also results.

Wet weather. Moist straw is much more flexible than dry straw, water acting, as we saw earlier, as a lubricant for the slippage of cellulose fibrils over one another. The greater degree of bending in response to wind leads to a greater likelihood of lodging, especially with a heavy wet ear.

Disease. This is perhaps the most important factor of all. Diseased branches of trees are always the first to break; diseases such as eyespot, resulting from fungi which attack the base of cereal stems, are important causes of lodging. Pathogenic fungi and bacteria cause structural failure by causing the death of cells which thus lose turgidity, by breaking down pectin, and also, to some extent, cellulose and even lignin.

4.5 Protective tissues of plants

Just as the plant has a 'skeleton' of lignified tissues for support, so it has a 'skin' to protect it against the environment. A 'skin' is especially necessary in a land, as opposed to an aquatic, plant as it forms the boundary between the aerial environment and the aqueous medium of life itself. It is a barrier, though by no means a perfect one, to the escape of water. An equally important function, however, is to keep excess water *out* of the plant, preventing water-logging of the tissues which would result in an insufficient supply of oxygen for respiration. The 'skin' of plants, like our own skin, also protects the plant against injury of the more delicate tissues underneath by impact or abrasion, and offers some defence against attack by pathogenic fungi and the depredations of insects.

The 'skin' of young roots, stems and leaves is the *epidermis*, a single or sometimes multiple layer of cells which, in above-ground parts of the plant, carries on its outer surface a more or less water-proof deposit – the *cuticle*. In older stems and roots which have thickened by secondary tissue development, and in swollen storage organs including some fruits, the epidermis is replaced by *cork*. This fulfils much the same functions but can be continuously regenerated as the older cork ruptures with the increase in girth of the organ.

4.5.1 Epidermis

In leaves and young stems the epidermis consists of rather tabular cells, often with thicker walls than the underlying parenchyma tissues. The thickening is of cellulose; only rarely is lignin present in epidermal cells. Many plants have some of their epidermal cells modified into hairs, which can perform a variety of functions. In some cases their major function is the secretion of unwanted substances (salt, for example, is secreted from the glandular hairs of many halophytes), but very

often they have a protective role, which will be considered in later sections of this chapter. Cotton fibres, as has been mentioned, are extremely long, thick-walled epidermal hairs from the seed coat. Root hairs, the absorptive function of which was dealt with in Section 2.3.1, are outgrowths from the epidermis of the young root.

Stomata – pores in the leaf or stem epidermis bordered by guard cells which control their opening and closing – have already been considered in detail and need not be further dealt with here. It may, however, be worth reminding the reader that the stomata are necessary to allow gas exchange for photosynthesis, since little gas exchange can take place through the epidermis with its outer cuticle, and that the guard cells are necessary to close the stomata when, in the short term, the need to prevent excessive water loss by transpiration is more important than the need to continue gas exchange for photosynthesis.

4.5.2 Cuticle

The outer-facing wall of epidermal cells in the leaf and stem sometimes contains some waterproofing, but most of the waterproofing is in a discrete layer lying outside the cell wall – that is, the cuticle. The walls of the mesophyll cells immediately beneath a stoma may also have a thin cuticle. Young roots must be able to absorb water; if a cuticle occurs here it is very rudimentary.

The cuticle is usually attached to the epidermal cell wall by a thin layer of pectin, continuous with the pectin which glues the epidermal cells together, as shown in Fig. 4.12. Immediately outside the pectin layer is the *inner cuticle*, consisting of cellulose fibrils embedded in a lipid substance known as *cutin*. Cutin is made up of long-chain fatty acids, with 16–18 carbon atoms in the chain, joined together by chemical bonds of various kinds. The *outer cuticle* is a crust of cutin with no cellulose, sometimes containing waxes and other protective materials. The outer cuticle of apple fruits, for example, contains tannins. On the surface of the outer cuticle there may be a further deposition of waxes, often in the form of rods, plates or crystals, giving a highly convoluted surface geography. It is this wax on top of the cuticle that gives rise to the familiar 'bloom', for example on cabbage leaves and plums, and which can be polished to put a shine on apples.

The cuticle is highly resistant to microbial decay and to animal digestion, although this resistance

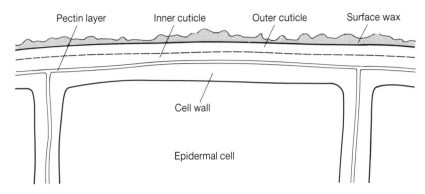

Fig. 4.12 The structure of a typical plant cuticle (thickness greatly exaggerated).

does vary with the detailed chemical composition of the cutin. For example, the fatty acids in the cuticle of apples of the variety Cox's Orange Pippin are more rapidly broken down than those of Golden Delicious, and this results in Cox's Orange Pippin having a shorter storage life.

4.5.3   Cork

Cork is a secondary tissue formed along with secondary xylem and phloem during the lateral growth of stems and roots. It does not occur in monocotyledons which, as has been noted, do not undergo secondary tissue development. Cork is produced by the cork cambium, a meristematic layer distinct from the vascular cambium. It develops in the cortex, and forms the outer bark of woody stems and roots.

The walls of cork cells are impregnated with a waterproofing lipid, *suberin* (the same material as forms the Casparian strip in the endodermis of roots – see Section 2.3.1). Suberin is chemically similar to cutin but with, in general, longer-chain fatty acids with 20–22 carbon atoms. If anything, it

is more water-repellent than cutin. Cork cell walls also contain varying amounts of lignin, waxes and tannins. Not unexpectedly, the cells of the cork are dead, and often they are air-filled – hence the buoyancy of commercial cork, which is harvested from the bark of the cork oak.

Just as the cuticularized epidermis requires stomata to allow gas exchange for photosynthesis in the underlying tissues, so the cork layer must also have a ventilation system, not for photosynthesis but to allow oxygen through for respiration in the living, working tissues of the phloem, the cork cambium and the vascular cambium. This ventilation system takes the form of areas in the cork layer known as *lenticels* consisting of thin-walled, non-suberized parenchyma, as shown in Fig. 4.13. The lenticels generally grow faster than the cork itself, so that the parenchyma cells of the lenticels tend to protrude through the gaps in the cork. The outermost cells are continually removed by sloughing but are replaced by new cells forming from the cork cambium. Substantial air spaces between the parenchyma cells provide a pathway for oxygen entry but also for water loss.

Fig. 4.13 Lenticel structure. Lenticels are not suberized and provide a site for entry of oxygen through the cork of the bark to the living tissues beneath.

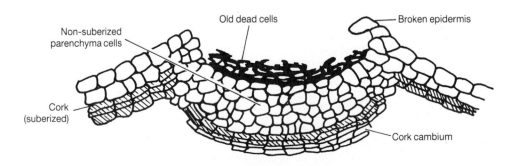

4.6 Protection from mechanical injury

The cuticle and cork protect the above-ground parts of the plant from abrasion by wind-blown sand or soil. Damage to wax extruded outside the cuticle, or to the outer layer of the cuticle itself, is soon repaired by further secretion of wax or cutin from the epidermis. Similarly, the cork (the outer bark) of trees protects the delicate phloem and vascular cambium (the inner bark) from damage, for example by animals.

The skin of potato tubers is a cork layer which protects the tuber from light knocks and scuffs, but not from heavy impact or abrasion, during handling. A similar role is fulfilled by apple skin, which is basically a cuticularized epidermis but with some development of cork and with lenticels replacing stomata.

The most delicate parts of plants are the meristems, which are essential for the further growth and development of the plant and so are given special protection. The root tip meristem is covered by the *root cap* (Fig. 4.14), which consists of expendable cells continuously regenerated from the meristem. As the root tip pushes through the soil, the root cap cells are removed by abrasion but the meristem itself remains undamaged.

The apical meristems of shoots, when in a dormant state, are surrounded by leaf primordia (leaves at an early stage of differentiation), which are in turn enclosed by older leaves forming an apical bud. Very often the outer leaves of the bud are modified as *bud scales* which have a purely protective function, helping to prevent desiccation, and never expand to become significant as photosynthesizing leaves. Sometimes, as in birch or horse-chestnut, the bud scales are gummy or resinous, further enhancing their protective effect. They are usually shed when the bud breaks in spring.

Growing shoots in which the apical meristem is not dormant still protect the meristem by young leaves as they grow (Fig. 4.15). In plants showing little elongation of the stem the meristem may be protected by specialized structures at the bases of the leaves, for example the leaf sheaths of grasses and cereals, the shorter tubular sheath or ochrea of rhubarb and docks, or the wing-like stipules of red

Fig. 4.14 The root cap protects the delicate root tip meristem from damage by abrasion as it grows through the soil, in transverse section as a photomicrograph and in diagram.

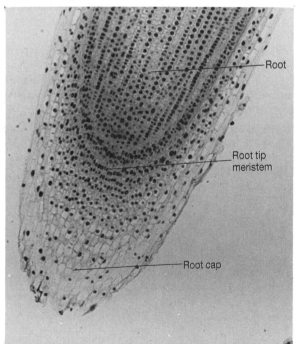

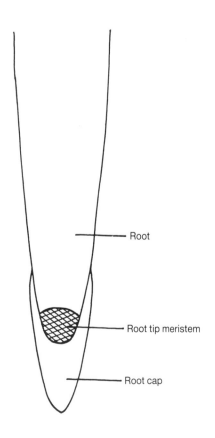

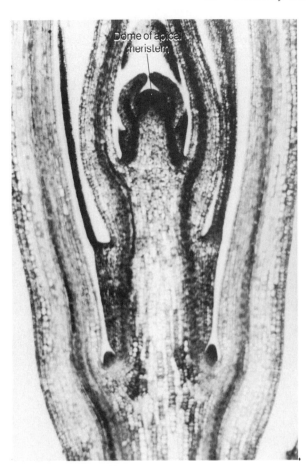

Fig. 4.15 A shoot apex in longitudinal section. The apical meristem of the shoot is protected by young leaves before they unfold.

clover. The axillary buds are similarly protected.

In the flower bud, the delicate inner parts of the flower are protected by the more robust outer part, the *calyx* (see Section 7.3.1). In grass and cereal flowers there is no calyx, but its function is taken over by highly modified scale-like leaves – the *lemma* and *palea*. These persist in barley and oats as the 'husk' of the grain. In addition, a further pair of scales, the *glumes*, may help protect a group of flowers.

Another important protective structure is the *seed coat*, enclosing the delicate, and at this stage wholly meristematic, embryo of a new plant. Many seeds remain within the fruit even after shedding, so that the fruit wall, or *pericarp* (often heavily lignified, as in nuts) gives further protection. A balance, however, must be struck between protection of the seed and the need for the embryo to burst out of its protective shell during germination. The influence of seed coat and pericarp on germination will be considered in Chapter 5.

4.7   Recovery from breakage or injury

Plants, then, are well provided with strengthening and protective tissues, but in spite of these some structural failure and mechanical damage are inevitable, especially in agricultural crops which have to bear heavy loads and which are subjected to mechanized harvesting. We must now look briefly at the ability of plants and harvested organs to recover from damage.

Breakage of a plant organ resulting from failure of the support system cannot normally be repaired. In the case of stem breakage, if the plant is perennial or if the damage occurs at an early stage in the life of the plant, the plant can usually recover by making fresh growth from the axillary buds below the break. Early lodging of cereals by buckling of the straw can usually be overcome by fresh growth of the upper internodes, bringing the ear up again, but this cannot happen with late lodging or where the stem is broken.

Wounds resulting from abrasion, breakage, animal grazing or other causes are a serious liability to the plant. They are sites of water loss, but more importantly they provide an easy site of entry for pathogenic (disease-causing) microbes and insects. Plants, like animals, need to be able to heal such wounds.

In leaves and young stems and roots, wounds are first sealed by the secretion of *gums* which give temporary protection while a new cuticularized epidermis develops. The wounds left by the shedding, or *abscission*, of leaves, flowers and fruits are similarly sealed by gums, and also by deposition of suberin in the cell walls. There is usually a well-defined *abscission zone* where the cleavage takes place; suberization may already have begun in this zone before the organ is shed. In woody species, cork develops beneath the abscission scar.

The 'skin' of potatoes, yams and some other root and tuber crops is a layer of cork which is vital in preventing the entry of pathogens and in controlling water loss both during storage and after planting in the soil. If this cork layer is damaged or the tuber is cut to expose the parenchyma tissue within, great losses can occur unless such wounds are rapidly sealed off. Harvesting inevitably damages tubers, thus post-harvest storage conditions must be such as to permit rapid wound healing. Newly harvested potato tubers are usually stored for up to two weeks at a temperature of 15–20°C to hasten the process of wound healing, which involves first the suberization of existing parenchyma cells near cut surfaces and later, the development of cork tissue to form a new 'skin' (Fig. 4.16). Cork barriers are known as *periderms* and those formed in response to wounding as *wound periderms*. The relatively high temperature treatment is known as curing, and is followed by storage at much lower temperatures to minimize metabolic activity, sprouting and microbial growth.

In woody organs there is no great mass of parenchyma in which cork can form, and in these the first response to wounding is the formation of a mass of undifferentiated parenchyma tissue from the vascular cambium and from living parenchyma cells in the xylem. This undifferentiated tissue is known as *wound callus* (Fig. 4.17) and in it a wound periderm develops, as in potato tubers, to seal off the wound. Trees have other well-developed systems of sealing off woody areas that have become exposed to infection.

Fig. 4.17 Wound callus forming around a wounded area on a tree base.

Fig. 4.16 During the post-harvest 'curing' of potato tubers, wounds are healed by the development of cork barriers under the cut surface.

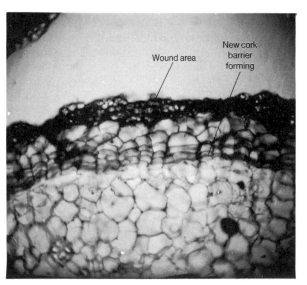

4.8 The cuticle as a barrier to water movement

4.8.1 Restriction of water loss

Much has been said about the plant's defences against excessive water loss in Section 2.4.1, but a few words on the role of the cuticle may usefully be added here.

The cuticle is not an absolute barrier to water movement. Some transpiration, perhaps up to 10% of the total, does occur through the cuticle. The thickness of the cuticle appears not to influence the amount of cuticular transpiration, but its chemical composition does have an influence. The more wax the cuticle contains, the less cuticular transpiration can take place. The detailed chemical nature of the wax is also important. However, since the great bulk of water loss is through the stomata, variations in cuticular transpiration are of minor significance. Indeed no great differences appear to exist between xerophytes and non-xerophytes in cuticle composition or thickness, adaptations to drought tending to lie in other areas such as sunken stomata, deep rooting or succulence (see Section 2.6.5).

The importance of the cuticle is as a barrier to water movement, which allows the stomata to exert maximum control over transpiration. The stomata of a leaf without an effective cuticle would be about as useful as fifteen-foot high gates in a prison with three-foot high walls.

4.8.2   *Restriction of surface retention and entry of water*

It is as important to allow free access of air to living cells as to prevent excessive loss of water. The importance of the internal air space of the leaf for the exchange of gases was seen in Section 1.6.9. Without a waterproof cuticle, the leaf might easily become saturated with water in wet weather, and this would be almost as damaging as desiccation in dry weather. A fact frequently overlooked is that the cuticle is waterproof in both directions.

The water repellence of the cuticle means that raindrops landing on it tend to remain spherical and roll off. The presence of wax on top of the cuticle increases its water-repellent character. Minute details of surface topography also influence the wettability of the leaf; projecting rods or plates of wax, or closely packed hairs, help further to prevent the wetting of the leaf surface.

If a rain droplet does remain on the leaf surface, the water may penetrate only slowly or not at all through the cuticle, depending, like cuticular transpiration, on the amount and composition of wax in the cuticle.

These characteristics of the cuticle are of great practical significance in the application of sprays containing chemicals, such as herbicides, fungicides, insecticides, growth regulators and foliar nutrients, in aqueous solution to plant leaves. If a leaf is very waxy, hairy or with particular patterns of corrugation at the microscopic or submicroscopic level, the chemical may not be retained on the leaf, or, if retained, may not be absorbed through the cuticle. It therefore becomes necessary to formulate the chemical either in a lipid-soluble form to dissolve its way through the cuticle, or to add wetting agents or other adjuvants to ensure better retention and penetration. Light and high humidity are known to aid penetration of water-soluble materials through the cuticle; it is believed that this is because of changes that take place in the arrangement of wax particles in the cuticle under these conditions.

As an example to show the importance of leaf surface wax in crop protection by herbicide application, many weeds in sugar beet crops can be controlled by the herbicide phenmedipham. The beet has a less easily wettable leaf than the susceptible weeds, and this largely accounts for the selectivity of the herbicide. As both crop and weeds get older, their leaves become progressively less readily wettable, mainly through further wax deposition. This means that to achieve the same level of weed control it is necessary to use higher rates of application of the herbicide or to add wetting agents or other adjuvants to the spray solution to assist retention of the herbicide on the leaves of the weeds. In the case of phenmedipham, mineral oil is a commonly used spray adjuvant, and is very effective in enhancing the activity of the herbicide. Sugar beet injury, however, results if mineral oil or increased phenmedipham rates are used at too early a stage when the leaf surface wax is less well developed.

4.9   Thermal injury and its avoidance

Both excessively high and low temperatures can kill or injure plants by causing *thermal injury*.

4.9.1   *High temperature injury and temperature control*

Plant leaves are very efficient interceptors of solar energy. Only a small proportion of the intercepted energy can be converted into chemical bond energy by photosynthesis, and all the remainder must be dissipated as heat. This is often difficult to achieve and leaves thus have a constant tendency to heat up. Unshaded leaves can heat up very quickly and in full sun can be up to 20°C above air temperature. In cooler areas, this can have an important influence in lengthening the growing season and increasing the growth rate of plants. Under warmer conditions, avoiding thermal injury and even death from overheating of the leaves is a considerable difficulty for the plant. This is a significant problem in agriculture, especially in the tropics.

A few plants simply tolerate high internal temperatures, the best known being prickly pear, which can tolerate a tissue temperature of 65°C, but most species die if their temperature reaches 45–50°C. This can be reached rapidly if leaf temperature is not regulated. The problem revolves around the balance between the gains and losses of energy by the leaf.

The first point to recognize is that not all the radiant energy falling on the leaf is absorbed. Some is transmitted, especially by thin leaves. Probably of greater significance is the reflection of radiant energy from leaf surfaces. Not all wavelengths are reflected equally; leaves tend to reflect less in those wavelengths that are useful in photosynthesis and more in photosynthetically inactive wavelengths. These are chiefly in the near infrared and are a substantial proportion of the sun's energy reaching the earth's surface. The cuticle plays an important part in this preferential reflection of infrared radiation. While it inevitably causes some reflection of photosynthetically useful radiation, it has been shown to reflect much more efficiently in the near infrared wavelengths which are responsible for much of the heating of leaves. Even on a single plant, leaves in full sun frequently have a thicker cuticle than those growing in the shade. This is to some extent an adaptation to restrict cuticular transpiration, but at the same time it increases the reflectivity of the leaf surface.

Not surprisingly, plants absorb photosynthetically useful wavelengths very efficiently. Since these wavelengths account for a sizeable fraction of the sun's energy and only a small part is finally fixed as chemical bond energy by photosynthesis, there still remains the problem of dissipating considerable energy. There are only four ways in which plants can dissipate unwanted heat – radiation, transpiration, conduction and the movement of warmed air away from the leaf surface on air currents (Fig. 4.18). They are considered in turn below.

Leaves are efficient at re-radiating large amounts of absorbed energy in the far infrared, and dispose of much excess energy in this way. They also absorb these wavelengths efficiently, but since only a small proportion of solar radiation at ground level is in the far infrared, this has little effect on the energy balance of the leaf. The higher their temperature, the more energy leaves radiate, and the more significant radiation becomes as a means of heat loss relative to the other processes. Leaves are, in fact, such efficient radiators of energy to the atmosphere that they can cool to well below ambient temperatures, especially at night. This happens particularly under clear skies when there is

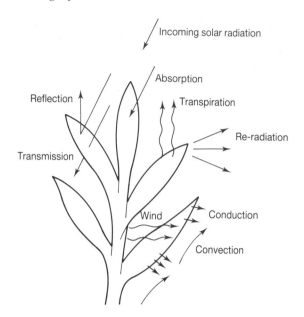

Fig. 4.18 Energy gains and losses by a plant. Of the incoming solar radiation received by the plant, some is reflected, some transmitted and the rest absorbed. Most of the energy absorbed is dissipated as heat and re-radiated to the atmosphere, or lost by transpiration, conduction and air currents caused by convection or wind.

no cloud cover to reflect or re-radiate energy back to earth. When cooler than the air, leaves collect dew by condensation of atmospheric water vapour upon them, and this can be a vital source of water to plants in arid regions. By virtue of the same cooling effect, however, leaves may suffer frost damage even when air temperatures around them are well above freezing point. On balance, radiation of energy by leaves is of great benefit to plants by protecting them from high temperature injury, especially when little or no transpiration is taking place.

This brings up the question of heat loss by transpiration. When water evaporates from any surface it absorbs considerable heat from its surroundings because of its high latent heat of evaporation. It follows that a transpiring leaf will be cooler than a non-transpiring one in the same environmental conditions. In one experiment, the temperature of tomato leaves rose by 4°C and of cotton leaves by 9°C when transpiration was suppressed by a chemical that induced stomatal closure. Well-watered plants with low resistance to transpiration can in fact hold their internal temperature close to ambient even when in full sun.

For plants growing in well-watered soils, therefore, transpiration can be an important means of heat loss.

It follows that if transpiration is impeded, for example by drought, leaf temperature may rise steeply with detrimental results such as reduced plant growth rate. It was seen in Section 2.5.1 that water stress can result from rapid transpiration even when moisture is freely available in the soil. In most agricultural crops the periods of peak solar radiation and therefore greatest risk of thermal injury usually coincide with periods of drought or of water stress induced by rapid transpiration, when further increase in transpiration to cool the leaves would lead to permanent wilting and death. Thus while transpiration is a major route of heat loss from plants, at critical times the other mechanisms of heat loss become more important. In arid and semi-arid regions in particular, unless adequate water is supplied by irrigation, transpiration cannot play a major cooling role.

The remaining routes of heat loss are by conduction and by air currents cooling the leaf. Conduction and air movements operate in tandem. Conduction transfers heat through the leaf tissue and from the leaf surface to the air at the leaf's outer surface, and air movement carries away this warmed air. There does not appear to be significant resistance to heat loss by conduction from leaves, except in the case of relatively massive tissues such as the thick leaves of certain succulents or the hearts of cabbages or lettuces. Air currents, which can be a major route of energy loss, can be caused by convection currents arising from the warmed leaf surface or by wind.

What decides the relative importance of the different routes of energy loss? This depends both on the design of the plant and on the environmental conditions in which it is growing. Under natural conditions there is a close link between the environment in which a species grows and the design it has evolved in adapting to it. This may not be true of crops, which are mainly introduced species.

The size and shape of the leaf, and its orientation to the sun, are the main design features affecting energy loss and gain. Large, flat leaves, such as those of brassicas, beets and sunflowers, present a large surface area per unit volume to the sun. By comparison with smaller or divided leaves they promote less turbulence in the air flow around them and this gives rise to a thicker boundary layer (see Section 2.4.2) of relatively static air close to the leaf surface. This thicker boundary layer restricts not only transpirational heat loss, but also the dissipation of heat by air currents. The internal air temperature of such leaves is thus closely coupled to the level of solar radiation, especially if they are held at right angles to the sun. Re-radiation is a major means of energy dissipation for large, flat leaves, but transpiration, restricted though it is by the boundary layer, retains an important role, as is witnessed by the thermal injury that can result in times of water stress when transpiration stops almost completely.

In contrast, small, needle-like or finely divided leaves, though they have a large surface area per unit volume, present a smaller fraction of this surface area to the sun; they also promote air turbulence around them, giving much thinner boundary layers. In well-watered situations, transpiration is probably a major means of cooling of such leaves, but conduction and air currents are also much more efficient mechanisms of heat loss than in large, flat leaves. Thus even when transpiration is severely restricted by lack of available moisture, the internal temperatures of these small or finely divided leaves tend to stay close to ambient temperature in nearly all conditions. Such leaves are therefore common on plants of arid or semi-arid climates.

Between the two extremes represented by large, flat, undivided leaves and small or finely divided leaves, many intermediate forms can be recognized. For example, lobed or serrated margins on flat leaves increase air turbulence and thereby accelerate cooling, and this is probably the main explanation for these leaf forms in many plants.

A vertical habit helps to diminish heating, mainly by reducing energy absorption rather than by increasing energy dissipation. Thus tea varieties with vertical leaves tolerate full sun much better than those with horizontal leaves, because of their angle of presentation to the sun. The vertical cylindrical habit of many cacti and other desert plants also has this advantage, coupled in many cases with a highly reflective cuticle. Nonetheless, cylindrical structures of poor reflective ability, such as tree trunks, may still overheat. Thermal injury to bark is therefore common in certain thin-barked tree species, and in areas of high solar radiation intensity the bark of trees may even be charred, as with larch in the European Alps.

The rates of heat loss through the routes favoured by these different factors in plant design will obviously vary rapidly with changes in environmental factors such as wind speed and moisture availability. Within limits, therefore, the relative contributions to temperature control of the different methods of energy loss are continuously fluctuating in response to constantly varying environmental influences.

4.9.2 Chilling injury

Many crops show a marked physiological disorder if exposed to temperatures below a certain critical level. This phenomenon, which always occurs above 0°C and without freezing in tissues, is called *chilling injury* and is entirely separate from frost injury. It occurs most commonly among tropical and subtropical crops, but also affects temperate crops. The threshold temperature tends to reflect the general climate of the crops' normal geographic distribution. Thus, in temperate crops such as apple, it occurs at 0–4°C, in subtropical crops such as citrus at about 8°C, and in tropical crops such as bananas at about 10–12°C. Cotton, peanuts, maize and rice are also susceptible. The injury is also time-dependent, and its extent increases as the period at the chilling temperature is extended beyond a certain minimum time required to induce it.

The entire physiology of the plant is disrupted, so that susceptibility to chilling injury imposes strict geographical limits on the distribution of many crops. In the case of tropical crops, this is further exacerbated by the frost sensitivity of many species, to which even brief, mild frosts are often lethal. In addition it complicates the marketing of many crops. Ripe bananas, for instance, cannot be stored for long, as they cannot really be stored at temperatures low enough to preserve them. The mechanism of chilling injury is still uncertain, and the symptoms vary with the organ and growth stage. If cotton seed receives chilling injury when it is hydrated, the radicle aborts, but at the seedling stage the same treatment produces a damaged root cortex. Fruit often shows a surface pitting and collapse of the underlying cells. It may require weeks of exposure to induce symptoms, but at the other extreme bananas suffer after just a few hours. Warming up the plant or organ before damage appears often prevents the development of symptoms.

4.9.3 Frost damage

Of all the factors limiting the distribution of plant species over the earth's surface, only low tempera-ture and the associated occurrence of frost ranks along with moisture availability as a major determining influence. Frost is a widespread and important cause of damage to crops, especially those dependent on seed or fruit developing from a sensitive flower, such as strawberries in temperate regions and coffee at high elevations in the tropics.

It is not clearly understood how frost damages plants since, even if plant tissue is frozen, it does not necessarily follow that it must be severely damaged. Examination shows that there are two ways in which freezing can take place in tissues. Ice crystals may form within the cells, that is *intracellular freezing*, or outside them, known as *extracellular freezing*. When water freezes within plants, if the ice crystals form extracellularly, frost-resistant species usually tolerate this, but if they form intracellularly this is almost invariably lethal in all species. Most frost-tolerant plants in fact achieve tolerance by mechanisms which encourage extracellular rather than intracellular freezing. One immediate effect of ice formation in tissues is the extraction of water from the cytoplasm to form the crystals. This dehydrates the cytoplasm and its contents and it is probably damage to membranes during this process that is mainly responsible for frost damage. During thawing water is released to rehydrate the tissue and more damage may be inflicted then.

There is a tendency to assume that frost acts primarily on the leaves, stems and flowers of affected plants, possibly because it is on these organs that symptoms are usually seen. But it is quite often the roots that are the more susceptible organs and which suffer most, at least during periods of cold that are prolonged enough to allow soil temperatures to fall below freezing point.

Frost hardiness is an important phenomenon in crops grown over the huge areas of the earth where frosts occur. Many herbaceous plants, including a wide range of agricultural crops such as spring wheat, peas and potatoes, withstand a few degrees of frost. This is chiefly due to the concentration of salts in their cells, which lowers the freezing point to −2°C or a little less. These species, however, can hardly be classified as truly frost-resistant.

There appear to be two major mechanisms involved in frost hardiness, and neither is present in a species all the year round, except in types locally adapted to regions where frost occurs during all seasons. Usually the mechanism for frost hardiness is induced in plants by patterns of temperature and daylength. Many species, including a wide range of important crops such as winter wheat, cabbage and various turf grasses, can acclimatize to tolerate temperatures down to $-25°C$. The mechanism involved here is the formation of ice crystals extracellularly. The most important way in which hardy plants avoid ice formation in the cells is by maintaining a high solute concentration in the cell sap, which acts as an antifreeze. If we compare, for example, swede or fodder beet varieties, we find those with the greatest frost hardiness are those with the highest dry matter content. Since much of the extra dry matter is sucrose, which is in solution, there is a relationship between solute concentration and frost hardiness.

Most deciduous forest species and fruit trees avoid damage by a second, and entirely different mechanism. Their tissues supercool down to $-40°C$. In some species there is in fact a mosaic of the two mechanisms with some tissues supercooling and others tolerating freezing by extracellular ice formation.

Very hardy woody plants do not supercool but induce extracellular freezing and may survive in this form down to exceptionally low temperatures, even to $-196°C$. Most conifers growing in the extremely low temperature areas near the poles, in the taiga, show this form of frost hardiness.

4.9.4 Frost hardening

As mentioned above, many plants, including a wide variety of deciduous and evergreen trees and crops such as cabbage and winter wheat, show a very marked ability to increase their frost hardiness. Under the correct stimuli, these plants show changes such as development of proteins more resistant to low temperature stress, reduced cell size, higher solute concentration in the vacuoles, more flexible cell walls, and changes in the permeability of membranes. Most of these changes probably confer increased frost hardiness. Frost hardiness, for a species, depends on the conditions under which the plant has been grown. Over-generous nitrogen fertilizer application to grass, for example, leads to more succulent (i.e. lower dry matter) herbage, which is more subject to frost damage and therefore to winter kill. This effect of high nitrogen fertilization is observed on most species. Nonetheless, adequate nutrition is important in conferring winter hardiness on plants.

Winter cereals which have been growing at $10°C$ or higher temperatures are less hardy than plants which have experienced colder conditions. A period of temperature around $2-4°C$ 'hardens' winter cereals, making them less liable to damage by subsequent frosts. Hardiness can only develop once the plant is well established; earlier in the life of a plant, the seedling is less capable of withstanding any kind of stress. The physiology of hardening is not well understood. One change very widely observed is a conversion at low temperatures of starch to sugar, effectively increasing the 'antifreeze' concentration in the cells. Thus, the harvesting of sugar beet is best delayed until after a period of cold weather, to maximize the extractable sugar content. Most plants of warm temperate climates, such as tomato or soya bean, do not harden under low temperatures and remain highly frost-susceptible.

In those species capable of maximum acclimatization, the whole process of hardening occurs in three stages and the extent to which it can develop depends on the species. Under the influence of either short days or low temperature, or both, a hardiness-promoting factor of unknown chemical structure is thought to be formed. In trees it is produced in the leaves and moves to the stems. Growth ceases, resting buds are formed, and there is an increase in the concentration of cell solutes. The second stage requires frost, is indifferent to daylength, and results in increased membrane permeability and other changes. Finally, at temperatures of -30 to $-50°C$ deeper resistance, which allows endurance of temperatures as low as $-196°C$ in some species, develops, but how is unknown. It is worth noting that the processes of development of resistance to heat, drought, and frost are closely related in plants and the ability of the tissue to endure desiccation and subsequent rehydration is a factor common to them all.

Seeds are highly frost hardy because of their low moisture content. In the 'seed banks' which are maintained for plant breeders to ensure that potentially useful breeding material does not become extinct, seeds are stored for many years at temperatures of $-20°C$ or lower, without significant damage.

4.10 Protection from invasion by pests and pathogens

Throughout their lives plants are attacked by many pests and microbes. Microbes which cause disease in other living things, usually by invading and parasitizing them, are called *pathogens*. The chief

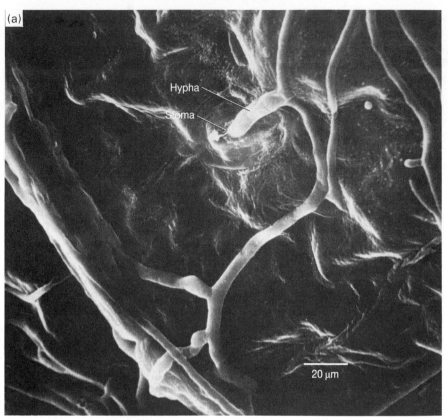

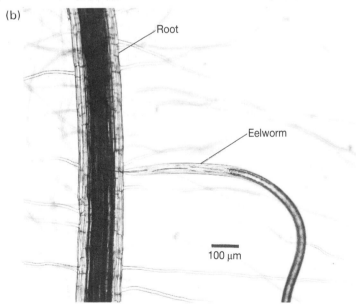

Fig. 4.19 (a) The fungal pathogen *Sclerotinia sclerotiorum* with its hypha penetrating a stoma of a potato (courtesy of D Jones, Macaulay Land Use Research Institute), and (b) the eelworm *Xiphinema diversicaudatum* with its mouthparts penetrating a root of rye (courtesy of W M Robertson, Scottish Crop Research Institute).

groups of pathogens in plants are *fungi* (Fig. 4.19), *bacteria* and *viruses*. Bacteria and viruses are discussed and illustrated in Section 1.2. Viruses are not regarded by most biologists as microbes or even as living things, and are best classified simply as infectious agents.

Animals which attack plants are usually referred to as pests. However, *nematodes* or eelworms (Fig. 4.19) are often considered pathogens rather than pests in American literature and they cause numerous disorders in plants. Undoubtedly, *insects* are the most important group of plant pests (Fig. 4.20). Other groups of pests that should also be mentioned because of the range of species that attack plants are *molluscs*, including slugs and snails, and *mites*.

A species susceptible to attack by a pest or pathogen is called its *host species*, and an affected individual of a host species is a *host*. (In American literature a susceptible plant is called a *suscept*, and is defined as a host only when it has been invaded.) Every plant species is susceptible to attack by a wide variety of pests and pathogens. But this range is small in comparison with the total range of pests and pathogens that exist, and it is evident that all plant species are highly resistant to attack by the vast majority of pests and pathogens, most of which are very restricted in the range of host species they can successfully attack. The defences of plants against this huge range of organisms can be dealt with only very generally in this book. Nonetheless, the main defensive measures employed by plants are briefly outlined here, since without them sustained growth by plants would not be possible.

4.10.1 Forms of plant defences

To invade a plant, a pest or pathogen must overcome a series of problems, not all of which are immediately apparent to the human observer. In many cases it must be able to sense the presence of a host at a distance, and in terms of the motility of microbes a few centimetres may be a considerable distance. Even when it reaches the surface of its host, there remain problems of recognition. The problems for a small organism such as a planthopper or aphid, even one with highly developed vision, in recognizing a host that is vast in comparison with itself, are not evident to a large organism like a human being, used to recognizing sizeable plants and animals by rapid visual scanning. Chemical diffusions, scent and tactile sensations at the plant surface play a major part in host detection and recognition. Thus the motile spores of some soil-borne fungi detect their host by sensing chemicals in root exudates, and they swim up the chemical concentration gradient towards the root.

Having arrived, the invading organism must penetrate the host, thrive on it, and later reproduce. To resist invasion, plants use diverse types of defence. These include mechanical barriers and chemical and physiological defences. These three cannot always be disentangled. The chemical defences include a range of toxins, which may be either present normally in the plants or produced as an active response to invasion. Plant behaviour may also confer resistance upon it. For instance, the airborne spores of *Claviceps purpurea*, a fungus which causes the disease ergot in cereals, readily invade rye but not barley. The explanation lies in the method of invasion and in differences in behaviour between the two hosts. Airborne spores of the fungus settle, by chance, on the open flowers of the plants and, like pollen, germinate on the feathery stigmas, later invading the embryo. Rye is cross-pollinating and opens its flowers to receive pollen carried by wind from other plants. It also receives *Claviceps* spores and is thus susceptible to ergot. Barley is self-pollinating and modern varieties keep their flowers closed almost all the time. Their stigmas therefore do not extrude from the flower to the same degree and thus escape invasion by the fungus.

Fig. 4.20 An insect pest – an aphid, showing its mouthpart penetrating the leaf it is feeding on.

4.10.2 Effectiveness of plant defences

Of the physical barriers employed by plants, the cuticle is not very effective in excluding fungi or insects. The mouthparts of insects penetrate it relatively easily. Many fungal pathogens exude enzymes that attack the epidermal cells and cuticle, and allow direct entry through the cuticle, while other species grow through the stomata or through wounds caused by abrasion of the cuticle.

An undamaged cuticle is a more effective mechanical defence against bacteria, and even more so against viruses. Viruses have no means by which they can penetrate the cuticle unaided, and are dependent on breaches created by mechanical damage resulting from wind or animal feeding. Often they have evolved elegant relationships with other pest or pathogen species in order to effect entry to the plant. Many are carried by plant-feeding organisms, especially insects, from host to host and rely on being introduced by them into the host cells through breaches in the plant's defences caused by feeding activity. Species which carry and introduce pathogens into hosts in this manner are called *vectors*. For example, aphids and nematodes act as vectors for many important viruses that cause crop diseases, often doing more damage by this indirect means than by their direct feeding on the plant.

A significant role of the cuticle lies in the manner in which its waxy surface encourages water droplets to roll off the leaf. Such water, if it were allowed to wet thoroughly the leaf surface, would encourage the growth and development of fungal spores and bacteria. The cuticle also reduces leaching from leaf cells of nutrients which would otherwise further stimulate fungal and bacterial growth.

Cork barriers are on the whole more effective than the cuticle in excluding pests and pathogens. Thick bark, which has this as one of its chief functions, excludes all pathogens and pests efficiently, except for a restricted range of wood-boring insects. When the cork barrier is very thin, as in the case of wound periderms on potato or yam tubers, it is less effective but still important. Lenticels can provide a major route of entry through such periderms. Periderms seem to owe their defensive properties to chemical as well as physical features, as certain of their chemical components are toxic to many pathogens.

Trees are vulnerable to insects and other animals which remove or tunnel into the bark and lay their woody parts open to attack by wood-rotting bacteria and especially fungi. They have evolved a sophisticated system of defences based on physical barriers in the structure of wood and on chemicals that block off infected xylem by plugging the vessels. Through these measures, they seal off compartments in the wood within which the invading pathogens are contained. Wound calluses (Section 4.7) also play an important part in these defence measures.

In some plant species, the hairs on leaves and stems have a defensive role, particularly against insects. In fact, it is becoming apparent from accumulating knowledge that almost all plant organs, tissues and functions have evolved in ways that improve the resistance of the plant to invasion by pests and pathogens.

Of the plant's physiological characteristics, its nutritional value and the pH and solute potential of its cell contents can all affect resistance to invading species. The increased susceptibility of many crops when high levels of nitrogen fertilizer are applied to them is in part due to changes in some of these factors. Physiological defences can be both subtle and effective. Some plants are able to detoxify toxins released by an invading pathogen and thus resist the pathogen's efforts. An opposite reaction is where plant cells die rapidly when invaded by a pathogen. This is called *hypersensitivity* and is a highly effective defence against obligate pathogens, that is, pathogens which are entirely dependent on the living tissue of their host for survival. Local death of host cells through the hypersensitive response is inevitably followed by death of the dependent pathogen, preventing its spread to other, healthy, tissues. Hypersensitivity can therefore confer immunity on the host.

Turning now to chemical defences, toxins play a major role in plant resistance. Some are always present in the plant at toxic levels. Coloured onions, for example, owe much of their resistance to certain fungal pathogens to the chemicals catechol and protocatechuic acid present in the bulb scales. Other toxins may be induced when invasion by pathogens occurs. These chemicals may already be present in small amounts in normal tissues, rapidly rising in concentration after invasion. Plant phenols are well-known examples of such toxins. A large group of chemicals called *phytoalexins* also function in this way in plants, and it is believed that most, if not all, species can produce one or more of them. In trees, substances such as terpenes, resins and rubbers increase in concentration following insect invasion. This improves resistance of the trees to further attack. For example, conifers produce large amounts of oleoresin when attacked by insects.

The range of chemical defences against insects is broad and sophisticated. Many of the toxic or unpalatable constituents of plants have an important function in this respect. For example, coumarin, found in many herbaceous species, acts as a powerful feeding deterrent for some insects. Similarly, the alkaloidal glycosides in potatoes and related plants discourage feeding by the colorado beetle. More subtly, some plants may simply fail to produce a volatile chemical that would normally act as an attractant to the insect or other pest and aid it in identifying its host. Others may be produced and act as repellents, or interfere with the pest's ability to recognize its host. Others within the plant may discourage initiation or continuance of feeding, or even inhibit insect development.

4.11 Protection from herbivorous animals

In this section the defences against larger grazing animals, whether wild or domesticated, are briefly considered. Not surprisingly, the range of defence mechanisms involved is again considerable. Chemical defences include bitter or toxic plant constituents, as discussed in Section 12.5, or others which decrease plant digestibility.

As we have seen, the woody growth habit is a method of conferring the strength and stiffness needed for growth to heights above about 2 m. It also reduces the digestibility and attractiveness of the plant to animals. Many plants, both woody and herbaceous, have evolved spines, thorns and prickles, in most cases heavily lignified, which discourage browsing.

With larger animals, grazing and trampling usually come together, and often the solutions to both problems come together also. An evolutionary solution to these twin problems has been to keep the growing points below or just above ground level so that when the plant is grazed only leaf material is removed. Leaves can then be rapidly regenerated from the growing points. This growth habit protects the growing point not only from trampling damage but also, often importantly, from fire.

Perhaps the best known example of this is provided by grasses, which are adapted to, rather than resistant to, grazing. In the vegetative condition grass shoots consist of a series of concentric cylinders, the leaf sheaths, formed from the basal portion of the leaves, with the shoot apex and axillary buds at or near ground level, inaccessible to even close grazers like sheep or rabbits. The leaf sheaths take over the function of the stem in carrying the leaf blades aloft for exposure to sunlight (Fig. 4.21). Additionally, grass leaves grow by extension of a meristematic area near the base of the leaf blade. Thus, if the upper parts of the young growing leaves are grazed, the leaves can continue extending and rapidly regenerate at least some of the lost photosynthetic area. The leaves are also highly resistant to fracture by trampling.

Another valuable pasture plant, white clover, has prostrate creeping stems, and hence all of its buds are near ground level. Long leaf stalks carry the blades up to the light. Many other broadleaved

Fig. 4.21 In grasses the apical meristem remains close to ground level for protection from grazing animals. The leaf blades are carried on leaf sheaths instead of an elongated stem.

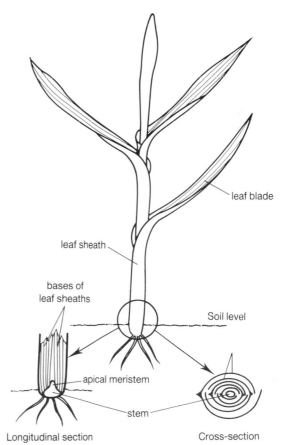

leaf blade

leaf sheath

bases of
leaf sheaths

Soil level

apical meristem

stem

Longitudinal section   Cross-section

plants show the *rosette habit* (Fig. 4.22). The stem, as in grasses, remains extremely short during the vegetative stage, so that the growing point at ground level is again protected from grazing, trampling and fire. In all these examples, the flowering head is formed near to ground level and is then carried up on a rapidly extending stem for flowering and seeding, thus shortening the period during which it is vulnerable to trampling. These growth habits will be considered in greater detail in Section 6.2.3.

Toxins and unpalatable or otherwise repellent substances in plants also play a major part in protecting them from grazing and browsing.

Summary

1. Plants use two basic structural materials – cellulose and water. Turgidity, the distension of cells with water, gives support to and maintains the shape of leaves, young stems, fruits and other fleshy organs. The cellulose cell wall is the major structural support. Cells are bound together into tissues by pectin cementing cell walls together.

When mechanical stress is applied to a plant or plant part it produces a deformation, the degree of which is called the strain. Different kinds of stresses such as tension or compression produce different kinds of strain such as stretching, compressing or shearing. The amount of stress required to produce a given strain depends on the stiffness of the material.

Cellulose confers on the plant tensile and shear strength, flexibility or rigidity (depending on the presence of other materials in the cell wall), and toughness. Plants vary the nature of cell walls to modify their structural properties as necessary by further deposits of cellulose, or by the addition of hemicellulose or lignin. Hemicellulose increases cross-bonding between cellulose molecules and stiffens the cell wall. Lignin stiffens cell walls in this way also, but chiefly acts by waterproofing cellulose and preventing shearing of the cross-bonds by hydrolysis.

Collenchyma is a non-lignified strengthening tissue allowing great flexibility, as in leaf stalks. Xylem cells are lignified to help them withstand the internal tensions created by the transpiration pull; this lignification also gives them a role in strengthening the plant. The narrower tracheids give more support than the broader vessels. The most

Fig. 4.22 The rosette habit of the dandelion.

important of the non-conducting lignified cell types, or sclerenchyma, are the long, thick-walled fibres.

2. The strengthening tissues of the leaf lie in the veins. In the herbaceous stem, support is given by the vascular bundles and their associated fibres, arranged around the periphery of the stem. This is inadequate for tall plants, which develop woody stems by the formation of secondary tissues from the vascular cambium. Wood is secondary xylem, combining the functions of water transport and mechanical support. Wood in tree trunks gains added strength from being pre-stressed, and possesses great toughness by effective crack stopping. The strengthening tissues of the root are located in the centre, enabling the root to withstand the tugging effects of wind. Roots, like stems, may undergo secondary tissue development.

3. Gravitational stress alone seldom causes structural failure of stems. When added to wind stress, however, it causes such effects as lodging of cereals and breaking or uprooting of trees. Structural failure of stems may also be induced by impact stress, as by trampling or wheel-tracking, and of roots by frost heaving.

4. Crop species and varieties differ in the strength of their support systems. Strength can also be influenced by the use of growth regulators, the application of nitrogen fertilizers, sowing density, weather factors and the presence of disease.

5. The main protective tissue of leaves and young stems is the epidermis, with its outer, more or less waterproof, cuticle. In woody stems a secondary protective tissue, cork, develops from the cork cambium outside the phloem. The cork is waterproofed by suberin, but unsuberized zones, or lenticels, remain to allow access of oxygen to the living tissues beneath the cork.

6. Cuticle and cork give some protection from mechanical injury. Especially delicate tissues such as meristems and the embryo of the seed are protected by special structures such as the root cap, bud scales, the calyx of the flower and the seed coat and pericarp of seeds and fruits.

7. Cuticle and cork also protect the plant from desiccation; the importance of the cuticle is as a barrier to water movement, thus allowing the stomata to exert maximum control over transpiration. Equally, cuticle and cork restrict the entry of water into the leaves and stems, preventing waterlogging but interfering with the retention and penetration of agricultural chemicals.

8. Plants in direct sunlight tend to heat up to above ambient temperatures, and avoid high temperature injury by reflection, transmission, re-radiation, conduction, transpiration and air currents caused by convection or wind. The relative importance of these routes of heat loss varies with the shape and orientation of the plant and its leaves, and the conditions under which it is growing.

9. Many plants, especially tropical and subtropical species, are susceptible to chilling injury at temperatures above freezing. The plant's defence against frost damage is largely a function of solute concentration in the cells. Plants which have experienced low temperatures are more frost-hardy than those which have been grown at high temperatures. Dehydrated tissues such as those of dormant seeds or tree buds are highly resistant to frost.

10. Plants are susceptible to attack by many pests, pathogens and large animals. The most important plant pathogens are fungi, but bacteria and viruses are also significant causes of disease. Insects are the most important group of pests attacking plants, other groups including nematodes, molluscs and mites. Plants resist invasion and attack by a wide range of devices including mechanical barriers such as cuticle and cork, chemical repellents and toxins, and physiological defences such as hypersensitivity. Grazing by large animals is resisted by the woody habit, by thorns and prickles, and by holding the growing point close to ground level to permit rapid regeneration of shoots after grazing. Toxic and unpalatable substances in the plant also discourage grazing.

5

Plant Growth and Development:
Seed and Seedling

5.1 The seed

When does a plant's life begin? To be accurate, it begins at the moment of fertilization in the parent flower. But it is simpler, and to the agriculturist more meaningful, to consider the life of a plant to begin with the fully formed seed.

The seed represents a pause in growth and development. The miniature plant, or *embryo*, which it contains developed while the seed was still attached to the mother plant, but growth has now ceased, the seed has lost most of its moisture and shows no outward sign of life.

If placed in moist conditions above a certain minimum temperature and with a plentiful supply of oxygen, many seeds will take up water and immediately begin to grow. Some seeds also need light for this to happen. The growth of a seed to produce a seedling is what we call *germination*.

Seeds which are in a state of constant readiness to take advantage of favourable conditions by germinating and which are prevented from doing so only by lack of some essential external factor, are said to be *quiescent*. Other seeds do not germinate even when growing conditions are ideal. They are held in check by some built-in mechanism that blocks germination; they are said to be *dormant*.

Both quiescent and dormant seeds should be distinguished from seeds which are not germinating because they lack *viability*. Such seeds may simply be dead, or they may be in some way imperfect. For instance, they may not contain an embryo, or they may be moribund and have insufficient energy reserves to fuel germination. Any sample of seeds is likely to include some which are non-viable.

5.1.1 Functions of seeds

Seeds are so familiar to us that we are inclined to overlook just what remarkable structures they are

and how many functions they perform. The seed, however small, contains all the genetic information necessary to guide the growth and development of the complete plant, however large. Through the medium of the seed, the parents transmit accurately all the characteristics with which they endow their progeny.

As well as being an organ of reproduction, the seed is an organ of multiplication and of dispersal. A single plant can produce many seeds and scatter them over a wide area, often with the aid of special dispersal mechanisms. The seed provides continuity of life over periods unfavourable to growth and is a time capsule programmed to recommence growth at an appropriate moment.

Alone of the parts of higher plants it can withstand desiccation to moisture contents below 10%, which for organisms normally existing at 80–90% moisture content is remarkable, to say the least. We know little of how this resistance to desiccation is achieved or of what goes on in the dehydrated seed. We do know that the quiescent or dormant seed has a very low, usually undetectable metabolic rate. Further, in most species, the seed is resistant to drought, to extremes of temperature and to microbial decay. Lastly, it requires no outside sources of nutrition of any kind to remain viable for long periods.

5.1.2 Structure of the seed

The structures of seeds of two representative crop species are illustrated in Figs 5.1 and 5.2. Both types of seed have three essential components: an *embryo* (the part that will give rise to a new plant), an *energy store* in the form of carbohydrate and often oil, sometimes with protein too, and a protective *seed coat*.

The embryo, like the plant it will eventually become, has a central *axis*. At one end of the axis is a rudimentary root, or *radicle*, and at the other

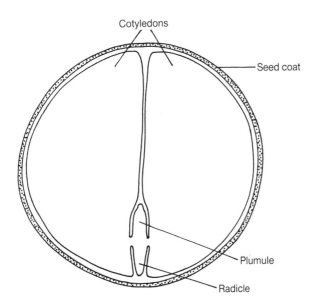

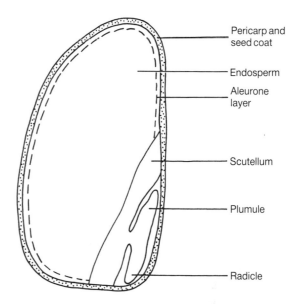

Fig. 5.1 Pea seed in longitudinal section (diagrammatic). The embryo stores carbohydrates, proteins and other nutrients in its two cotyledons, which occupy most of the volume of the seed.

Fig. 5.2 Wheat seed in longitudinal section (diagrammatic). Most of the carbohydrates and other nutrients are stored in the endosperm, rather than in the embryo itself. The single cotyledon forms the scutellum, which is adjacent to the endosperm and absorbs nutrients from it during germination.

the beginnings of a shoot, the *plumule*. During germination this axis has to grow and break out of the seed coat and for this it must be nourished from the energy store.

In some seeds, including for example those of cereals, this energy store is located in a tissue called *endosperm* which is not part of the embryo itself but lies alongside or around it. In all seeds, however, a central role in nutrition of the embryo axis is played by certain structures which are part of the embryo. These structures are attached to the axis at the base of the plumule and are called *cotyledons*.

The number of cotyledons possessed by the embryo is used to divide all angiosperm plants into two subclasses: *dicotyledons* such as beans, brassicas, beets and beeches have two, and *monocotyledons* such as grasses and cereals have only one. It may seem an insignificant difference on which to base such a major division of flowering plants, but it is accompanied by a host of other differences in root, stem, leaf and flower structure that show it is indicative of a real and fundamental division.

What, then, is the central role that the cotyledons play in nutrition of the embryo during germination? First, in seeds such as those of pea, field bean and cabbage which have no endosperm, the cotyledons are the energy storage organs. They are large relative to the rest of the embryo, in many cases occupying almost the entire volume of the seed (Fig. 5.1).

Second, in seeds with a well-developed endosperm, the cotyledons absorb nutrients from the endosperm and transfer them to the embryo axis. This is so in many monocotyledons such as onion, where the single cotyledon is embedded in the endosperm, and cereals, where the cotyledon forms a shield-shaped structure, the *scutellum*, adjacent to the endosperm (Fig. 5.2) and absorbs nutrients from it during germination.

Third, in most dicotyledons, as we shall see, the cotyledons take on another role after germination of the seed – that of the first functional leaves. They continue to nourish the seedling, but by photosynthesis rather than from reserves laid down by the parent plant.

The seeds of some plants such as broad bean and maize contain a plumule already well formed, with several rudimentary leaves and a central portion recognizable as a stem. In other seeds, such as those of brassicas, the plumule is little more than a tiny group of cells, giving rise to stem and leaves by cell division only after germination begins.

The seed coat (or testa, as it is sometimes called) surrounds and protects all the other tissues of the seed. We shall have more to say about it when we consider dormancy and germination.

In many species, the 'seed' of commerce is not just the seed but the seed plus all or part of its enclosing fruit. In cereals, for example, the fruit wall, or *pericarp*, becomes inseparably fused to the outer surface of the seed coat, and the whole functions as if it were simply a single seed coat. The cereal grain is thus, botanically speaking, a fruit, but for all practical purposes it can be thought of as a seed.

There is a similar arrangement in lettuce and sunflower, except that the seed coat and pericarp remain distinct. The fruit of these plants is called an *achene*. In carrot and parsnip the organ of dissemination is a half fruit containing a single seed. Each half of the fruit, which breaks in two when it is mature, resembles an achene.

In these and other examples, the pericarp gives added protection to the dormant or quiescent seed. Other structures such as outer husks, wings and barbs occur widely in seeds and have a variety of additional functions in seed formation, dispersal, dormancy and germination, which we shall mention when we come to consider these topics.

5.1.3 Imbibition

So much, then, for the structure of the seed. We must now go on to consider what happens to it when it falls on or is sown into moist soil.

When harvested or shed from the parent plant, seeds contain remarkably little water. As conventionally stored, crop seeds have a moisture content around 15%. In moist soil, indeed in any situation where water is available, such seeds absorb water, and do so very powerfully. This uptake of water occurs irrespective of whether the seed is dead or alive and is thus a purely physical process. We call it *imbibition*.

Imbibition is sometimes described as the first stage in germination, and certainly no seed can germinate unless it has imbibed water. Some seeds have coats which are impermeable to water; they cannot imbibe, therefore they cannot germinate. This, as we discuss below, is one cause of dormancy.

But a seed which imbibes water does not necessarily germinate. Imbibition occurs irrespective of whether or not external conditions favour germination, and whether the seed is quiescent or dormant, unless dormancy is due to an impermeable seed coat that prevents the entry of water. Most dormant weed seeds in the soil are partially or fully imbibed, ready to germinate once their dormancy is broken.

What are the forces giving rise to imbibition by seeds? The answer lies in the structure of protoplasm, which consists of large molecules of protein and other substances suspended in water. Such a colloidal structure, when partially dehydrated, has a strong affinity for water. In other words, it exerts a negative water potential (see Section 2.2.2). In a seed at around 15% moisture content this water potential is sufficient to draw in water from soil which is well below field capacity. Indeed a seed will imbibe water from humid air when no free water is available.

Seeds typically imbibe 1–2 times their own weight of water. Imbibition in dormant seeds is usually completely reversible; the seeds can dry out again and suffer no damage. In seeds which are merely quiescent, however, only the early stages of imbibition are reversible. As we shall see, once a certain point is reached various life systems are 'switched on' and a series of irreversible changes occurs, leading to germination. Drying after this point generally kills the seed. Crop seed, which is seldom dormant at sowing, must therefore be sown only when soil conditions are expected to stay moist for some time after germination has begun.

5.1.4 Functions of seed dormancy

What exactly is this condition called dormancy? It is a condition observed not only in seeds but in vegetative organs such as buds, tubers and bulbs. At this stage we shall confine our attention to dormancy in seeds.

A dormant seed may be defined as one which is viable – that is, with the ability to germinate – but which does not do so even when germinating conditions are ideal. It may be dry, or partially or fully imbibed. The period for which dormancy persists in a seed may be a matter of days, weeks, months or years, depending on the species and the external conditions to which the seed is exposed. Individual seeds within a species or even within a crop variety may show a wide range of dormancy periods.

Dormancy is basically a device to allow the seed some control over when and where it germinates.

Without this control germination is dictated entirely by the soil or other environment in which the seed happens to find itself. Let us look at three ways in which dormancy prevents a seed germinating at the wrong *time*, and then at how it prevents germination in the wrong *place*.

First, ripe seed on the parent plant before it is shed or harvested is often exposed to mild, wet weather, particularly in temperate climates where the harvest season is autumn and in the humid tropics. It is only dormancy that prevents this seed from germinating on the parent plant. A very short dormant period will suffice, and most modern cereal varieties, for example, show just enough dormancy to prevent germination of ripe seed on the cereal head during a damp period before harvest.

Second, seed production by most plants is completed just before the onset of winter or a dry season. Some plants are quite winter-hardy in the seedling stage, but for many it would be disastrous for seed to germinate in the autumn and for plants to start the winter as delicate seedlings. Germination just before a dry season would be even worse. Dormancy is a delaying device, assisting in the timing of germination to coincide with the start of a new growing season, not with temporarily favourable conditions at the end of an old season.

Third, dormancy may distribute the germination of the progeny of one growing season over many subsequent ones. If one season does not permit good seed production or good seedling establishment, there will still be seeds available to germinate in the following season. Dormancy spreads the risks; it is, in part, an insurance policy.

Finally, dormancy may prevent the seed from germinating at too great a depth in the soil. This is particularly important in small seeds with limited energy reserves for seedling growth, such as those of fine bent grass or foxglove. Many of these require light to break dormancy; in the case of some woodland species there is insufficient light for this in the shade of trees, and seeds will germinate only in clearings.

In farming and horticulture, man has taken over the tasks of timing, insurance and placement. Thus, though some form of dormancy is the norm in wild species, when a crop is sown, the grower wants no delay but an even, rapid, full germination. In many garden ornamentals that have been comparatively little altered from the wild state and in many crops that are traditionally propagated by vegetative means, such as apples and pears, the seed retains its primitive dormancy characteristics and artificial dormancy-breaking techniques have to be used. By way of contrast, in most agricultural seed crops such as cereals, peas and beans, dormancy has been reduced by centuries of selection to a very short period after harvest.

It would not be desirable to lose this dormancy altogether; as we have seen it prevents crop seeds from germinating prematurely while still on the parent plant during a wet harvest. Dormancy lasting more than a few weeks can, however, be a serious hindrance, particularly where the seeds are to be sown shortly after harvest, as in the case of winter cereals. Barley for malting must also have a short dormancy period, as the malting process requires rapid and uniform germination of the grains.

5.1.5 Types of dormancy in seeds

Dormancy, then, is more than simply a blockage of germination. The removal of the blockage in response to the right environmental signal is the other half of the picture. In other words, the *breaking* of dormancy is as important in the control of germination as is dormancy itself.

In true dormancy the blockage is within the seed itself, and in many instances the passage of time is all that is needed to break dormancy. The barrier to germination, whatever form it may take, is progressively lowered as the days or weeks pass, just like the dribbling of sand through an hourglass. Eventually the dormancy 'runs out' and the seed becomes merely quiescent or, if the conditions are favourable, it germinates.

In other instances of true dormancy, some cue from the seed's environment removes the barrier. The cue is commonly a temperature signal, but other cues such as light or even the length of day are involved in certain species. The cue tells the seed either that favourable conditions have returned, in which case the seed germinates immediately, or that the unfavourable period (normally winter) has come, in which case the seed remains quiescent until the new growing season begins.

So far we have considered only dormancy which is possessed by the seed when it is first shed or harvested. This is known as *innate dormancy*. Most species, including agricultural crops, show it to a greater or lesser degree. In many wild species, a seed which has lost its innate dormancy and is merely quiescent may become dormant again, usually as a result of exposure to certain environmental conditions such as high temperature or deep burial. This is called *induced dormancy* and is an important characteristic of many weed seeds. In knotgrass, for example, it restricts germination of seeds in the soil to the early part of the growing season when the plants have a good chance of completing their life cycle before the onset of winter. Induced dormancy is a form of true dormancy in that when the inducing conditions are removed the dormancy remains. Some internal barrier has been set up in the seed which has to be lowered either by the passage of time or by an environmental cue as in the case of innate dormancy.

One consequence of the interplay of innate and induced dormancy is the marked seasonal periodicity seen in the germination of many weed seeds (Fig. 5.3). To take one example of this, freshly shed seeds of knotgrass do not germinate in the autumn because their innate dormancy needs to be broken by low temperatures in winter. At the same time, seeds in the soil surviving from previous seasons cannot germinate because of induced dormancy brought on by high temperatures in the summer. They too must wait for winter to break their dormancy. By early spring seeds of all ages are merely quiescent, ready to germinate in a flush as soon as the soil temperature rises high enough.

We have said that in true dormancy the barrier to germination is within the seed itself. The situation frequently arises, however, where germination is prevented by some external barrier, yet conditions are what would normally be considered favourable for germination. For example, buried weed seeds in the soil may be in a warm, moist environment with plenty of oxygen, yet fail to germinate. The moment they are brought to the surface by cultivation they germinate, showing that the barrier cannot be an internal one. This failure to germinate for purely external reasons is akin to quiescence but, because it is part of the way in which the seed controls its germination by sensing external cues, it is usually considered to be a form of dormancy and

Fig. 5.3 Seasonal periodicity of germination of weed seeds. From H A Roberts (ed) *Weed control handbook*, vol. 1, 7th ed. Blackwell Scientific Publications, Oxford, with permission.

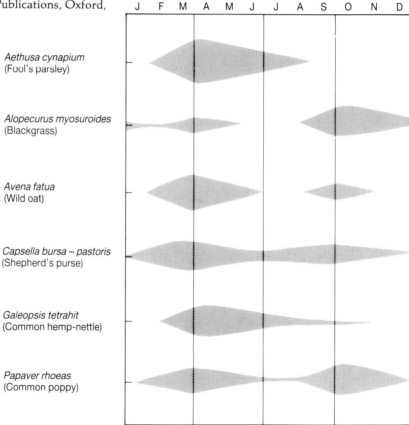

the term *enforced dormancy* is used to describe it. The enforcing factor may be darkness or high carbon dioxide level, as is found at depth in the soil, or a combination of both.

It is enforced dormancy that enables weed seeds to survive in undisturbed soil for very long periods. Most farmers will at some time or other have been surprised by the appearance of large numbers of weeds on the ploughing up of grass which has not allowed these weeds to grow for many years. Some weed seeds survive in a state of enforced dormancy for fifty years or more. They form a 'seed bank' which is drawn upon every time the soil is cultivated, and to which a deposit is made every time weeds are allowed to set seed.

5.1.6 *Mechanisms of dormancy and dormancy breaking*

Leaving aside the special case of enforced dormancy, there are two basic ways in which dormancy can be achieved in a seed. The simplest we have already touched upon, namely an impermeable seed coat. The coat of legume seeds such as peas and beans is not readily permeable by water. These seeds, however, have a pronounced scar where they were attached to the pod, and this scar, known as the *hilum*, is usually more permeable. In addition a tiny pore in the seed coat called the *micropyle* provides a pathway for the entry of water. A large proportion of the seeds of some legume crops, including clovers and lucerne (alfalfa), have even these routes for water uptake blocked by waterproof plugs and are known as *hard seeds*. They cannot imbibe and are therefore dormant.

In some cases it is impermeability to oxygen rather than to water that causes seed dormancy. The dormancy of wild oat seeds has been shown to be partly due to the husk that persists around the seed and restricts the supply of oxygen to the embryo.

Where dormancy is due to a seed coat which is impermeable to water or oxygen or both, it may be broken in the wild by the gradual decay of the seed coat by soil microbes. Alternatively, the seed coat may be damaged by frost. Another important dormancy-breaking agent for such seeds is fire. In many natural and semi-natural grasslands and forests, fire is so frequent a catastrophe that the species living there have evolved and adapted to it, and indeed many now depend on it for their continued existence. The seeds of such species may require the heat of fire to damage the seed coat before germination can take place.

Agricultural and horticultural seeds with impermeable coats, including the hard seeds of legumes, have this barrier to imbibition removed by a process which damages the seed coat and is called *scarifying*. Depending on the species, the most appropriate method of damaging the seed coat may be abrasion, for example with sand, corrosion with chemicals such as sulphuric acid, or in some cases heating in warm water.

The second basic form of dormancy is not physical but physiological, and is a property of the embryo rather than of the seed coat. Perhaps the simplest of such mechanisms is that seen, for example, in anemone, the embryo of which is not fully formed at the time the seed in shed. In such cases a period of time is necessary to allow the embryo to mature. This is known as *after-ripening*, but the term is frequently used for all situations in which the passage of time is all that is needed to break dormancy, regardless of the mechanism involved.

Much more common than an immature embryo is the action of one or more chemical *growth inhibitors* within it. We shall have more to say about these in Section 8.5.5, but suffice it to say at present that whether a seed is dormant or not depends in many species on the balance between growth inhibitors and growth stimulating chemicals. If the stimulators increase relative to the inhibitors, dormancy is broken; if the reverse happens, dormancy is induced.

The balance can be tilted away from dormancy towards germination by various environmental influences. Some seeds, for example, require to be held fully imbibed at some particular temperature for a period of time. In horticulture this is achieved by storage between layers of moist sand, soil, peat or other medium and the process is called *stratification*. This term has been extended to include the natural exposure of seeds of wild plants to low temperatures in winter, resulting in the breaking of dormancy. Artificial stratification at temperatures above about $7\,^{\circ}$C is known as *warm stratification*, and at low temperatures, normally around 0–$5\,^{\circ}$C, as *cold stratification*.

The apple is an example of a species the seeds of which require stratification, natural or artificial, to break their dormancy. This is clearly an adaptation for preventing germination just before a cold winter, and indeed the apple is native to the cool temperate zone where cold winters are a major climatic feature. Like most timing devices in plants the cold requirement is prolonged and the effects of successive cold periods are additive. Thus a brief spell of atypically mild weather after a short cold period does not lure such seeds into premature germination.

It is worth remembering that many crops now grown in temperate latitudes originated in climates with a hot dry season; it was this rather than a cold winter that had to be survived. The seeds of such crops, including barley, may require a period of dry storage to break dormancy.

Other timing devices which affect the balance of inhibitory and stimulatory substances in the seed, and thereby break dormancy in particular species, include alternating cold and warm periods, the washing of a soluble inhibitor out of the seed by rain or soil moisture, and a mechanism that many plants have for measuring the length of day and night. The effects of daylength on plant growth and development will be examined in Chapter 8. Dormancy is broken artificially in many horticultural seeds by soaking them in a solution of a growth stimulator such as gibberellic acid (see Section 8.7.2).

Species also occur with *double dormancy*, possessing both an impermeable seed coat and a dormant embryo. The physical barrier has to be removed first so that the seed can imbibe. In the wild the seed coat may be breached in the first year after shedding and the physiological barrier overcome in the second. In cultivation, the seed has to be first scarified, then probably stratified.

5.2 Germination

The breaking of dormancy is followed by a period of quiescence or, if conditions permit, by immediate germination of the seed. During germination, the plant moves suddenly from a state of being able to tolerate extremely harsh treatment, to the most vulnerable stage in its whole life – the seedling stage. Good crop management during this crucial early period is one of the most important skills of cultivation, and we shall therefore examine germination and seedling establishment in some detail.

5.2.1 Patterns of growth and development during germination

The first area of the embryo where growth recommences following quiescence or dormancy is the radicle. This extends, mainly by the elongation rather than the multiplication of cells, until it ruptures the water-softened seed coat. The radicle quickly begins to function as a root, absorbing water and minerals and soon providing firm anchorage for the seedling in the soil.

Fig. 5.4 shows the pattern of water uptake by an initially dry seed. The initial rapid uptake in the period up to point A on the graph represents imbibition. This phase lasts for several hours to several days depending on the species and the availability of water in the soil. If the seed is dormant, there is no further water uptake after point A, but a simply quiescent seed, after a plateau lasting until point B, continues to take up water again. At first this is slow, and is associated with growth of the radicle within the seed coat. After point C, when the radicle bursts through the seed coat, water is absorbed at an accelerating rate through the rapidly expanding surface area of root.

In cereals, the tip of the radicle remains protected by a small cylindrical sheath called the *coleorhiza* until this has penetrated through the seed coat. The coleorhiza grows no further, but the radicle emerges from a hole in its tip to produce not one but 3–7 slender roots. Roots like this derived from the radicle are known as *seminal roots* (literally 'seed roots').

Growth and development of the plumule, following emergence of the radicle, has to contend with the difficulties of pushing a delicate young

Fig. 5.4 The pattern of water uptake by dormant (---) and quiescent (—) seeds. Points A, B, and C as described in the text.

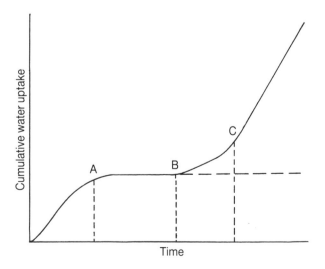

shoot up through the soil, and yet it must create leaf area for photosynthesis in a very short space of time. The pattern of plumule growth, however, depends very much on the plant species. Let us begin by considering a cereal, for example wheat. Carbohydrate reserves are supplied by the endosperm via the scutellum to nourish the growth of radicle and plumule. Just as the radicle is enclosed in the coleorhiza, the plumule is similarly enclosed and protected by a sheath called the *coleoptile*. After emergence of the radicle, the plumule, still in its ensheathing coleoptile, elongates to emerge from the seed and grows upwards. Once it appears above the soil surface, the coleoptile stops growing, its tip ruptures and the first leaf breaks through into the light (Fig. 5.5). At this stage there

Fig. 5.5 A wheat seedling.

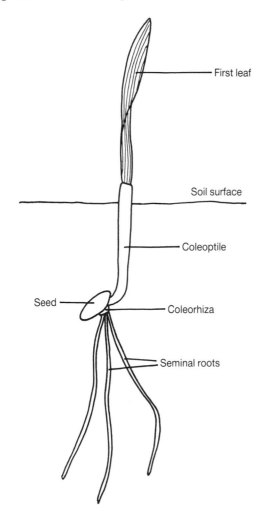

First leaf

Soil surface

Coleoptile

Seed

Coleorhiza

Seminal roots

has been little elongation of the stem; the shoot consists almost entirely of leaves. A little later the lowest one or two internodes of the stem may extend to bring the tip of the stem up to ground level; this portion of stem is known as the *epicotyl*.

In an onion seed, the tip of the cotyledon remains embedded in the endosperm during germination. Here it continues to act as an absorbing organ, but its middle portion elongates greatly to form a loop which comes above ground, turns green and begins photosynthesis (Fig. 5.6).

In dicotyledons, we find two basic patterns of germination. The first of these is seen, for example, in broad bean and pea. The plumule, formerly lying between the two large cotyledons, emerges from the seed coat and grows upwards bent over in the form of a hook. This protects the delicate apical meristem from damage as it is pushed up through the soil. Once into daylight, the hook straightens, the plumule turns green and rapidly grows into a photosynthesizing shoot system of stem and leaves. The cotyledons remain below ground, gradually shrivelling as their store of energy is used up (Fig. 5.7). This type of germination is known as *hypogeal*, meaning simply 'below ground'.

More commonly, as in turnip, beet, carrot and lettuce, there is little growth of the plumule immediately following emergence of the radicle. Instead it is a portion of the embryo axis just below

Fig. 5.6 An onion seedling.

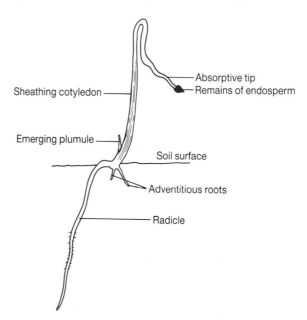

Absorptive tip
Remains of endosperm

Sheathing cotyledon

Emerging plumule

Soil surface

Adventitious roots

Radicle

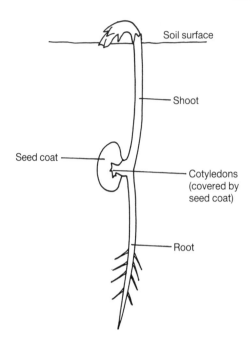

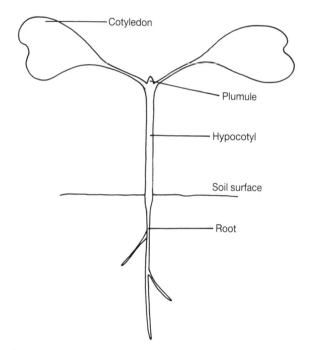

Fig. 5.7 A broad bean seedling showing hypogeal germination, in which the cotyledons remain below ground.

Fig. 5.8 A turnip seedling showing epigeal germination, in which the cotyledons are carried above ground on an elongated hypocotyl.

the cotyledons that elongates, carrying not only the entire plumule but also the cotyledons above ground. This stem-like structure is known as the *hypocotyl*. Once above ground the cotyledons expand greatly to become the first functional leaves of the new plant. Only at this stage does the plumule itself begin to grow, eventually taking over from the cotyledons as the photosynthetic system (Fig. 5.8). This is described as *epigeal* ('above ground') germination.

A form of germination intermediate between hypogeal and epigeal is seen in the French or haricot bean, the plumule of which is well formed in the seed and is carried above ground together with the cotyledons on an elongating hypocotyl. In this case, however, the cotyledons never contribute significantly to photosynthesis in the seedling; once their energy reserves are exhausted they wither and fall off.

5.2.2 *Patterns of metabolism during germination*

Prior to imbibition, the metabolic activity of a quiescent seed is effectively zero. The first signs of metabolic activity in a seed about to germinate are

not those of respiratory enzyme systems but of enzymes whose role is to *mobilize* energy reserves, whether in the cotyledons or in the endosperm.

In a cereal seed virtually all the energy store is in the form of starch, which is insoluble and therefore cannot be moved from cell to cell. The seed needs two enzymes to convert starch to sugars, in which form it can be transported via the scutellum to the embryo axis. These enzymes are α-amylase and β-amylase. It appears that β-amylase is present in the quiescent seed, needing only minor alteration to its structure to activate it, but α-amylase is synthesized from its constituent amino acids after metabolic activity in the seed resumes. Most of the amylase is produced in the *aleurone layer*, a zone of the endosperm immediately inside the seed coat (see Fig. 5.2), and is secreted from here into the rest of the endosperm.

Activation and synthesis of amylase and the mobilization of starch as sugars underlie the commercial process of malting, in which barley grains are allowed to germinate and are then killed by drying. The dried germinated grain is known as *malt* and is rich not only in easily fermented sugars for the production of alcohol, but also in amylase. Because of its high amylase activity, malt is frequently added to other starchy materials such as ungerminated maize grain to produce fermentable sugars.

In some crop seeds, oxygen may not enter fast enough through the seed coat to permit fully aerobic respiration in the earliest stages of germination. There may therefore be a certain amount of anaerobic respiration until the radicle has ruptured the seed coat and oxygen can enter freely.

Once this stage is reached the seedling respires extremely rapidly, causing rapid depletion of stored energy reserves and thereby a loss in dry weight of the seedling. Much of the sugar resulting from the mobilization of reserves is not, however, respired away but provides carbon skeletons for a wide range of compounds required by the growing plant – amino acids (and hence proteins), lipids, nucleic acids and so on. A large proportion of the sugar is assembled into cellulose for the building of cell walls.

When the cotyledons or plumule come above ground and photosynthesis gets under way, the rate of dry weight loss declines. Eventually sufficient photosynthetic area has been produced to allow a positive net assimilation rate. The seedling is now no longer dependent on energy reserves bequeathed by the parent plant, and can be said to be *established*.

5.3   Germination and establishment in the field

The period of germination and seedling growth prior to establishment is, as mentioned, the most vulnerable stage in a plant's life. Since achieving good crop establishment is one of the most important skills of husbandry, let us consider some of the hazards the plant has to face during this period.

1. Germination takes place in the upper layers of the soil, which are prone to rapid and wide variations in moisture content and temperature. The surface layers of most soils dry out rapidly during even the short dry spells encountered in moist, temperate climates. At the other extreme, they may become waterlogged during heavy rains. Before the seedling emerges above ground its root system receives no oxygen via the shoot; it is therefore entirely dependent on a supply of oxygen from the soil and is more susceptible to waterlogging than an older plant. Ground frost may cause chilling or other injury. For example, as water in moist soil freezes, it expands, causing the soil also to expand. This may uproot seedlings gripped in the ice. The process is called *frost heave* (see Section 4.3.4) and the uprooted seedlings frequently die from later desiccation. Direct sunshine causes higher temperatures to occur at the surface of soils than at a short height above them, and this may be equally damaging.

2. The zone of the atmosphere immediately above the soil, into which the delicate shoot of the seedling emerges, may be a particularly hostile environment, not only because of temperature extremes but also because of sandblasting by wind-blown soil particles.

3. Because of their small size, seedlings are seriously damaged if attacked by pests and diseases. Seedlings showing epigeal germination which have their growing points removed by insects have no powers of regeneration, since they lack axillary buds below ground. Many soil fungi, including *Pythium*, *Phytophthora* and *Corticium*, attack and often kill seedlings, causing the disease known as *damping off*. Infections may also be carried in the seed itself.

4. Again because of their small size, seedlings are particularly easily choked by weeds, many of which are adapted to grow extremely rapidly and establish quickly from seed.

Plant mortality is thus usually highest at the seedling stage and the resulting reduction in plant population density has major influences on crop yield, as we shall discuss in Section 11.2.

The hazards outlined above are especially serious in modern precision-seeded crops, where there is little margin for error (see Section 11.2.5). As an example, in the traditional method of turnip growing, seeds were sown thickly in rows, so that a 'hedge' of seedlings arose which had to be thinned. Predation by turnip flea beetles, a common insect pest of this crop, could remove a considerable proportion of seedlings without affecting final yield. Nowadays turnips are almost exclusively sown by machines which deposit single seeds at exact spacings (precision drills). This removes the need for laborious thinning, but means that any seedlings which fail to establish leave sizeable gaps, causing yield loss. Gaps in a row crop cause variability in size of produce, which reduces quality as well as yield, especially in vegetables grown for processing. This problem is examined in more detail in Section 11.2.5.

The establishment of crop seedlings must be

achieved within a short time – in fact, a very few weeks from sowing. The longer establishment is delayed, the more likely the seedling is to run out of essential nutrients, the store of which in the seed is distinctly limited, or fall prey to some of the hazards just described.

Another advantage of quick establishment is that in effect it lengthens the growing period of the crop within a season and thereby allows bigger yields to accumulate. The practice of *fluid drilling*, whereby seed is partially germinated and then sown in a gel to keep it moist, is a new technique designed to shorten the vulnerable period before establishment.

5.3.1 Seed germinability and vigour

From the foregoing it is obvious that the *speed* of germination and establishment is at least as important as the final *percentage* of seeds which germinate. Both percentage and speed of germination are affected by external, or seed-bed, factors and by internal, or seed, factors. If we remove the effects of seed-bed factors by placing seeds in an artificial medium with ideal growing conditions, we can measure the germination percentage as influenced only by seed factors such as age, storage conditions, fungal infection and dormancy. This is a measure of the inherent *germinability* of the seed sample. Routine laboratory tests as performed in seed-testing stations can very quickly detect seed lots of poor germinability (Fig. 5.9).

Seed, as opposed to seed-bed, qualities which affect the speed of germination and establishment are collectively expressed as *seed vigour*. This is far more difficult to determine in the laboratory than germinability. Sometimes a measure of germination percentage after a very short time, say four days, is used to estimate seed vigour but this has not always related well to field performance of the seed lot. A new technique, known as controlled deterioration testing, is proving successful in identifying vegetable seed lots with low vigour and poor field performance, thus helping to improve seed quality for the grower. In this technique, the extent to which seed vigour has deteriorated by aging (see Section 5.3.5) is determined by measuring the effect of a precise period of artificially induced and accelerated aging on germination.

The quality of seed, in particular its germinability and freedom from impurities, is in most developed countries subject to legislative control and certification schemes to guarantee minimum standards of purity and health. These have so greatly improved the reliability of seed for the grower that the availability of good seed is almost taken for granted in these countries, although there may still be problems of low vigour. In systems of 'subsistence' agriculture, however, especially in the developing world, farmers still rely heavily on home-saved seed and for them poor seed quality in all its aspects is a major limitation to productivity.

Fig. 5.9 A laboratory seed test, showing turnip seeds of good (right) and bad (left) germinability.

Although we talk separately about germinability and vigour, there is of course no absolute distinction between these two aspects of seed quality. At the extreme, seeds with very low vigour simply fail to germinate. Nonetheless, vigour seems to be a highly significant factor in crop establishment even when germinability is high, especially in legumes such as peas and beans and in small-seeded vegetables. Vigour is even more important where a crop is being grown near the limits of its climatic tolerance range or sown into adverse soil conditions (two factors that often coincide), further slowing down seedling growth. In other words, poor vigour has its greatest potential negative impact in the very situations where good vigour is most needed. The pea crop in Scotland is a good example of this.

Since these two aspects of seed quality are so important for crop establishment, the main factors influencing them are considered in Sections 5.3.2–5.3.5. They include pre-harvest growing conditions, harvesting and drying, seed size, age and storage conditions. Let us first briefly mention two other factors, seed dormancy and microbial infection.

The importance and mechanisms of seed dormancy have been examined in Sections 5.1.4–5.1.6. Among agricultural crops it is seldom of importance in the field except when newly harvested seed is being sown.

Microbial pathogens, especially fungi, infecting or contaminating crop seed have a major effect on seed quality, influencing not only germinability and vigour of the seed but the later health of the plant if it succeeds in establishing itself. The existence of such seed-borne infection and contamination is a major reason for the application of fungicide dressings to seeds.

5.3.2   Effects of pre-harvest growing conditions on seed quality

Almost any kind of stress on the parent plant can damage the germinability and vigour of its seed. This includes stress both from inadequate nutrition and from unfavourable weather conditions.

Let us consider first the question of nutrition. Parent plants subject to stresses such as drought or which carry heavy microbial infections, for example cereals infected with rust fungi, have less carbohydrate and protein available for seed filling and tend to produce small shrivelled seeds with poor germinability and vigour. A more subtle

example is the condition known as 'marsh spot' in pea seeds. Affected seeds, when split open, show a dark brown circular patch in the centre of each cotyledon. The cause is manganese deficiency in the parent plant; the effect is lowered seed vigour. The condition was formerly common in seed from Romney Marsh in south-east England, hence the name, where high soil pH rendered the manganese unavailable for uptake by the plant (see Section 3.5.2).

Temperature is one of the most significant of pre-harvest weather conditions affecting seed vigour. For example, in the disorder of pea seeds known as 'hollow heart', each cotyledon has a hemispherical hollow in its inner surface. This appears to result from faulty maturation of the ripening seed in hot weather conditions and causes reduced vigour. A warm, dry ripening season can reduce germinability of barley seeds by causing prolonged dormancy. At the other extreme, freezing of seed on the mother plant, before the seed has dried out during ripening, can also damage its germinability and vigour. This can be a problem with maize in some parts of North America. Even the wetting and drying of some seeds on the parent plant in showery conditions has been shown to be detrimental.

The effects of pre-harvest weather and nutrition on seed quality can thus be important and this may be an area in which there is still much to be learned.

5.3.3   Effects of harvesting and drying on seed quality

A first requirement for obtaining good quality seed is that the seed should be fully matured. The longevity and vigour of seed are often reduced if it is harvested when immature. This is partly due to increased mechanical damage when seed moisture content is high, as shown below, but other factors may be involved. If seed is harvested too early and dried artificially it can show a dramatic loss in germinability. In one experiment peas harvested and then dried 37 days after full bloom had only 5% germinability, whereas delaying harvest for just 4 more days and then drying produced seeds of 79% germinability. Seeds evidently may acquire the ability to withstand drying in a very short period during ripening.

Hand threshing, mechanical harvesting and subsequent handling such as grading and dressing, introduce seeds to processes they have not evolved to withstand, and the result is physical damage to

content in the soil thus reduces both the speed and the percentage of germination.

Crop species differ in the minimum moisture content of the seed which will permit germination. Thus maize and sugar beet seed can germinate at moisture contents as low as 30%, whereas soya bean seed must reach 50% moisture content before it will begin to germinate. The moisture content reached by the seed depends not only on soil moisture content but also on other physical characteristics of the soil. Soil texture, for example, affects the ease with which water moves through the soil, this being easier in light, sandy soils than in heavy, clay soils. A fine tilth, achieved by raking or harrowing, and a firm seed-bed, achieved by rolling, are essential for good seed–soil contact to facilitate the transfer of moisture from soil to seed. Germination can be poor in a cloddy or fluffy seed-bed even at adequate levels of soil moisture.

Increasing soil moisture content, up to an optimum when the capillary spaces are full of water but adequate air is present, increases the speed of germination and seedling growth. At higher moisture contents, however, waterlogging becomes a danger to the seedling. The effect of oxygen deprivation through waterlogging becomes more acute with increase in soil temperature because the increased metabolic rate leads to faster accumulation of ethanol, and there are considerable differences in tolerance between species. Rice and lettuce seed, for instance, will germinate at very low oxygen supply levels, but broad beans, peas and sugar beet will not. The importance of these differences in practice is, however, uncertain. High soil moisture content can also encourage attack on seedlings by pathogenic microbes in the seed or soil.

Soil compaction. Compacted soil can delay or prevent seedling emergence by impeding physical penetration by the radicle and plumule. The effects of compaction on root growth are dealt with in Chapter 6, but the plumule is very limited in its ability to push through soil. Clay soils, in particular, as they dry out, can form a strong barrier to plumule growth, and the commonest form of this is *capping*, that is the formation of a surface crust on the soil either by heavy rain or by the passage of machinery.

Fertilizer application. Adequate mineral (including nitrogen) nutrition is important in ensuring rapid seedling establishment, and fertilizer

practices are partly intended to achieve this. The supply of phosphate is, as was seen in Chapter 3, particularly important because of its immobility in the soil, especially at low temperatures. The modern practice of combine-drilling, which places fertilizer in a band below the seed at sowing time, not only increases the efficiency of fertilizer use by reducing the necessary application rate but places phosphate in a readily accessible position. This has been found to be particularly effective in the sowing of certain forage legumes. Placement of fertilizers, as opposed to surface spreading, has comparatively little effect on nitrogen and potassium nutrition of seedlings because of the higher mobility in the soil of these nutrients.

Fertilizer placement can lead to problems by creating osmotic forces around the seed which may inhibit water uptake. These forces may also cause direct damage to seedling roots, especially in dry conditions.

5.3.7 Germination and establishment in relation to sowing depth

In the wild, seeds simply fall on the soil surface and eventually must germinate there, unless they are small enough to be washed into cracks in the soil or are buried by the activities of earthworms and other soil animals. Only a few species, including wild oats (Fig. 5.11), have devices for self-burying.

The burial of crop seed by drilling is a man-made practice imposed on seed which in nature would normally germinate on or near the surface. The practice has the advantages of sheltering the seed from extremes of temperature and moisture variation and from predation by birds and mammals. However, the emergence of buried seed requires the expenditure of more energy from the seed's reserves, the extent of which is related to the size of the seed. Thus in general large seeds can emerge from deeper levels than smaller ones and the larger the seed the deeper it is usually sown.

With small seeds, such as those of carrot and flax, depth of sowing has finer margins of tolerance and optimum sowing depth may vary between different sized seeds in the same batch. Small-seeded perennial forage species must be sown shallowly for this same reason, and are often simply scattered (broadcast) on the soil surface, followed by some form of raking to cover them lightly. The finest seed is broadcast without raking, a practice which is also necessary for seeds such as

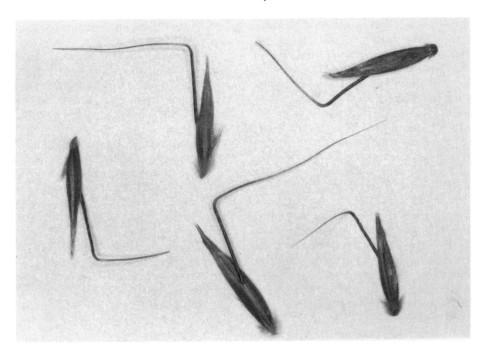

Fig. 5.11 The seeds of wild oat. The long bent awn or bristle on the wild oat seed turns like a corkscrew in response to changes in humidity. This turning action helps to bury the seed in soil.

those of tobacco or fine bent grass, which require light for germination.

Shallow-sown seed, whether broadcast or drilled, is very vulnerable to drying out of the soil. Many wild species exude mucilages from the seed to improve contact with the soil in this situation and to retain a film of water around the seed. A few cultivated species, including brassicas, preserve this ability to some extent. Good seed–soil contact is, however, very important with surface or shallow sowing to ensure efficient water uptake, and is generally improved by heavy rolling to press the seed on to and into the soil.

In most species the hypocotyl or plumule continues to elongate either until it emerges above the soil surface or until the energy reserves in the seed are exhausted. In some plants, including dwarf varieties of soya bean, however, the length of the hypocotyl is genetically limited and the plumule above the cotyledons may be left with the task of penetrating the top layers of the soil, a task for which it is less well equipped than the hypocotyl. Often it fails to reach the surface at all.

Similarly, in some cereals, especially dwarf and semi-dwarf varieties of barley and wheat, the lengths of coleoptile and epicotyl are genetically determined. These organs stop growing regardless of whether the tip of the coleoptile has reached the soil surface. A leaf emerging from the coleoptile tip below the surface is often unable to push through the soil, especially if capping has occurred. In unevenly consolidated seed-beds, seed drills often sow to variable depths and under these circumstances a proportion of seed may be sown too deep and fail to emerge.

Deeper sowing is sometimes advocated as a method of encouraging deeper rooting. In cereals at least this practice cannot work since the downward growth of seminal roots is not affected by the seed's position in the soil and the adventitious roots develop from stem nodes which, regardless of sowing depth, are located by the plant just beneath the soil surface.

Summary

1. The first phase of growth and development is germination of the seed and establishment of the seedling. Viable seeds which are not germinating are said to be quiescent if growth is prevented by lack of an essential external factor such as water, or dormant if they will not germinate even in favourable external conditions because of some internal barrier to growth.

The seed is the vehicle by which genetic information passes from parent to progeny. It is also an organ of multiplication, dispersal and survival during unfavourable periods. It contains an embryo, an energy store (either in the cotyledons of the embryo or in the endosperm outside the embryo) and a protective seed coat. The embryo consists of one or two cotyledons with a plumule, or rudimentary shoot, above, and a radicle, or rudimentary root, below them. In seeds containing endosperm the cotyledons are usually involved in the transfer of carbohydrates from endosperm to radicle and plumule during germination.

Seeds, as conventionally stored, contain very little water. Before they can germinate they must imbibe water. This is a purely physical process resulting from the strong negative water potential of partially dehydrated protoplasmic colloids. The rate of imbibition may be moderated by the resistance of the seed coat to the entry of water. In some seeds the seed coat is completely impermeable. This is one of the main mechanisms of dormancy, which is not broken in the wild until the seed coat is damaged by microbial decay, frost or fire. In cultivation, such seed can be released from its dormancy by scarifying.

The other basic mechanism of dormancy is a property of the embryo, and this form persists even when the seed is fully imbibed. In some cases the embryo is not fully formed and needs time to mature. More commonly this type of dormancy results from the action of chemical growth inhibitors. The level of these, or the balance of inhibitory and stimulatory substances, in the seed is affected by external conditions, which can therefore lead to the breaking of dormancy. The most important dormancy-breaking mechanism is stratification, the natural or artificial exposure of the imbibed seed to particular, usually low, temperatures.

2. The first part of the embryo where growth recommences following a period of quiescence or dormancy is the radicle, which emerges from the seed coat to begin functioning as a root. Shortly afterwards, in grass and cereal seeds, the plumule emerges from the seed coat and grows upwards within a protective sheath, the coleoptile; the first leaf breaks through its tip once this is above the soil surface. In dicotyledonous seeds germination may be hypogeal, with the plumule emerging and the cotyledons remaining below ground, or more commonly epigeal, where the hypocotyl elongates to bring the cotyledons above ground and the plumule generally does not begin to develop until later.

Germination and seedling growth is characterized by great metabolic activity, first the mobilization of energy reserves, then respiration in the embryo. It is some time before photosynthesis in the cotyledons or plumule can balance the dry weight loss due to respiration, but once this stage is reached the seedling can be considered to be established.

3. The period up to this point is the most vulnerable in the life of the plant. Germination takes place in the upper layers of the soil, where extremes of temperature and moisture are experienced. The seedling, because of its small size, is very susceptible to damage by pests and diseases and choking by weeds. It is therefore essential that germination and establishment are completed in as short a space of time as possible. The speed and evenness of germination are at least as important as the germination percentage.

4. The inherent germinability of seed and its vigour depend on dormancy, the existence of microbial infection, pre-harvest conditions (especially disease and mineral deficiency), damage during harvesting and handling, seed size, age and storage conditions. Seed vigour is particularly important where the crop is being grown close to the limit of its climatic tolerance range or sown into adverse soil conditions.

The longevity of seed ranges from a few days to a few years, depending on species. Seed stored dry ages because normal cellular repair mechanisms do not appear to function in dried seed, and damage due to the deterioration of many systems within tissues accumulates. In the early stages of aging seeds show reduced vigour, decreased ability to germinate at extremes of their environmental tolerance range, and increased susceptibility to pathogens. In the later stages they cannot germinate at all. The rate of aging is influenced by the growing conditions of the parent plant, the species, variety, and above all the temperature and moisture content of the seed in store, except for some species with very hard-coated seed.

5. The main seed-bed factors affecting the speed and percentage of germination are soil temperature and moisture content. Drainage helps to prevent waterlogging and hence oxygen deprivation of seed and seedling; it also accelerates the warming of the soil in spring. The preparation of a fine tilth is essential for good seed–soil contact and hence for rapid water uptake. Rolling consolidates the seed-bed and further improves seed–soil contact, but excessive compaction can cause a barrier to pene-

tration of the plumule through the soil, leading to reduced establishment. Mineral nutrition, influenced by fertilizer application, is also important, but fertilizer placement can result in high osmotic pressures around the seed, inhibiting water uptake.

6. Burial of seed at sowing places it in a zone of the soil where there is less extreme fluctuation in temperature and moisture content and where it is protected from birds and mammals. The larger the seed, the more energy reserves it possesses and the deeper it can safely be sown. Very small seeds and light-requiring seeds must be broadcast or surface-sown. Care must be taken not to sow too deeply seeds of species or varieties whose capacity for plumule or hypocotyl elongation is genetically limited.

6

Plant Growth and Development:
The Vegetative Plant

6.1 Growth and development of the root system

The underground parts of a crop are out of sight and too frequently out of mind, yet more often than not they account for more than half of the weight and bulk of the crop. It is through these parts that the great majority of cultivations have their effect on crop performance.

While it is known that poorly timed or carelessly applied inter-row cultivations can reduce the yield of row crops by cutting roots, experiments indicate that much of the time plants have a considerably larger root system than they need to support them. Perhaps this is because if they are to survive, plants must have sufficient roots to carry them through the period of maximum stress from drought or other causes they are likely to experience in their life, not just the average level of stress.

Both shoots (in particular, their leaves) and roots must have large surface areas to perform their respective functions of photosynthesis and absorption efficiently. Leaf design, especially regarding surface area, is limited by the need to conserve water, as was seen in Section 2.4.1, but root surface area has no such constraint placed upon it. In one study of a mature rye plant the surface area of the roots was calculated to be 639 m² or 130 times the surface area of the shoot.

This massive surface area is achieved by repeated branching into fine rootlets. The rye plant mentioned above had over 13 million roots, measuring more than 500 km in total length. Such a system clearly explores the soil very thoroughly, yet it is doubtful if a root system ever occupies more than 5% of the volume of the soil, and the usual figure is probably nearer 1%. Only a small proportion of the soil is therefore in direct contact with the roots, hence the importance of mass flow and diffusion of water and minerals through the soil and the problems of nutrition that arise with

relatively immobile ions such as phosphate (see Section 3.2.4).

To understand how this remarkable system develops it is easiest to look first at the growing root tip, which spearheads the advance of the root through the soil.

6.1.1 Growth and development at the root tip

After the radicle has emerged from the seed coat, it relies for its continued growth not on cell elongation alone but on cell division followed by elongation. Now cell division in plants is localized in small regions called meristems. The meristem by which a young root grows is to be found at the extreme tip of the root, and is known as the *root tip meristem* (see Fig. 4.14). Elongation of cells, originally occurring throughout the radicle, becomes restricted to a zone a few millimetres long immediately behind the root tip.

The root tip meristem, like all meristems, consists of small, thin-walled, unvacuolated cells with large nuclei. These meristematic cells retain the capacity for division throughout the life of the root. They are very delicate, and the root tip is probably exposed to more physical abrasion than any other part of the plant as it pushes forward through the soil.

The meristem is protected from this abrasion by a shield of loosely packed cells, the *root cap*, which constantly sheds cells from its surface. Root cap cells are renewed from behind by division in the root tip meristem. Separate groups of cells can be identified within the meristem, one group giving rise to the root cap, and another to tissues of the root itself. The root cap exudes mucilage which forms a sheath around it, acting as a lubricant for the forward thrusting of the root tip.

At the tip of the meristem, just beneath the root cap, is a small group of meristematic cells called the *quiescent zone*, where relatively little cell division occurs. Its function is uncertain but it appears to regulate division in the remainder of the meristem. Occasional cells divide off from the quiescent zone and these then divide repeatedly and very actively, forming the main body of the root tip meristem.

The cells towards the rear of the meristem stop dividing and their volume enlarges 30–150 times, mainly through *vacuolation*, the formation of water-filled vacuoles which fill most of the volume of the cell. This is also a zone of rapid cell wall deposition, as the greatly enlarged cells must be accommodated, protected and strengthened by more extensive and thicker walls than the cells of the meristem. The cells do not enlarge equally in all directions, but become elongated in the direction of the root axis. It is this elongation that provides the forward thrust of root growth. In good soil conditions the rate of root extension can be rapid, for example 2 cm day^{-1} for a main root of wheat. Up to 6 cm day^{-1} has been recorded in maize.

Immediately behind the zone of cell enlargement, but overlapping with it, is a zone of cell differentiation. Here the outermost layer of cells becomes identifiable as epidermis, with some of them growing out as root hairs in some species. Beneath the epidermis the cells continue to enlarge and begin to round off from one another forming the characteristic loosely packed parenchyma tissue of the cortex.

It is in this same zone that suberin is deposited in the Casparian strips of the endodermis, allowing the root to select and accumulate mineral ions, as we discussed in Section 3.4.1. Differentiation of xylem tissue in the centre of the root involves lignification and thickening of the cell walls and death and loss of the cell contents. Differentiation of sieve tubes in the phloem requires dismantling of most of the cell organelles, including the nucleus, although large numbers of mitochondria remain to provide the energy for translocation. Only in the narrow cylindrical zone surrounding the vascular tissue but within the endodermis do the cells remain relatively undifferentiated. This zone, it may be recalled from Section 2.3.1, is the *pericycle* of the root.

6.1.2   Branching and maturing of the root

Some distance behind the root tip, small groups of cells in the pericycle become meristematic again, dividing rapidly to produce branch roots. Originating as they do deep within the tissues of the root, branch roots at early stages of development are well protected from the soil environment. Soon, however, they break out through the endodermis, cortex and epidermis, rupturing these tissues as they do so. Usually they arise opposite the points of the xylem star (Fig. 6.1), so that if there are only

Fig. 6.1 Origin and development of a branch root. A branch root originates in the pericycle of the main root, its xylem and phloem linking up with those of the main root. As it grows out through the endodermis, cortex and epidermis, these tissues are ruptured.

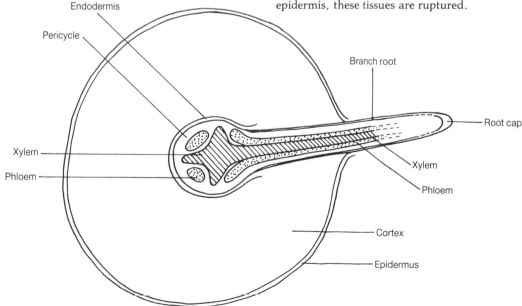

two such points, as in sugar beet or swedes, the branches appear to be borne in two opposite ranks. Xylem and phloem quickly differentiate in a branch, linking up with those in the main root. At the tip of the branch is a meristem protected by a root cap exactly like that of the main root.

The shedding of root cap cells is accelerated by their death and degeneration following invasion by soil bacteria. As the root matures, the root hairs, epidermis and cortex are soon similarly invaded; these entire structures are sloughed off, often before the root is very old. The root is now said to be *decorticated*, leaving the endodermis as the outermost layer. The Casparian strips and other deposits of suberin in the endodermis appear to prevent the bacteria from entering the pericycle.

Quite early in the life of many dicotyledonous roots a cylinder of meristematic cells arises between the phloem and xylem. This is the *vascular cambium*, which gives rise by division and differentiation inwards to secondary xylem and outwards to secondary phloem, exactly as in dicotyledon stems (see Section 4.2.2). It allows a large increase in girth, and therefore in strength and water-carrying capacity of the root. Around the same time, a second zone of meristematic cells forms in the pericycle. This *cork cambium* produces the protective cork tissue which surrounds

the mature dicotyledon root. (Monocotyledon roots do not form vascular or cork cambia.)

The roots of neighbouring plants intermingle closely in the soil. As their girth increases by secondary tissue development they may come in contact with one another. In forests, timber plantations and orchards the roots of adjacent trees of the same species as they come in contact often form natural grafts one to the other. Their vascular tissues unite just like those of an artificial graft between two stems. These root grafts may provide pathways for transfer of chemicals, viruses and other pathogens from tree to tree throughout the forest or plantation.

6.1.3 Patterns of root development

The typical dicotyledon root system, such as that of sunflower (Fig. 6.2), centres on a more or less vertically growing main root derived directly from the radicle and called a *tap root*. Numerous more slender *lateral roots*, themselves much branched, grow out semi-horizontally from it. In annuals such as sunflower and lettuce the tap root is primarily an anchoring, absorbing and transporting organ but in biennials and perennials it often becomes much thickened and serves as a carbohydrate storage organ (Fig. 6.3). We shall examine some of the

Fig. 6.2 Root system of sunflower, typical of many dicotyledons, showing vertical tap root and more slender semi-horizontal lateral roots.

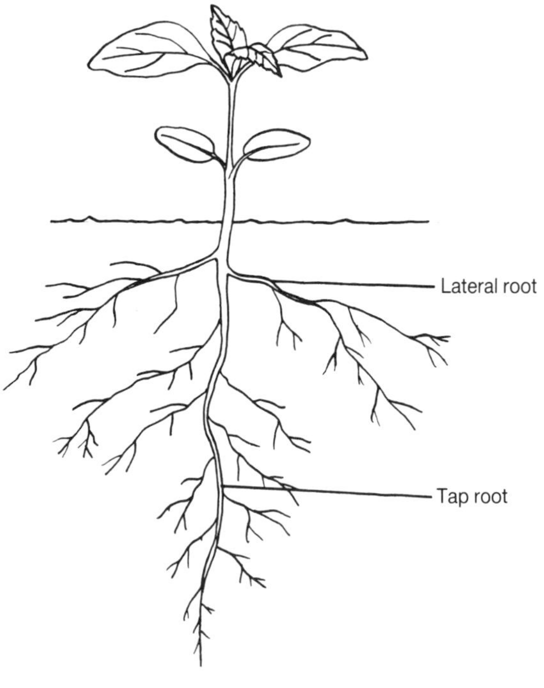

— Lateral root

— Tap root

Fig. 6.3 Storage organ of sugar beet, derived from the tap root.

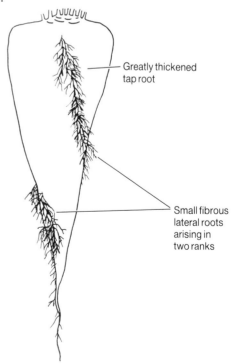

— Greatly thickened tap root

— Small fibrous lateral roots arising in two ranks

ways in which such storage organs develop in Section 6.1.9.

Tap roots often go particularly deep into the soil and literally 'tap' reserves of water that are out of reach of shallower root systems. They therefore give resistance to drought. It is the tap root of lucerne (alfalfa) that enables this crop to be grown as forage in areas too dry to support grass. Trees often have tap roots reaching down many metres into the soil, as well as a system of lateral roots ramifying nearer the surface. In some tree species such as birch, the deep tap roots have an important indirect effect on the associated vegetation. Through the leaching (washing out of soluble materials) and decay of birch leaves, minerals taken from deep in the soil are returned to the surface where they are accessible to the much shallower roots of grasses and other herbs.

In some dicotyledons, such as the common annual weed groundsel (Fig. 6.4), the tap root aborts at an early stage, and the entire root system is derived from a profusion of lateral roots. Known as a *fibrous root system*, it seldom penetrates as deep into the soil as a tap root, but often produces a greater concentration of roots just below the surface.

Commonly, new roots arise from the lower nodes of the stem, and these may supplement or replace the primary root system. Roots such as these, which develop from stem tissue rather than

as branches of existing roots, are called *adventitious roots*. Creeping stems such as the stolons of strawberry or creeping buttercup, or the rhizomes of coltsfoot, produce adventitious roots from nodes all along their length. In plants propagated vegetatively (whether by natural or artificial means) from organs such as the tubers of potato, which are modified stems, all roots must be adventitious in origin.

In cereals and grasses (in other monocotyledons the situation is similar), the seminal roots derived from the radicle are often not much branched. They grow to considerable depths in the soil and, in the cereals at least, persist for most of the life of the plant. Quite early, however, they are supplemented by adventitious roots from the short stem; these become the main anchoring and absorbing roots of the mature plant (Fig. 6.5). In perennial grasses the seminal roots tend to die off once the adventitious root system is formed.

Part of the significance of adventitious roots arises from the fact that where a shoot branches at ground level, as in cereals and grasses, each branch shoot (*tiller*) develops its own roots. It can thus become largely independent of other shoots which originally formed part of the same plant.

In adventitious root systems no one root is more prominent than the others and in this respect they resemble fibrous root systems. In general they do not penetrate as deeply into the soil as tap roots, although in dry conditions the adventitious roots of wheat or rye have been observed to go to great depths.

This is an example of a more general phenomenon, namely that the root system is highly flexible

Fig. 6.4 A fibrous root system, produced by the early death of the tap root and profuse branching of lateral roots in the upper layers of the soil.

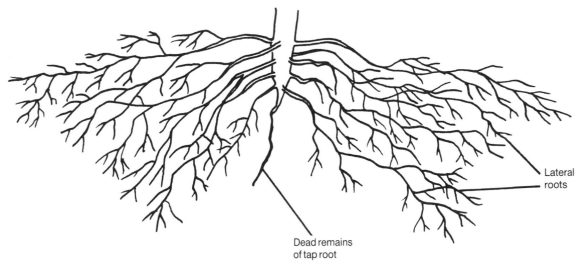

Lateral roots

Dead remains of tap root

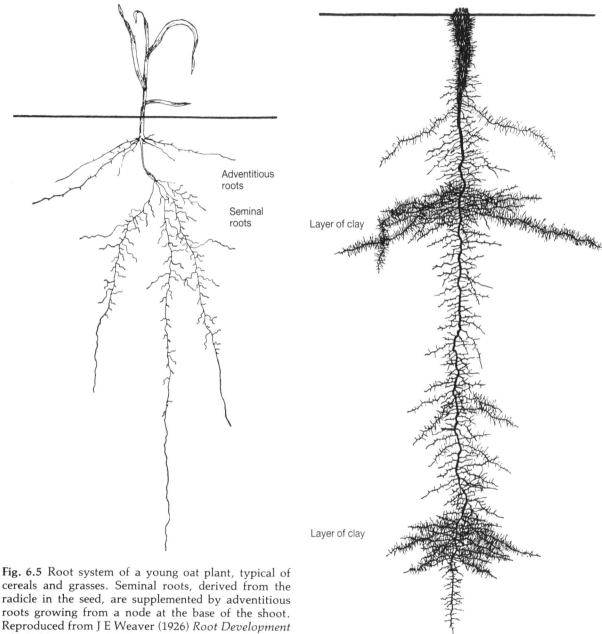

Adventitious roots

Seminal roots

Layer of clay

Layer of clay

Fig. 6.5 Root system of a young oat plant, typical of cereals and grasses. Seminal roots, derived from the radicle in the seed, are supplemented by adventitious roots growing from a node at the base of the shoot. Reproduced from J E Weaver (1926) *Root Development of Field Crops*. McGraw-Hill, New York, with permission of the publisher.

in its response to soil conditions during its development. Roots develop preferentially wherever soil conditions are more suitable, and the general pattern of root development of a species can be greatly modified by cultural conditions (Fig. 6.6). For example, roots proliferate around sources of moisture, whether seeking the water table deep in the

Fig. 6.6 Root system of sugar beet in sandy soil with two layers rich in clay. Lateral root development has been more profuse in the layers of clay, where water and mineral nutrient availability are greater. Reproduced from J E Weaver (1926) *Root Development of Field Crops*. McGraw-Hill, New York, with permission of the publisher.

soil in dry conditions, or ramifying just under the surface in a regularly watered garden. Too frequent watering can, by encouraging surface rooting, lead to increased drought susceptibility if watering is discontinued. Roots also proliferate in areas of increased mineral availability, such as result from fertilizer placement by combine seed drills.

6.1.4   Transplanting

Root development can thus be seen as a continuous process from emergence of the radicle from the seed through to the establishment of the very extensive root system of the mature plant. In the husbandry of certain crops, however, the continuity of root growth is interrupted by the uprooting of the seedling or young plant and its replanting elsewhere. This is the practice of *transplanting*, a common feature of the growing of timber crops, many temperate horticultural crops and ornamentals, and a large proportion of tropical agricultural crops.

The seeds of these crops are sown not in the field in which the crop is ultimately to be grown to maturity but in a special area set aside for the purpose. The term *nursery* is often used for such an area if it is outdoors, but many horticultural seedlings are raised in glasshouses or other protective structures before being transplanted outside.

The nursery or glasshouse offers several advantages over sowing the crop direct into the field.

1. The plants are grown at much higher density than in the field, so that the thousands of plants which will eventually occupy each hectare of ground can be accommodated in a few square metres. Forest trees need particularly wide spacing to grow to maturity and a small nursery can provide a succession of young trees for huge areas of plantation over many years. Because of the small area involved, it is possible to protect the plants during their most vulnerable stage from pests, diseases, drought, wind and extremes of temperature much more easily and effectively than in the field.

2. The best possible site can be selected for the nursery. Plants are moved to less favourable situations only when they have reached a more tolerant stage of growth.

3. It is possible to start the plants off from seed before the beginning of the growing season, either under glass where field sowing is prevented by low temperature, or in an irrigated nursery where a dry season limits field sowing. Once the spring or rainy season arrives the seedlings can be transplanted in

the field. The growing season for the crop has thus been extended, with obvious benefits for yield or for earliness of harvesting.

4. Seedlings are extremely susceptible to damaging competition from weeds (see Section 11.4). The nursery culture of seedlings allows close attention to weed control and the very fact that they are grown *en masse* rather than widely spaced as in the field probably affords them some protection from weeds. Transplanting into newly cultivated soil gives the crop a head start on the weeds of the field. The choking of seedlings by weeds is a particularly serious problem in the humid tropics and is one of the main reasons for the traditional husbandry system for rice, which involves the laborious transplanting of nursery-grown seedlings. Now that selective herbicides are available for weed control in the young rice crop, nursery culture and transplanting are less widely practised.

5. The raising of seedlings in a nursery allows selection of the strongest, healthiest, most promising individuals for planting out. This is especially valuable in the case of forest trees which are to occupy the land for many years or for perennial plantation crops such as tea which are to be harvested repeatedly over a long period.

6. For crops which are propagated vegetatively by artificial means such as cuttings (see Chapter 10), it is usually necessary to go through a nursery stage while they develop roots and become established.

These, then, are some of the reasons for transplanting. But the practice is not without its difficulties. Transplanting inevitably causes some damage to roots, particularly the delicate young roots through which the plant absorbs most of its water. Furthermore, the intimate contact that exists between root and soil and which, as we have explained, is so necessary for water uptake is broken and has to be restored. On the other hand, the shoot is relatively undamaged and continues to act as a large surface for water loss by transpiration. The effect is therefore to put the newly transplanted plant under considerable water stress, resulting in a marked check to growth.

The nub of the problem for successful transplanting is to restore the plant's water supply before it dies of drought. The speed of this depends on three main factors: the species, since this largely decides the ease with which roots can regenerate, the root/shoot ratio of the plant after it has been lifted, and the soil moisture supply.

The aim during transplanting must be to get the plant established in its new site as quickly as possible in order to restore rapidly its water supply and reduce the growth check to an economically acceptable level. To achieve this, root damage at lifting must be minimized, new roots must be regenerated rapidly to restore root–soil contact in the new site, and excessive water loss must be avoided or made good.

To minimize root damage, horticultural crops, ornamentals and young trees are often transplanted with a large block of soil around the roots. In theory any plant can be moved from one site to another if a sufficiently large soil ball is taken along with its roots. Horticultural vegetable, fruit and ornamental plants are increasingly commonly *container grown*, very often in containers made of peat or other materials which decompose in the soil, in which case the entire container and its contents are simply bedded into the soil. This permits transplanting at any season.

For mechanized transplanting, however, as in field-scale growing of cabbage or Brussels sprouts, seedlings have to be transplanted with bare roots. Considerable root damage is unavoidable, and the success of transplanting depends very much on the rate at which the damaged roots can be replaced and root–soil contact restored.

Fine fibrous root systems tend to regenerate much more rapidly than the thicker, more brittle roots typical for example of peas or onions. Thus species with more fibrous roots, including tomatoes and brassicas, show rapid root regeneration and therefore normally transplant more readily. Others, such as beans and maize, with thicker roots are slow to regenerate roots and cannot be successfully transplanted on a commercial scale. At best, the check to their growth is so prolonged as to nullify all the advantages of transplanting.

Even species which are readily transplanted as seedlings gradually lose their capacity for root regeneration as they get older. When an older plant is transplanted most of the vigorous young roots are lost or damaged and those that remain are mostly suberized and less able to initiate new branch roots. In addition, as a plant matures its root/shoot ratio tends to decline, and inevitably a smaller proportion of the root system is dug up for transplanting. Thus a relatively larger shoot system has to be supported on the damaged root system, increasing the rate of water loss. However, if properly prepared and lifted with a large enough ball of roots and soil, even sizeable trees can be transplanted. This technique is now widely practised in commercial landscaping.

Transplanting at the seedling stage usually damages tap roots severely. The roots which replace it are densely branched laterals or in some crops a profusion of adventitious roots arising from the base of the stem. Crops such as sugar beet or carrot, in which the tap root is the harvestable part, cannot therefore be successfully transplanted on a commercial scale. It is necessary to prevent the establishment of a deep tap root by seedling forest trees in nurseries to avoid unacceptable damage at transplanting. This may be done by pruning the roots with a machine that undercuts them, thus encouraging the development of a more branched root system that is more suitable for transplanting.

Considering now the question of soil moisture supply, transplanting of annual or biennial crop seedlings often has to take place during the summer months when transpiration rates tend to be high. It is essential that the soil receiving these seedlings is well watered and it is best that transplanting is done in cool, humid weather. Soil around transplants must always be made firm to help restore close root–soil contact. The seedlings themselves should be drought-resistant; glasshouse grown plants may need to be 'hardened' by withholding water for some time before transplanting. This encourages development of a thicker waxy cuticle on the leaves. Commonly some of the leaves are removed before transplanting to reduce the surface area for transpiration but it should be noted that this practice also reduces the photosynthetic area and restricts the production of carbohydrate to fuel new root growth.

Perennials are most commonly transplanted in winter, when they are dormant and, in most species, leafless and when weather conditions are less conducive to transpiration. These features combine to minimize water loss. Herbaceous perennials present no problem, as during the winter they have little or no top growth through which water can be lost. The root/shoot ratio is so large that considerable root damage can be tolerated without endangering the survival of the plant.

Woody perennials present rather more difficulty, as they have a smaller root/shoot ratio. Deciduous trees and shrubs, however, lose very little water during their leafless period in winter and most of them transplant readily at this time. They are best transplanted in autumn when soil temperatures are still high enough to allow some root growth and thus establish a water supply before winter sets in. Evergreens, such as coniferous forest trees, are more of a problem because they present a large surface area for water loss even during the winter. It is particularly important to transplant these at a young stage when their root/shoot ratio is still fairly high.

The success of transplanting dormant perennials is due to the fact that the dormancy does not extend to the root system. However, sufficient time must be allowed for root growth to restore the moisture supply before the shoot resumes growth. To allow this, transplanting should be done only at soil temperatures that permit root growth, and not during cold, frosty periods.

The period immediately after transplanting is a vulnerable one for any species and the grower must do all he can to avoid other stresses being placed on the plant at this time. Young transplanted forest trees, for example, are very sensitive to competition from grass and other natural vegetation, which must usually be controlled. All transplanted crops are vulnerable to attack by pests and diseases because until a new vigorous root system is established their ability to replace lost or damaged tissues is limited.

6.1.5   The rhizosphere

Returning now to the root system itself, the sloughing off of root cap, epidermal and cortical cells previously mentioned contributes to a flow of materials from the root to the soil which are collectively known as *root exudate*. Other contributions to this flow include the mucilage sheath around the root cap as well as a range of chemical substances released by intact cortical cells and root hairs or leaking from ruptures created by the emergence of branch roots. The total production of root exudate is considerable, probably accounting for 1% or more of a plant's dry matter production in the course of a year. This substantial quantity of material has important agricultural implications, if only as a significant addition to soil organic matter.

Some substances which have been shown to occur in root exudate are toxic to the roots of other species. Where toxic substances released by one plant inhibit the growth of others, this is known as *allelopathy*. It has been demonstrated, for example, that barley roots produce substances inhibitory to the growth of lettuce. The full significance of allelopathy in the field, however, is not yet known.

Numerous bacteria, fungi and nematodes that cause diseases of plants are able to recognize the root exudate of their particular host species, and their dormant spores or cysts in the soil can respond by germinating when a susceptible root grows nearby. The pathogen then grows or moves, depending on its form, towards its host by sensing the direction of increasing concentration of exudate in the soil solution. An example of this behaviour is furnished by the fungus *Plasmodiophora brassicae*, the causal organism of club-root disease of brassicas.

Sometimes quite unrelated plants produce root exudates so similar that a pathogen may be misled by a non-host species into germinating when its real host is not present. This may be put to use. For example, roots of the marigold *Tagetes* stimulate germination of cysts of the potato cyst nematode and thereby reduce soil populations of this pathogen.

The zone of soil extending for a few millimetres around a root in which the root exerts an influence on microbial life, chiefly through the production of root exudate, is called the *rhizosphere*. Since root exudate contains a range of organic nutrients such as carbohydrates and amino acids, the rhizosphere is characterized by the proliferation of microbes, especially bacteria, whose growth is stimulated by these nutrients.

Rhizosphere microbes are known to have effects on the uptake of minerals by plant roots but their impact has yet to be properly assessed. They also play a role in protecting roots from certain pathogens.

6.1.6   Symbiotic associations between roots and microbes

As mentioned in Section 3.2.4, certain non-pathogenic soil fungi invade the cortical cells or intercellular spaces of roots without causing the death of these cells. They receive carbohydrates and other organic materials from the root cells, but at the same time assist the plant to take up minerals. The association is therefore of mutual benefit to both plant and fungus. Any such mutually beneficial association between two living organisms is said to be *symbiotic*; this particular association between fungus and plant root is called a *mycorrhiza*.

Mycorrhizas fall into two distinct types (Fig. 6.7). In an *ectomycorrhiza* the fungus forms a dense weft of filaments or *hyphae* around a young root, with some of these hyphae penetrating the root and for-

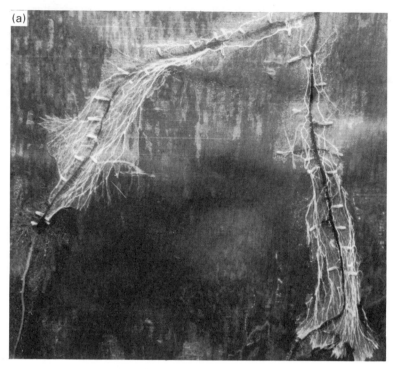

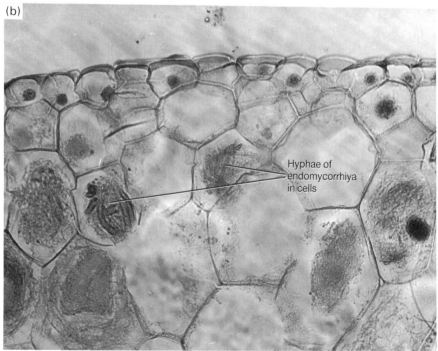

Fig. 6.7 Mycorrhizas, formed by association of a fungus with the roots of a plant. (a) Ectomycorrhiza composed of hyphae of fly agaric in association with roots of Sitka spruce. (courtesy of R Jones, University of Aberdeen.) (b) Endomycorrhiza composed of fungal hyphae inside cells of the root cortex of a crop plant.

ming a network in the intercellular spaces of the cortex. Extension growth of the root itself is often stunted, sometimes with the development of small club-shaped swellings at the tip. Ectomycorrhizas occur in most trees, including nearly all commercially important hardwood and softwood timber species. They unquestionably enhance mineral, especially phosphate, uptake by the plant and are often vital to the successful establishment and early growth of forestry plantations or to the establishment of seedlings in the wild.

In the other type, an *endomycorrhiza*, there is no weft surrounding the root but instead the fungus invades the cortical cells, with hyphae extending up to 2 cm out into the soil. The existence of this type of mycorrhiza has been shown to enhance phosphate and in some cases micronutrient uptake, probably because of the more thorough exploration of the soil permitted by the fungal hyphae. Endomycorrhizas occur widely on many crop species including cereals, grasses, legumes, potatoes, cotton, tobacco and various fruits. They are probably seldom essential to the nutrition of the crop but may often be beneficial. This is a subject about which much more needs to be known.

Another important symbiotic association is that between nitrogen-fixing bacteria and the roots of legumes and certain other plants, which was discussed in Section 3.2.1. *Nodules* are formed on the roots in response to the invasion of cortical cells by the bacteria. The bacteria stimulate renewed cell division in the cortex to form the nodule, which is supplied with vascular tissue to link it to the stele just as in the case of a developing branch root.

6.1.7   Soil conditions and root growth

Soil factors which influence root growth and development and thereby affect the yield of a crop can be classified as chemical (mainly nutritional), biological, such as attack by root pathogens, or physical, including soil temperature, aeration and resistance to penetration by roots. Relatively early in the investigations by soil scientists of this subject it was realized that the nutritional factor was the one most commonly limiting yield, and the result was the development of the use of chemical fertilizers, giving big yield increases. We have already discussed the nutritional factor in Chapter 3.

Soil physical factors are now of increasing significance, firstly because on good land the nutritional limitations to crop growth have been largely removed by the use of fertilizers and lime, and sec-

ondly because agriculture is moving on to poorer land where physical limitations are more severe. These limitations cannot be removed, but they can be alleviated with an understanding of root growth. Some physical problems, such as poor aeration or drainage and low soil temperature, have been dealt with in earlier chapters, but here we shall consider one of the major limitations to root growth, namely the resistance of the soil to root penetration.

The root system is confined entirely to the soil pores, which occupy some 40–60% of total soil volume. All life in the soil is totally dependent on this pore system, allowing as it does the movement of water and oxygen and the entry of roots.

Simple experiments measuring the forces exerted by growing root tips have revealed that these forces are not great. Except in very loose or friable soils, advancing root tips cannot widen pores. (Much later, as root girth increases with secondary tissue development, soil particles are easily pushed aside but this has no bearing on the initial penetration of the soil.) Roots cannot force their way through most soils, nor as it turns out, can they shrink themselves to squeeze through pores smaller than their diameter. From this central fact, much else flows. It means that roots are entirely dependent for soil penetration on the existence of pores in the soil which are both wide enough to accept them and continuous enough to allow lengthy growth. Now most roots, even fine fibrous rootlets, have a diameter of 60 μm or more. Many are much larger than this – for example, the main adventitious roots of cereals and their immediate branches are about 500–750 μm in diameter, and those of onions are even larger.

How do these diameters compare with the pore sizes occurring in different soils? In clay soils, which have the smallest soil particles and therefore the finest pores, relatively few pores are larger than 60 μm in diameter, probably accounting for not more than 3% of total soil volume. This proportion declines rapidly with depth, even in the top 40 cm or so. Sandy loam soils have much larger particles and therefore a much larger average pore size, but even here root-sized pores probably account for less than 10% of soil volume. Since many pores of suitable size do not lead to continuous spaces the apparent restrictions on the development of root systems are considerable, and the question arises as to how clay soils, in particular, are penetrated satisfactorily.

The answer lies very largely in the development

of larger spaces in the soil. The chief example of this is the result of the binding together of aggregates of small soil particles, as described in Section 2.6.3, to form a crumb structure in the soil. Crumbs store available moisture within them and, especially in fine-textured, small-pored clay soils, open up larger pores between them through which roots may penetrate and air and water may freely drain. They are therefore of great importance to soil fertility.

The main binding material forming crumbs is soil organic matter. Where this declines below a certain limit, crumb structure begins to break down and with it the larger pores it created. The resulting impedance to root growth is one of the major reasons for the decline in soil fertility that is often observed with a decline in organic matter following, for example, continuous cropping with cereals, especially if straw and stubble are burned rather than ploughed in.

Certain quick-growing crops are sometimes grown purely for ploughing in to increase soil organic matter content and thereby improve crumb structure. They are known as 'green manure' or *soilage* crops; mustard is one commonly used. It is, however, generally more economic to grow crops which give a harvestable yield as well as improving soil structure; legumes such as peas and beans are particularly valuable in this respect because of the ease with which they can be accommodated into cereal husbandry systems and because their roots and straw are relatively rapidly decomposed in the soil.

Enhanced soil structure is also observed after a few years under grass. This enhancement appears to be due not simply to the increase in soil organic matter the grass brings about, but to some part of it, possibly polysaccharides originating from bacteria associated with the grass roots, binding soil particles and increasing the stability of soil aggregates. This is another example of the importance of rhizosphere microbes.

There are at least three other mechanisms by which even larger spaces are generated in the soil:

1. Earthworms, which occur in very large numbers in most agricultural soils in temperate areas, leave numerous long continuous tunnels.
2. Old roots die and decay, leaving vacated tracks for exploitation by new roots.
3. Wetting and drying and freezing and thawing cause expansions and contractions in the soil which open up fissures that roots can explore.

From the foregoing discussion of the importance of soil pores for root penetration it is obvious that even a thin layer of non-porous, or even very fine-pored material at any depth in the soil will deflect root growth and confine it to the layers above. Such layers, of varying degrees of thickness and hardness, occur widely in soils and can severely limit their usefulness for cultivation. The most common cause of their development is the deposition of dissolved salts of soil minerals, usually principally of iron and aluminium in insoluble forms, filling the soil pores like a cement and forming layers impenetrable to roots.

In many wet, acid soils in temperate areas, for instance, dissolved oxides of iron are leached out of the upper layers but are then deposited in the anaerobic conditions found often less than half a metre down. There they form a thin layer of material made of insoluble iron salts and soil particles and known as an *iron pan*. This is often hard and brittle, but even when relatively soft it is impermeable to roots and water alike and not only interferes with surface drainage, thereby creating problems of waterlogging, but confines root development to surface layers.

Soils with an iron pan often occur in areas used for forestry plantations. The pan clearly limits the volume of soil available for exploitation of minerals and water, and therefore causes increased susceptibility to drought. In addition, as the trees grow larger on the resulting shallow root systems lacking deep anchorage, they become liable to windthrow as we saw in Section 4.3.2. Before planting, therefore, the iron pan must often be broken up by one of a number of techniques such as deep ploughing to improve drainage and root distribution.

In tropical areas, deposition of iron and aluminium, although under different soil and climatic conditions, can also produce impenetrable layers often of considerable depth, especially on certain soil types known as laterites. Another widespread example occurs where land has been glaciated in the past. The weight of the overlying ice compresses layers of soil, which then become overlaid by detritus left by the retreating ice as it melts and which becomes the upper layers of the soil. Subsequent deposition of salts, in the manner described above for pans, increases the impermeability of these compressed layers, which completely resist root penetration.

6.1.8   Cultivation, soil compaction and root penetration

Ploughing of agricultural soils serves many purposes, one of which is to increase the friability of the soil so as to ease root penetration. On occasion, however, ploughing can have the opposite result, especially if performed in wet conditions, when the smearing effect of the plough base creates an impenetrable layer in the soil. This layer is known as the *plough sole* and is especially marked in clay soils or where ploughing is repeated to the same depth year after year. Unless it is broken, the plough sole prevents root penetration and causes shallow rooting, which in turn produces poorly nourished and drought-susceptible plants.

While ploughing in general eases root penetration, subsequent operations may have the opposite effect by over-compacting the soil. Every passage by tractor or other machinery through a crop, for the application of fertilizers, the spraying of chemicals, inter-row cultivation or other purposes, compacts the soil further, decreasing average pore size.

In recent years, this problem has been greatly aggravated by the introduction of farm machinery too heavy for soils to support without considerable compaction. Even one season's cultivation can cause compaction problems with such machinery if the soil is in a wet and therefore plastic condition. As might be expected, crops with relatively thick roots, such as peas or onions, are more susceptible to this form of damage than those with finer roots, such as most grasses. The other common cause of soil compaction is trampling by livestock, especially cattle kept for instance in grass fields during wet weather. This problem also has tended to increase where agricultural intensification has produced denser stocking rates or heavier animals or both. Similar effects can occur following heavy rain or over-generous irrigation, when a compacted condition known as *soil slump* is observed.

Compaction, like other soil conditions causing reduced pore size, is most damaging in fine-textured soils. In sandy soils a moderate degree of compaction may even be beneficial by improving water retention. The smaller the pore size the better is the soil's ability to retain water, but the poorer is its aeration and the greater is its resistance to root penetration. A balance between these effects is therefore needed to ensure good root growth.

In recent years, various systems of crop husbandry known collectively as *minimum tillage* have become possible, largely through the replacement of cultivation, as a means of weed control, by the use of herbicides. The object of minimum tillage systems is often to reduce cultivation costs and to shorten the interval between the harvesting of one crop and the sowing of the next. In low rainfall areas such as the northern Great Plains of North America and southern Australia, the prime aim is to limit soil erosion and to avoid excessive loss of soil moisture, as happens when moist layers of soil are brought to the surface by ploughing. In situations like these, the term *conservation tillage* is often used.

One system of minimum tillage involves *direct drilling* whereby the seed is sown into soil which has not been ploughed since harvest of the previous crop. Cereals, for example, can be direct-drilled into the stubble bed of the recently harvested cereal crop; turnips are often direct-drilled into stubble or into a chemically-killed grass sward.

Minimum tillage, by comparison with systems involving ploughing, leads to greater compaction of the surface layers of the soil, with a corresponding decrease in pore size. This might be expected to restrict root growth over a period of years and result in a gradual lowering of crop yield. In practice, however, this tends not to happen. The problem of compaction is offset by an increase in organic matter in the surface layers of the soil, because this is not being mixed throughout the soil volume. Organic matter, as we have seen, stabilizes soil aggregates and therefore maintains larger pores. Meanwhile, at greater depth, fissures opened by soil shrinkage or left by the decay of old roots persist, as they are not destroyed by ploughing. Finally, there tends to be an increase in earthworm population. Thus root growth in minimum tillage systems is often no more hampered by physical obstruction in the soil than in conventional ploughing systems.

6.1.9   Development of the root as a storage organ

So far we have been concerned primarily with the root as an absorbing and anchoring organ. But the roots of many plants, particularly those of biennials and herbaceous perennials, as a major part of their function store carbohydrates and other nutrients during unfavourable periods such as a winter or a dry season. It is largely root storage, for example, that fuels the spectacular growth of the stems of giant hogweed from nothing to 3 m or more in a period of about six weeks in the far from tropical

Fig. 6.8 Giant hogweed. The enormous shoots of this plant grow very rapidly, mainly at the expense of carbohydrate reserves in the root.

climate of northern Scotland (Fig. 6.8).

On a more modest scale, root storage is essential for the early spring growth of grass. If grass roots are prevented from accumulating nutrient reserves before winter, for example by hard grazing late in the season, the commencement of leaf growth in the following season will be considerably delayed.

Several plant species that lay down abundant carbohydrate reserves in their roots have in the course of history been domesticated by man as root crops. Some of these, particularly in the tropics, are perennials, such as cassava (the crop from which we obtain tapioca), but in cool temperate climates most of them are biennials. Artificial selection, both intentional and unintentional, over the centuries has greatly enlarged their roots and thereby enhanced their capacity to store carbohydrate, but the differences between these crops and their wild ancestors are not fundamental.

The storage of carbohydrate, whether in the form of sugar, starch or inulin, requires large, living parenchyma cells and the capacity for continuous tissue growth to accommodate more and more carbohydrate from the leaves. In all biennial root crops, the storage parenchyma is produced by secondary tissue development, that is, from the vascular cambium (Fig. 6.9).

In carrot and parsnip the vascular cambium is of the standard form, producing a massive body of secondary phloem outwards and rather less secondary xylem inwards. Both tissues contain abundant storage parenchyma in addition to the functional conducting cells normally associated with phloem and xylem; conducting cells in fact account for only a very small proportion of the tissue. The sugar content is higher in the phloem than in the xylem 'core', reflecting the fact that the phloem is in the direct translocation pathway from the leaves.

Beets and mangels form many concentric vascular cambia, each producing xylem inwards and phloem outwards. Groups of lignified cells in the xylem give rise to the familiar rings seen in sliced beetroot. Again the sugar content is higher in the phloem than in the xylem; the highest-yielding sugar beet varieties containing most sugar tend to

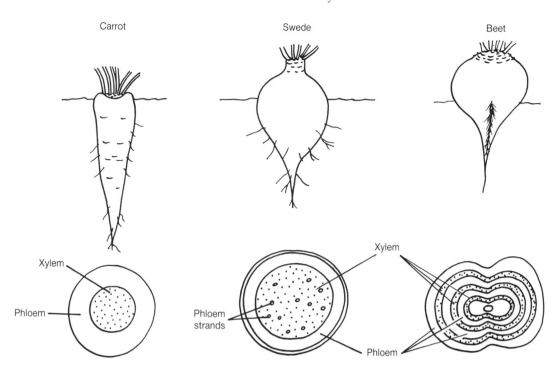

Fig. 6.9 Storage tissues of biennial root crops. These tissues form by proliferation of secondary phloem and xylem from vascular cambium, the pattern varying with the species.

be those with the broadest zones of phloem relative to the xylem.

In turnips and swedes only a very narrow outer ring of secondary phloem forms from the single vascular cambium. The secondary xylem, when it first differentiates from the cambium, is quite regular in appearance. Soon, however, small secondary cambia arise in the xylem parenchyma. Division of cells in these cambia soon obliterates the regular pattern of the xylem. They give rise to strands of phloem tissue scattered throughout the xylem, and these are probably important in distributing sugar throughout the mass of the storage organ.

In all of these root crops, the storage organ is derived partly from tap root and partly from hypocotyl. Carrot, parsnip and sugar beet show little extension of the hypocotyl beyond the seedling stage, so that the great bulk of the organ is true root. It tends, therefore, to lie almost entirely underground. In mangold, beetroot, swede and turnip a significant proportion of the storage organ arises by renewed extension and swelling of the

hypocotyl, although this proportion varies greatly from variety to variety. Much of the 'root' of these crops lies above ground, making for easier harvesting. In many horticultural radishes, the entire storage organ forms from the hypocotyl.

6.2 Growth and development of the shoot system

Just as the root grows by meristematic activity at its tip, so does the shoot, but in rather a different way. The *apical meristem* of the shoot has to do two things simultaneously: generate stem tissue, and initiate leaves.

The stem is produced in a similar fashion to the root. Cell division takes place in the meristem itself, elongation of cells occurs just below the meristem, and overlapping the elongation zone is a zone of cell differentiation. The stems of many dicotyledons, like their roots, form a vascular cambium for the production of secondary xylem and phloem, which leads to girth increase.

6.2.1 Leaf initiation and development

Leaves appear first as microscopic bumps or *primordia* on the dome of the apical meristem;

143

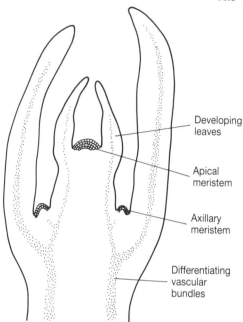

Fig. 6.10 Shoot apex of a dicotyledon plant in longitudinal section. Leaves arise as primordia on the apical meristem and rapidly elongate.

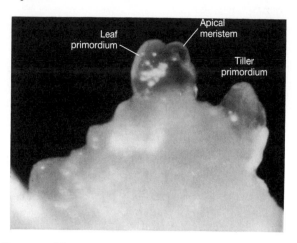

Fig. 6.11 The apical meristem of wheat revealed by peeling back successive layers of leaves.

these primordia rapidly elongate (Fig. 6.10). At first they are cylindrical or conical in shape, but then by continued cell division along two opposite sides they acquire their characteristic flattened shape. At this stage young but fully differentiated leaves enfold the shoot apex, protecting the apical meristem and leaf primordia rather as the root tip meristem is shielded by the root cap. It may be necessary to peel away many progressively smaller leaves to reveal the meristem itself; the meristem of wheat, revealed in this way, is illustrated in Fig. 6.11.

Different rates of meristematic activity at different levels along the sides of the leaf primordia give rise to lobed leaves or, in extreme cases, to compound leaves such as those of the potato or pea. In general, leaf shape becomes more complex as the plant gets older. The first leaf of red clover, for example, has only a single blade, whereas the second and subsequent leaves have a blade divided into three quite separate leaflets. Progressive increase in the complexity of leaf shape is well illustrated by the leaves of a swede as it gets older (Fig. 6.12).

Cell division in the leaf is virtually complete while the leaf is still in the primordial stage, later growth to full size being mainly by expansion of already existing cells. Often there is particularly marked elongation of cells at the leaf base, giving rise to a stalk. No portion of meristematic tissue is left in the mature leaf – it is thus an organ with no further capacity for growth and the limits of its size are determined by the number of cells initially present. Such organs are said to be *determinate*. In contrast, stems and roots retain meristems and with them the capacity for indefinite further growth throughout their lives and are said to be *indeterminate*.

Leaf primordia do not arise at random on the surface of the apical meristem but in a regular pattern (Fig. 6.13). In some species two primordia arise simultaneously on opposite sides of the meristem. The next pair arise again on opposite sides but at right angles to the previous pair. This leads to the *opposite decussate* leaf arrangement characteristic, for example, of mint and sycamore. More commonly, however, only one primordium is initiated at a time, so that each node on the stem carries just one leaf – the *alternate* arrangement.

In grasses and cereals leaves are initiated alternately on two sides of the apex, giving two ranks of leaves on the shoot. The leaves of most dicotyledons are arranged in a spiral, each leaf being initiated at a more or less constant angle (typically around 137°) to the previous one. This ensures the minimum of overlapping, and thus more effective light interception by mature leaves.

6.2.2 Branching in the shoot

At the same time as a leaf is initiated, a small group of meristematic cells is left behind by the advancing

Fig. 6.12 The cotyledon and the first and second leaves of a swede plant, illustrating the progressive increase in complexity of leaf shape as a plant develops.

Fig. 6.13 Arrangement of leaves on a stem. Left: opposite decussate; right: alternate spiral arrangement.

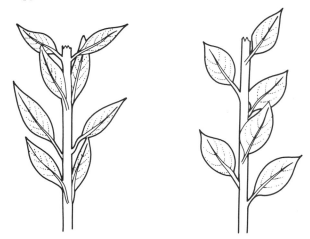

apical meristem. These cells are always located in the *axil* of the leaf, that is, in the angle formed by the junction of the leaf and the stem above it. The meristem may remain no more than a small cluster of cells, or it may develop into a miniature shoot apex complete with primordial leaves – an *axillary bud*.

Usually its development is arrested at this stage, but it retains the capacity for later growth to form a branch shoot of similar structure to the main shoot. Branching of the shoot system thus arises in quite a different way from that of the root system. Firstly, the branch shoot arises from a superficial meristem, not from deep within. Secondly, in shoots only meristems specifically laid down for the purpose can give rise to branches, whereas branch roots can arise anywhere in the pericycle opposite any one of the points of the xylem star.

The shoot produced by any axillary bud possesses its own axillary buds which in turn can potentially give rise to yet more axillary buds in the same way. A bud may even develop a cluster of new buds without significant growth of stem or leaves. This proliferation of buds provides a large number of potential growth points over the shoot system.

This is of great importance to the plant because the shoot is always at risk from damage by grazing, weather, disease and other causes. It must also be able to adapt its growth pattern to changing conditions such as shading and competition from other plants. The shoot therefore requires considerable powers of recuperation and adaptability – it must be flexible in its growth. The axillary bud system, by providing an indefinite number of potential growth points distributed over the shoot, confers this flexibility.

It should be noted that in plants with compound leaves, axillary buds occur only at the base of the whole leaf, never at the bases of individual leaflets. Any structure bearing axillary buds must be a stem, never a leaf stalk or a root.

Normally axillary buds close to the shoot apex do not grow into branches, whereas those lower down the stem are more likely to do so. One explanation for this might be that axillary buds need to have reached a certain age before they can grow into branches. Alternatively, the apical meristem may have an influence, which declines with distance, tending to suppress axillary bud development.

Familiar experience of the effects of pruning tells us that the latter explanation must be the correct one, for removal of the apical meristem by pruning releases axillary buds from its suppressive influence, stimulates their growth and produces a more bushy habit of growth through prolific branching. The suppression of axillary bud growth by the apical meristem is known as *apical dominance* and occurs to a greater or lesser extent in almost all plants. The nature of the suppressive influence will be discussed in Section 8.5.2.

6.2.3 Patterns of shoot development

As shoot growth in biennials and perennials slows down and stops at the end of the growing season, leaves tightly enclose the meristem and leaf primordia, forming an *apical bud*. Similar changes often take place in the axillary buds too, forming structures which, as we saw in Chapter 4, are extremely resistant to drought and frost.

Another important change that occurs at the shoot apex is the switch from leaf to flower initiation, which we shall examine in Section 7.2. Once this happens, the apex never switches back.

Eventually that apex must die after completion of the reproductive sequence of developmental events.

In this chapter, however, we shall confine our attention to patterns of shoot development during the purely vegetative phase of the plant's life cycle. These patterns determine the *growth habit* of the plant, which in turn determines to a very great extent the type of habitat in which the plant can live. The patterns we shall be examining are largely genetically controlled and are therefore characteristic of particular species or varieties. We should not forget however, that growth habit is also influenced by environment (see Fig. 1.5) and by damage to the plant. We have seen, for example, that pruning can convert a relatively unbranched plant into a very bushy one, by the removal of apical dominance.

In the typical dicotyledon, for example kale, sunflower or broad bean, leaf initiation and development is accompanied by continuous stem elongation. The stem is self-supporting and has distinct nodes, marked by leaves and axillary buds, separated by internodes. Such a plant is said to have an *erect* growth habit.

The degree of branching of the erect stem depends on the strength of apical dominance. Thus marrowstem kale with its strong apical dominance shows little or no branching, whereas thousand-head kale, a form of the same species with less pronounced apical dominance, is profusely branched (Fig. 6.14).

Fig. 6.14 Two forms of kale differing in the strength of apical dominance. Left: marrowstem kale with strong apical dominance, suppressing axillary bud development. Right: thousand-head kale with weak apical dominance, allowing development of branch shoots from axillary buds.

Fig. 6.15 Chickweed, a plant showing the trailing growth habit, growing in a strawberry bed.

Some species develop a growth habit in which they simply straggle over the ground or other vegetation to give a *trailing* growth habit, for example chickweed (Fig. 6.15), which by virtue of this growth habit can quickly smother young crop plants. Such species do not need stems strong enough to stand upright, carrying their own weight plus that of the leaves.

Other weak-stemmed plants have developed various ways of clinging on to vertical supports to enable them to grow tall. These plants have a *climbing* growth habit (Fig. 6.16). Most commonly in nature they use the stems of other plants for support, but when grown horticulturally they may be trained up stakes, canes, trellises or walls. In some climbing plants such as runner bean and clematis the slender stem, as it grows, twists itself around any suitable support. The same method is used by climbing weeds such as field bindweed and black bindweed, which can be very damaging to cereal crops by increasing the load on their stems and causing lodging, especially in wet weather (see Section 4.3.2). Another weed which uses the same means of support is morning-glory, probably the most damaging weed of soya bean and cotton crops in the southern United States. A similarly troublesome weed of winter cereals in Europe, cleavers, uses a different method of clinging to its support, namely a dense covering of tiny hooked bristles on its stem and leaves.

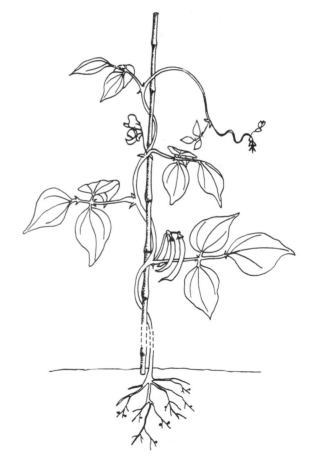

Fig. 6.16 Runner bean, illustrating the climbing growth habit.

147

Peas have leaflets modified into whip-like structures called *tendrils*, which coil tightly around any available support. The development of very short-stemmed varieties of pea has, however, enabled this crop to be grown extensively on a field scale with no support other than that provided mutually by the mass of stems of the crop itself.

Woodland plants which have used the climbing habit as an economical means of growing very tall include various species of ivy. These produce a mass of short adventitious roots which exude a glue allowing them to adhere to tree trunks (and to walls). In tropical forests lianas are climbing plants which ramify through the forest canopy. Their stems may be hundreds of metres long and provide local people with 'ropes' for a wide range of purposes.

Another type of growth habit is produced in many plants by internodes which show little or no elongation, except in some cases for a few internodes near the base of the stem. Such plants have all their leaves borne in a cluster around the shoot apex at or just above ground level and are said to show the *rosette* habit. In some (e.g. sugar beet, carrot) the long leaf stalks have taken over the function of the stem, namely to carry the leaf blades aloft for effective light interception. In others such as dandelion the leaves lie prostrate on the ground.

Fig. 6.17 Tussocks of common rush, a plant showing the tufted growth habit.

A variant of the rosette habit arises when the short stem branches repeatedly to create a mass of shoots and axillary buds on a single *crown*. This gives the characteristic *tufted* growth habit, for example of grasses in which the branch shoots are known as *tillers*. Tufted plants tend to become more so with every year that they grow; witness, for example, the huge tussocks of common rush that develop in poorly drained grassland (Fig. 6.17).

The growth habit of cereals is basically a tufted one, and it is only the fact that they are annuals that prevents them from tillering quite as profusely as perennial grasses. A single plant of barley, however, grown on its own with no competition for light and nutrients from other plants, can in the course of a growing season produce a hundred or more tillers. In a normal barley crop each plant produces on average no more than four to six tillers (Fig. 6.18). This is a dramatic illustration of the flexibility that the axillary bud system permits.

Some plants show characteristics both of erect and of tufted growth habits. Good examples are red clover and lucerne (alfalfa). Early in the year they generate a profusion of shoots at ground level and later many of these shoots become erect. They retain, however, a large stock of axillary buds close to the crown at ground level.

What is the significance of the rosette and tufted growth habits, which are so common among crop plants and agricultural weeds? It lies in the point just made, namely that the axillary buds are in a

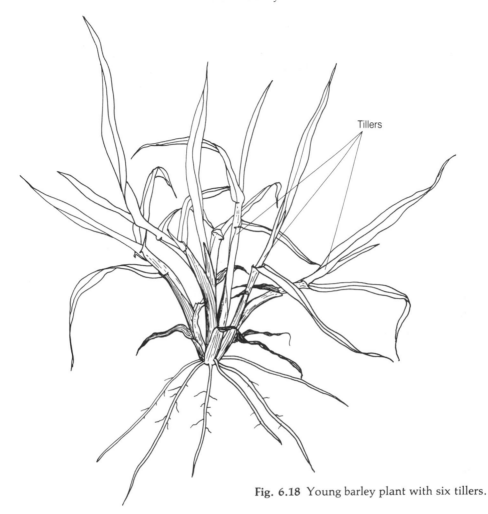

Tillers

Fig. 6.18 Young barley plant with six tillers.

very sheltered position, at, just above, or even below, ground level. In many plants, including dandelion, the tap root repeatedly contracts to draw the crown down below the soil surface. This protects the buds from grazing and from fire, two factors which dominate the ecology of the grasslands in which these plants evolved and which many of them still inhabit. Thus after fire or grazing the plant rapidly re-establishes its leaf area from its buds. The same growth habit has made grasses ideal for cutting as forage, not only because of the rapid regeneration afterwards but because most of the mass of herbage is easily digestible leaf material, not fibrous stems.

In most rosette and tufted plants, including grasses and cereals, the flowering head, or inflorescence, forms at the shoot apex while this is still in its very sheltered position. Once it is fully formed, there is rapid stem elongation, permitted by meri-

stems at the base of each internode, all of which grow simultaneously. These are known as *intercalary* meristems. They allow flowering and seed dispersal to be accomplished with as short as possible a period of exposure to the dangers of grazing and fire.

Another advantage possessed by tufted plants is the microclimate created within the dense mass of shoots. The delicate meristems at the crown are well protected from frost by the shield of dead and living leaves that forms around them.

Finally, the prostrate leaves of many rosette plants are a means of keeping other plants at a distance and preventing shading. Weeds such as dandelion, broadleaved plantain and cat's-ear which have this habit of growth can be very damaging to lawn or sports turf by shading out grass from sizeable patches.

Many plants have specialized, horizontally

growing stems which put down adventitious roots at the nodes. These *creeping* stems serve a number of clear functions and should be distinguished from simply trailing stems such as those of chickweed. Two main categories can be recognized, although, as with all attempts to classify plant structures in a rigid manner, there are many cases of creeping stems which show elements of both categories.

The first type are relatively short-lived and function solely as a means of short-range vegetative spread. They do not accumulate carbohydrate reserves in any quantity and are most commonly found above ground. They are called *stolons*, and plants possessing them are said to be *stoloniferous*. In white clover, for example, the stolons are branched and carry long-stalked leaves at nodes throughout their length. In strawberry and creeping buttercup the stolons form 'runners' which establish new rosettes or tufts at some distance from the original plant and serve as a means of vegetative propagation (see Fig. 10.1).

The other type of creeping stem is the *rhizome*. A long rhizome may, like a stolon, act as a means of spread or vegetative propagation but all rhizomes,

whether long or short, are essentially overwintering organs. They are long-lived, perennial structures which accumulate carbohydrate reserves and are almost always below ground. The leaves borne at the nodes on a rhizome are usually reduced to small scales. Examples of plants showing the *rhizomatous* growth habit are couch grass (Fig. 6.19) with long, relatively slender rhizomes and yellow flag with very thick, short rhizomes.

The full range of ways in which the forces of evolution have adapted the basic pattern of stem, leaves and buds that form the vegetative shoot would be outside the scope of this book, so only the more outstanding examples will be mentioned here. One of the most important is the massive development of heavily lignified secondary xylem forming the wood of trees, which was considered in Chapter 4. Of note, too, are the succulent stems and spine-like leaves of cacti, which enable them to survive long periods of drought; the leaves of insectivorous plants that trap and digest insects as a source of nitrogen; the floating leaves of water lilies and other aquatic plants; and the twining, leafless, chlorophyll-less stems of dodder which clasp tightly around clover or other stems and send out modified adventitious roots ('suckers') to parasitize the living cells of the host.

6.2.4 Shoot-derived storage organs

The stem of many species, including some important crops, becomes enlarged to act as a carbohydrate storage organ in a fashion similar to the root of turnip or sugar beet. Examples are marrowstem kale (see Fig. 6.14) or the more grotesquely swollen kohlrabi (Fig. 6.20), in both cases the main storage tissue being derived from the pith in the centre of the stem.

The *tuber* of the potato is the immensely swollen tip of an underground stolon (see Fig. 10.3); the 'eyes' represent axillary buds and the one at the end is the apical bud. In cross-section the tuber is seen to have somewhat unusual tissue composition. Just beneath the skin (which is a layer of cork derived from a cork cambium) is a narrow zone of phloem and within that an even narrower zone of xylem containing a few lignified cells. Inside the xylem is the main starchy storage tissue, formed from internal phloem. In the centre is the pith, which tends to be translucent as it is not so densely packed with starch grains as the parenchyma of the phloem (Fig. 6.21).

Some crops lay down carbohydrate reserves for

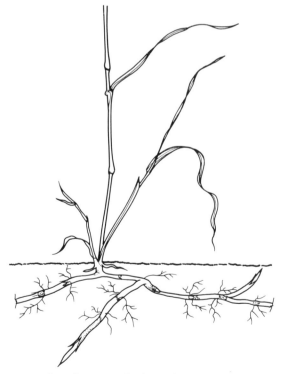

Fig. 6.19 Couch grass. The long rhizomes of this plant serve as undeground storage organs and as a means of vegetative spread.

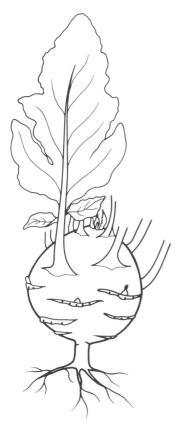

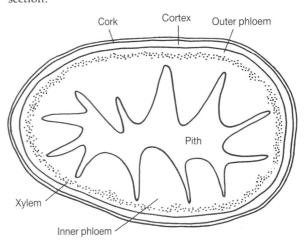

Fig. 6.20 Kohlrabi with its stem swollen for adaptation as a storage organ.

Fig. 6.21 The tissues of a potato tuber as seen in cross-section.

the winter in a greatly expanded apical bud with fleshy leaf bases. Perhaps the best example is the cabbage (Fig. 6.22). The *bulb* of onion or narcissus (Fig. 6.23) is a swollen apical bud at or below ground level. In Brussels sprouts (Fig. 6.22), the axillary buds become enlarged like miniature cabbages, while in other brassica crops, such as

Fig. 6.22 Modification of buds as storage organs. Left: cabbage (apical bud); right: Brussels sprouts (axillary buds).

151

vegetation. Other phylloplane organisms produce substances akin to the plant's own growth regulating substances and may thereby influence the pattern of growth and development in the plant. By their presence and activities some organisms of the phylloplane increase the resistance of plants to certain disease-causing organisms that may attempt to invade the leaves. Lastly, some of the phylloplane substances exuded by the plant or produced by microbes are washed down by rainfall and may influence the rhizosphere. We still have much to learn about the significance of phylloplane organisms in plant life.

Not all these organisms, however, are saprophytes. Leaves and stems, like roots, may be invaded by a wide range of parasitic bacteria and fungi which induce disease, as also may the flowers and fruits. As plant tissue senesces it becomes liable to invasion not only by these parasites but by phylloplane organisms which up to this time have lived purely saprophytically. The fungus *Botrytis cinerea*, for example, grows on dead and senescent tissues of a very wide range of species but in suitable circumstances it will also invade healthy tissue. Tissue which is damaged, for example by frost, is also liable to such invasion.

The fungi *Cladosporium* and *Alternaria* are prominent among the saprophytes which invade dying leaves and stems. They grow profusely on the straw of ripening cereals, often causing it to darken in colour. Their significance in the diet of farm animals fed such straw is unknown. Their spores, however, contribute massively to the population of microbes found in the air and are a common cause of allergy in human beings, particularly farmworkers who handle infected products and inhale large numbers of spores in the process.

6.3 Senescence, dormancy and death

We have looked, then, at patterns of growth and development in root and shoot systems, but the picture would be incomplete without a look at the other side of the growth and development coin, namely the decline of roots and shoots towards their death. All plant organs and whole plants eventually die, but it would be a mistake to assume that this is simply due to the remorseless march of time, occurring when the cells of which they are made up have inevitably reached the limit of their life span. Death may, of course, result from factors beyond the plant's control (for example prolonged

drought, frost or the ravages of disease), but more commonly death in the plant is a highly organized affair.

The changes taking place in plants or parts of plants in preparation for this planned death are collectively known as *senescence*. As we shall see, senescence serves a number of important functions, one of which is preparation for winter or other unfavourable season. In this it is closely linked to the dormancy of those organs which must survive the unfavourable period, whether those organs be seeds (whose dormancy we discussed in Section 5.1) or vegetative organs such as roots, stems, buds, tubers or bulbs.

It is important to understand the patterns of senescence, dormancy and death which are observed in plants. Nearly all our knowledge of senescence and vegetative dormancy concerns shoot systems, but undoubtedly root senescence and dormancy are equally organized affairs of which we as yet know little.

6.3.1 Patterns of senescence in shoots

In discussing patterns of senescence it is useful to draw a distinction between *woody* plants such as trees and shrubs, all of which are by their nature perennial, and non-woody or *herbaceous* plants, which may be annual, biennial or perennial.

Most herbaceous plants, including virtually all agricultural crops, show a pattern of leaf senescence in which old leaves die off one at a time as they are superseded or shaded by younger, more efficiently photosynthesizing leaves. The same sort of pattern is seen in woody evergreens such as holly and rhododendron and is known as *sequential leaf senescence*.

The components of the protoplasm in senescent leaves are systematically dismantled and incorporated into new growth elsewhere in the plant. Sequential leaf senescence is thus a device for reusing redundant materials. In a mature vegetative plant the rate of leaf senescence may be exactly matched by the rate at which new leaves appear, so that the total leaf area of the plant remains more or less constant. A similar sequential pattern of senescence is seen during flowering as successive flowers senesce in turn after their useful life is terminated.

In deciduous woody plants, all the leaves senesce around the same time regardless of age, a pattern known as *synchronous leaf senescence*. It is part of a programme of events through which the plant prepares for the onset of winter or other season unfavourable to growth, when leaves become an embarrassment to it. In winter, leaves continue to act as evaporation surfaces at a time when the availability of water for uptake by roots is limited by the soil being frozen or when roots are not functioning efficiently because of low soil temperatures. The same pattern of senescence occurs in many species in climates with a hot dry season, when again it is a means of minimizing water stress.

In both sequential and synchronous patterns of senescence, the leaves are first pillaged of all transportable materials which can be used in the organic and mineral nutrition of other parts of the plant. Many non-transportable materials such as proteins and nucleic acids are mobilized by being broken down into smaller molecules and then exported. This is all part of a general property of plants whereby they constantly redistribute materials in response to changes in supply and demand. We shall examine the role of this redistribution in the control of growth and development in Section 8.2.

As cells die, their membranes lose their functions. Soluble nutrients remaining in a leaf as it senesces thus become more liable to being leached out by rain. In a study of mineral cycling in a Scottish birchwood, the concentration of potassium ions in rain falling through the canopy in October was found to be 0.34 mM, as compared with only 0.05 mM in June. Mineral nutrients may be reabsorbed by the roots, but organic nutrients washed out of leaves are lost to the plant altogether. Hence the more that can be exported before senescence is well advanced, the better.

One of the most obvious signs of leaf senescence is the loss of green colour as a result of chlorophyll breakdown. In some cases this breakdown simply unmasks yellow pigments, but in other cases, particularly in certain deciduous trees such as cherries and maples, the intermediate breakdown products of chlorophyll give rise to the vivid reds and purples of autumn leaves. The process of fruit ripening also involves colour changes; as we shall see in Section 7.5.3 ripening is in many ways akin to senescence.

Leaf senescence in dicotyledons is accompanied by renewed cell division in a zone at the base of the leaf stalk. This results in a line of weakness at which the leaf readily becomes detached (Fig. 6.25). The controlled detachment of leaves in this way is called *abscission*. The wound is rapidly sealed by the deposition of cellulose and gums but remains visible as an *abscission scar*.

In most grasses dead leaves do not become

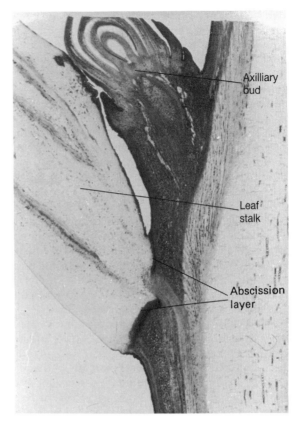

Fig. 6.25 Abscission layer at the base of a leaf stalk. This is a zone of renewed cell division, which becomes a line of weakness at which a senescent leaf detaches from the stem.

detached in this way, but simply flop on to the ground and decay where they lie. An exception is purple moor grass, in which the leaf blades are shed from the leaf sheaths in winter.

In many herbaceous perennials the entire aboveground portion of the plant dies back annually, leaving a below-ground carbohydrate storage organ. This pattern is called *top senescence* and an excellent illustration of it is provided by the potato plant. Again the shoot exports much of its useful and transportable material to the storage organ before it dies, and the physiological changes in the senescing organs are similar to those described above, although there is not necessarily any well-defined abscission. Potato tubers do form an abscission layer at the base, and this allows them to detach readily from the stolon at harvest.

A fourth and final pattern of senescence is seen in plants which flower and set seed only once in their life – the *monocarpic* plants. The most important of these are the annuals such as cereals, peas and mustard, and the biennials such as carrots and cabbages. Here the whole plant senesces with the maturation of seed. Such plants show a kind of 'big bang' reproduction in which the entire resources of the plant are poured into seed production and all the transportable resources of every other organ are utilized to achieve it, even at the expense of the plant's life. This is certainly not an inevitable result of ageing since if flowers are removed as they are formed senescence can be prevented.

6.3.2 Factors inducing senescence

In monocarpic plants senescence of the vegetative organs is induced by reproductive development elsewhere in the plant. To a large extent the creation of a strong sink for carbohydrates and other nutrients is responsible. Similarly, sequential leaf senescence may be induced by the formation of new sinks in the younger areas of the plant.

Senescence and abscission are also a useful defence mechanism against disease. Leaves infected with microbial pathogens may be detached, thus preventing the infection from spreading through the plant. Another use by plants of their capacity to senesce and shed leaves is the shunting of toxic minerals into older leaves, which are then discarded.

Synchronous leaf senescence in deciduous trees and the senescence of above-ground parts of herbaceous perennials are attuned to environmental cues such as falling temperature, the first occurrence of frost or shortening daylength. How the plant senses such cues and how this leads to control of senescence will be dealt with in Section 8.6.2.

6.3.3 Vegetative quiescence and dormancy

Other aspects of preparation for winter or a dry season are the development of quiescence and dormancy. We have encountered these states in seeds (Section 5.1) but they can occur also in entire vegetative plants or parts of plants.

It appears that the roots of trees are seldom dormant but in winter are merely quiescent, ready to recommence growth as soon as soil temperature rises in the spring. In truth, however, the subject has been little studied.

By contrast, the shoots and particularly the buds of trees show pronounced dormancy patterns akin to those in seeds. Chemical inhibitors, probably identical in many cases to those in dormant seeds,

are responsible, and awakening from dormancy requires the level of these to be reduced or the level of growth stimulatory substances to be increased. We shall discuss how the induction and breaking of vegetative dormancy is achieved in Sections 8.5.5 and 8.6.2.

However dormancy is induced, the pattern of events tends to be similar in most woody perennials. Stem elongation ceases and there is increased deposition of lignin in the cell walls of the current year's growth. Special small protective leaves called *bud scales* are produced instead of foliage leaves and these clasp tightly around apical and axillary buds. Cell division in these buds slows down and in the vascular cambium it stops. The moisture content of buds declines but never falls as low as in seeds. The later stages of the process are accompanied by senescence of the leaves, with most of their transportable nutrients being transferred to the stems before leaf abscission. Cell division is able to continue at a slow rate in the dormant bud, slowly initiating leaves or flowers for the following season. Low temperatures break the dormancy and by about midwinter the buds are merely quiescent, remaining so until rising spring temperatures induce budbreak.

Dormancy also occurs in underground storage organs such as tubers and bulbs. Potato tubers, for instance, will not grow for several months after harvest. These organs, however, have a much higher moisture content and metabolic rate than seeds or even dormant tree shoots, and do not have the same level of resistance to extremes of temperature and drought. They are, to a large extent, devices for avoiding rather than enduring these conditions, by retiring below ground. Considerable physiological and metabolic activity takes place within them even when dormant.

In tulip and narcissus bulbs, dormancy is induced by high temperatures, but during the dormant period flower formation takes place in the bulb. Dormancy is broken by cool temperatures, and thereafter root growth and shoot emergence require a fairly prolonged period below about 10°C. Subsequent growth can then take place at higher temperatures. This pattern of requirements reflects seasonal changes in the natural habitat of these plants, but the knowledge is used in 'forcing' bulbs to produce saleable flowers in time for Christmas rather than leaving them to flower at their natural time in the spring.

Summary

1. Roots account for a larger proportion of a plant's dry weight than is commonly realized. They achieve a massive absorptive surface area by repeated branching. Roots grow by the activity of the root tip meristem and the zones of elongation and differentiation behind it. Branch roots arise in the pericycle some distance behind the root tip and erupt through the cortex and epidermis. As the root matures its epidermis and cortex are sloughed off, leaving the stele protected by the endodermis and later often by a layer of cork.

The main types of root system are a tap root with laterals, a fibrous root system arising by abortion of the tap root, and adventitious roots arising from stem nodes. Tap roots in general penetrate deeper into the soil than other types. All roots are extremely flexible in their growth; for example, they tend to proliferate in regions of enhanced water or mineral availability in the soil.

The practice of transplanting interrupts the continuous process of root growth. Water uptake is reduced because of root damage and the breakage of root–soil contact, and this leads to water stress which is alleviated only when new roots grow and root–soil contact is restored.

Death of the root cap, epidermis and cortex contributes to a flow of root exudate which has several important effects. It may cause allelopathy, whereby root growth of other species growing nearby is inhibited, and it may attract pathogenic microbes and nematodes. It also acts as a source of nutrients for non-pathogenic fungi and bacteria inhabiting the rhizosphere, some of which may influence mineral uptake. Mycorrhizal fungi and nodulating bacteria live symbiotically within plant roots.

Root growth is restricted to existing pores in the soil. Only a small proportion of the pores in a clay soil are large enough to accommodate roots, unless a good crumb structure forms by aggregation of soil particles with organic matter. Soil structure can be improved by crop rotation, especially if this involves several years of grass. Pores are also created by earthworms, the death of old roots and fissuring caused by moisture and temperature fluctuations. Impenetrable layers in soil result from development of an iron pan in acid soils or a plough sole following repeated ploughing to the same depth. Compaction caused by the passage of machinery or trampling also reduces root penetra-tion, but minimum tillage systems, although giving more compacted soil, are not necessarily associated with poor root growth, partly because more organic matter tends to accumulate near the surface.

In roots modified as storage organs, sugar is stored in parenchyma tissue derived from secondary xylem and phloem. A substantial part of the storage organ may be formed by expansion of the hypocotyl.

2. At the shoot apex, the apical meristem generates stem tissue and initiates leaves. Leaf primordia arise in a regular pattern, developing the final leaf shape by differential cell division and expansion. Leaves are normally fully determinate structures, unlike stems and roots which retain meristems throughout their life. Branching of shoots arises from axillary buds, the growth of which may, however, be suppressed by the influence of the apical meristem. This apical dominance disappears if the apex is removed, for example by pruning.

In most plants the initiation of leaves is accompanied by elongation of stem internodes, giving rise to an erect growth habit with a greater or lesser degree of branching depending on the strength of apical dominance. Stems of climbing plants cannot support their own weight but have various devices for clinging on to supports. In rosette plants there is little or no internode elongation; the situation is similar in tufted plants which produce a mass of axillary buds on a crown at or below ground level. Such plants regenerate rapidly after cutting, grazing or burning. Other plants have horizontally creeping stems, either stolons or rhizomes. Shoot-derived carbohydrate storage organs include rhizomes, tubers and bulbs.

The surface of shoots, particularly leaves, provides an environment suitable for the growth of saprophytic microbes. These have various effects on plant growth and health, and some may become parasitic, particularly during senescence.

3. Senescence is the programmed run-down of cell, tissue and organ function, leading ultimately to death. Four main patterns can be discerned in vegetative shoot systems. Leaf senescence may be sequential or synchronous; in either case the process involves mobilization and export of nutrients from the leaf, pigment changes and abscission. In many herbaceous perennials the whole shoot senesces, leaving an underground storage organ to survive the winter or dry season. Annuals and biennials during flowering and seed formation show senescence of the whole of the rest of the plant.

Shoots, especially their buds, and storage organs such as tubers and bulbs show both quiescence and dormancy, the latter being imposed by inhibitors as in many dormant seeds. Metabolism and very slow meristematic activity may continue in these organs during the dormant phase. Roots appear to show only limited dormancy.

7

Plant Growth and Development:
Flower, Fruit and Seed

Having looked in some detail at patterns of growth and development during the vegetative phase of a plant's life cycle, we must now turn to growth and development during the reproductive phase. Reproduction in angiosperms may be accomplished with or without the formation of special male and female sex cells (*gametes*), and subsequent fusion of a male and a female gamete (*fertilization*). Reproduction involving gametes, that is sexual reproduction, takes place in the flower and results in the formation of fruits containing seeds. Asexual reproduction, which does not involve fusion of gametes, may mimic sexual reproduction in taking place through the medium of flower, fruit and seed, as explained in Section 7.5.1. Many plant species, however, have evolved systems of asexual reproduction that depend solely on vegetative organs, often strongly modified to function in this way. These systems are known as *vegetative propagation* and are dealt with in Chapter 10.

7.1 Sexual reproduction in the angiosperm plant

Before going into some of the intricacies of sexual reproduction in angiosperms, let us begin by examining the basic structure and function of a typical flower. We will then go on to outline the sequence of events leading up to, during and following flowering, and conclude this section by considering the importance of sexual reproduction to plant species.

7.1.1 The flower as a means of sexual reproduction

A flower is a complex assemblage of organs. Fig. 7.1 shows the various organs that make up the flower of strawberry.

The first thing to note is that the flower is borne at a shoot apex. The stem bearing the flower at its tip is called the *pedicel* and the tip itself, which bears the various organs of the flower, is called the *receptacle*.

On the centre of the receptacle, surrounded by the other floral parts, are found one or more organs called *carpels*. These are often described as the female parts of the flower because, as we shall see, the female gametes are produced here. Each carpel has at its base a completely enclosed chamber, the *ovary*, containing one to several small structures called *ovules* (in strawberry there is only one ovule in the ovary of each carpel). Inside each ovule is borne a single female gamete. The portion of the carpel above the ovary consists of a short stalk, the *style*, surmounted by a sticky and slightly enlarged region called the *stigma*.

Arranged around the carpels are the *stamens*, generally thought of as the male organs of the flower because they produce the male gametes. In strawberry, each stamen consists of a slender stalk called the *filament* on which is borne a body called the *anther*. Inside the anther a dust-like mass of tiny grains, each containing two minute male gametes, is formed. These grains, collectively known as *pollen*, are exposed when the mature anther splits open.

One of the key events in the sexual reproduction process in angiosperms is the transfer of pollen

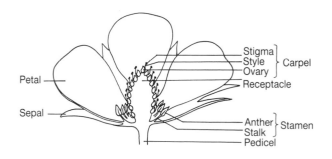

Fig. 7.1 Flower of strawberry, in half-section.

from anthers to stigmas. This transfer is known as *pollination* and in strawberry is effected by insects. An insect visiting one flower may pick up some pollen on its body and later deposit it on the sticky stigma of another flower (*cross-pollination*) or of the same flower (*self-pollination*). As we shall see in Section 7.3.1, the flower evolved as a device to promote cross-pollination by insects, but in many species it has become adapted for pollination by wind (Section 7.3.2) or in some cases by birds, bats and other agencies.

Following pollination, each pollen grain on the stigma produces a slender outgrowth called a *pollen tube* which grows down through the tissues of the style to the ovary. The tip of the pollen tube eventually reaches a tiny pore in the ovule called the *micropyle* (see Fig. 7.25) where the two male gametes are released. One of these fuses with the female gamete inside the ovule and the resulting single cell subsequently divides repeatedly to become the embryo of a new plant. As the embryo develops, the ovule enlarges and goes through other changes to become the seed. We shall consider the events following pollination in greater detail in Sections 7.4 and 7.5.

Returning now to the strawberry flower (Fig. 7.1), arranged around the stamens are the most conspicuous parts of the flower, the *petals*. These have the function of attracting insects for pollination. In many species the petals have special glands at the base that secrete a sugary solution called *nectar*, as an inducement to insects to visit the flower.

One further set of organs outside the petals remains to be considered. These are the *sepals*, green, leaf-like structures whose function is to protect the more delicate inner parts of the flower during its early development. They play little part in pollination.

The basic structure of the flower which we have just reviewed is subject to a bewildering array of variations on the theme of carpels, stamens, petals and sepals in the quarter-million or so species of angiosperms. We shall look at some of the more important of these variations in Section 7.3.

7.1.2 The sequence of events

Sexual reproduction in the angiosperm can, as we have just seen, be viewed as a sequence of events. The sequence, however, begins much earlier than the formation of gametes in the flower. The true beginning is when a shoot apex suddenly switches

from purely vegetative development to the production of flowers. This point we can call the *initiation* of flowering.

Once flowering has been initiated, existing leaf primordia around the apical meristem continue to develop as leaves, but new primordia arise on the meristem and these rapidly develop as floral organs. In some species such as poppy and tulip an apex produces only a single flower; such a flower is said to be *solitary*. In most plants, however, several flowers, often a large number, arise in succession from a single apex, and the resulting assemblage of flowers is called an *inflorescence* (Fig. 7.2).

When the flowers first become visible they are tightly enclosed either by the sepals or by specially adapted small leaves called *bracts*. We refer to these unopened flowers as *flower-buds*. During the flower-bud stage there is often rapid stem elongation below the inflorescence, especially in plants whose stems have remained short throughout the vegetative stage. In most cases differentiation of floral organs such as carpels and stamens is complete by the time the flower-buds become visible, but growth of organs, mainly by cell enlargement, and maturation of pollen and ovules continue inside the flower-bud until quite suddenly the bud opens and the flower is revealed. The increase in size of organs during opening and final maturation of the flower is achieved largely by intake of water, not by true growth.

Pollination then takes place and is followed by fertilization. The flower, now having completed its allotted task, senesces and dies, leaving only the carpels, sometimes together with other parts, to develop as the fruit and their enclosed ovules to become seeds. Fruit and seeds then undergo a series of changes, collectively known as *ripening*, before the final stage of reproduction, the dispersal (or, in agriculture, the harvest) of the ripe seed.

The sequence of events for any one flower in an inflorescence is therefore: formation of primordium, differentiation and growth of floral organs, appearance of flower bud, opening of flower, pollination, fertilization of ovules, development of fruit and seeds, ripening of fruit and seeds, and dispersal.

The inflorescence of many species is, like the vegetative shoot, indeterminate in its development, continuing to produce new floral primordia at the apex or from lateral meristems over a long period of time. Once formed, each primordium is launched into the sequence of events just outlined, with the result that flowers open, fruits ripen and

Fig. 7.2 Part of an oilseed rape inflorescence, showing the sequence of reproductive development. At the tip new flowers are opening; at the base fruits are fully formed and beginning to ripen.

seeds are shed in the order that the primordia arise. On an oilseed rape inflorescence, for example, seed is being shed from ripe fruits at the base while new floral primordia are still being generated at the tip. Between the tip and the base are flowers and fruits at every intermediate stage of development (Fig. 7.2).

Other species such as peas and tomatoes produce

small inflorescences in which all the flowers develop, and produce and ripen fruit within a limited period of time, but because the initiation of such inflorescences continues indefinitely flowering and fruiting of the whole plant are again spread over a long period of time.

In these examples of *staggered flowering*, then, there is an indefinite number of flowers in the inflorescence or an indefinite number of inflorescences. This means that the number of harvestable fruits or seeds is not fixed at an early stage. Damage to early flowers, for example by frost, may delay and reduce yield but is not devastating in its effect since later flowers may bear fruit. Staggered flowering, however, means staggered ripening and therefore the need for repeated harvests from the same crop to obtain its full yield. This is not suited to situations where the crop is harvested in a single operation, often mechanically. It is more acceptable in horticultural enterprises or in economies where labour is cheap and plentiful and harvest is by repeated hand picking. Plant breeders have developed pea varieties for field-scale growing in which all seeds come to maturity at around the same time, but for horticultural purposes and in many third world situations varieties with staggered flowering and ripening are still preferred.

In cereals, a large but determinate number of floral primordia form at the shoot apex in a short space of time after flowering has been initiated (Fig. 7.3). The number of primordia formed is determined partly genetically and partly by past and present growing conditions. Once that number has been reached, the formation of new primordia stops. As the inflorescence develops, the earlier-formed primordia develop more slowly than those formed later, so that by the time the flowers open to release and receive pollen the discrepancy in timing between youngest and oldest has been reduced to a matter of hours. Cereals thus show *synchronized flowering*. All the seeds ripen more or less together and a full yield can be obtained in a single harvest. However, because a limit is set during inflorescence initiation to the number of seeds which develop, yield is very sensitive to growing conditions around that time.

Superimposed on the development of the inflorescence are changes in the vegetative organs produced before the initiation of flowering. Early in the sequence of events the leaves must be fully functional, but later, as noted in Section 6.3, they go through a programme of senescence during which they are pillaged of all exportable organic

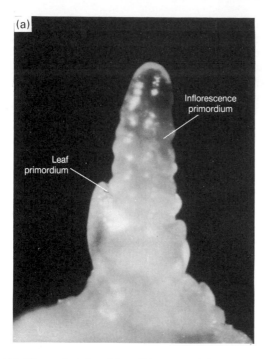

Fig. 7.3 The early (a) and late (b) stages of the development of the floral primordia in the shoot apex of wheat.

materials and these go to nourish the developing fruits and seeds.

Reproduction and vegetative senescence are intimately connected. Treatments which delay one tend to delay the other; similarly, acceleration of one tends to promote the other. Continual removal of flower-buds, for example, delays the senescence of leaves. Thus annual meadow grass, which if undisturbed normally dies after flowering, often behaves as a perennial in frequently mown lawns where its reproductive sequence is constantly interrupted by removal of its inflorescences.

Heavy applications of nitrogen fertilizer to a crop have the effect of encouraging vigorous vegetative growth. This tends to depress and delay flowering and seed production. Thus optimum nitrogen application rates for the yield of grass, where vegetative production is of paramount importance, are in the range 400–500 kg ha^{-1} year^{-1}, but in wheat or barley, where seed production is the important thing, the highest yields, even in the absence of lodging, are obtained with only 100–150 kg nitrogen ha^{-1} year^{-1}.

7.1.3 Sexual reproduction and genetic variability

Flowering and the development of fruits and seeds thus clearly represent a major investment of energy and other resources by the plant. Such intricate structures and such complexity of organization as are involved in the flowering process would lead us to suppose that some major benefit must accrue to species that reproduce sexually by this means.

What the flower does is to provide an efficient system for transfer of male gametes to fertilize the female gametes contained in the ovules. If the flower is cross-pollinated, the two gametes that fuse may come from different parent plants and in this case will contain different sets of genetic information. For reasons to be explained in Section 9.3 even gametes originating in the same parent plant can be quite different from one another in the genetic information they carry. When two gametes fuse, the resulting cell, and ultimately the plant derived from it, will contain and respond to a genetic blueprint which is a resynthesis of information received from the parents, via the two gametes, and which is different from that of either parent. This resynthesizing of genetic information in sexual reproduction is known as *recombination*. Because each cell contains many thousands of separate bits of genetic information, recombination gives almost limitless scope for variation.

The progeny of sexual reproduction, therefore, resemble both their parents but are not identical to either of them or, with certain exceptions, to each other. Recombination is thus important for the maintenance of *variability* in a population.

Why is genetic variability important? Wherever there are differences between individuals in a population there are bound to be some individuals better suited to their environment than others. A biologist would say that these are better *adapted* individuals. For example, in a population of a low-growing weed species in a cereal crop, those more tolerant of shade and able to photosynthesize efficiently at low light intensities will be better adapted to their environment. The better adapted an individual is, the more likely it is to survive to the reproductive stage. More gametes and, later, more seeds will be produced by well-adapted than by poorly adapted members of the population, so that the next generation should show a higher proportion of well-adapted individuals.

Plants possessing unfavourable characteristics, for example a high light compensation point in a shaded environment, will be gradually but remorselessly eliminated with the passing generations, while plants possessing any advantageous characteristics will tend to become more prevalent. This is the basis of *natural selection*, which is regarded as the chief mechanism of evolution.

A fine example of natural selection in operation has been observed in groundsel, a common weed of nurseries, orchards and vineyards which produces up to three generations of seed per year. For many years herbicides of the triazine group, principally simazine, have been used regularly for weed control in such situations. Initially these herbicides gave good control of groundsel, but where they have been applied repeatedly groundsel is no longer as susceptible as it was and many plants tolerant of triazines are surviving to return seed to the soil. Clearly these populations of groundsel are now better adapted to an environment containing triazines than they were, and this has arisen through natural selection of individuals which have some genetically determined characteristic conferring resistance to triazines. Triazine resis-

tance in a wide range of weed species is becoming a major problem in France and elsewhere through the continued use of atrazine, another herbicide of the triazine group, for weed control in maize. Fears have been expressed that the indiscriminate use of other types of herbicide in agriculture, horticulture and forestry may similarly result in the evolution of herbicide resistance in many undesirable species.

For evolution to function, there must be genetic differences between individuals on which natural selection can work. These differences, as shown in Section 9.2.5, are produced at a slow rate by mutation. An asexually reproducing population could evolve by selection of individuals acquiring advantageous mutations, but this would be an extremely slow process compared with general rates of evolution observed in sexually reproducing species. Also, populations of genetically diverse individuals, reproducing asexually, show a form of limited evolution if environmental influences select for better adapted individuals within it. However, it is the recombination of genes into new groupings that is the main and most fruitful means of producing the genetic variety of individuals on which evolutionary influences work. Fungi and bacteria have evolved mechanisms whereby recombination can occur without sexual reproduction, but plants and animals are entirely dependent on it for this function. Hence, only sexually reproducing species of plants can continue to evolve, other than in a very limited way.

Let us make clear, however, that the use of the word 'adapted' in this context does not imply foresight or purpose on the part of evolving species. Species are not adapted for future conditions, in that sense. Their adaptation simply reflects their response to their history. In so far as past influences that have induced evolutionary changes in species continue into the future, individuals of these species are more fitted to survive and reproduce.

Just as natural selection directs the course of evolution in wild plant populations, artificial selection by plant breeders (see Section 9.4) is the basis for the production of new, improved varieties of crops. Selection by breeders is equally dependent on genetic variability and therefore on sexual reproduction. In crops such as potatoes, apples or sugar cane, which are normally propagated vegetatively, new varieties can be produced only from true seed formed as a result of the sexual reproduction process.

7.2 Initiation of flowering

The transition from vegetative development to flower or inflorescence formation at a shoot apex may or may not involve a major change in the developmental pattern of the apical meristem. In peas and beans, for example, the apical meristem continues to generate leaves which differ little from the leaves produced by the meristem of a purely vegetative shoot apex. In these plants it is the behaviour of the axillary meristems that differs between a vegetative shoot and one which has initiated flowering. On a vegetative shoot the axillary meristems develop as axillary buds, which may later grow into branch shoots. On a flowering shoot, however, they produce small inflorescences in the axils of the leaves (Fig. 7.4).

At the other extreme, as, for example, in cereals and grasses, the shoot apex becomes fully committed to the production of an inflorescence and generates no more leaves. In such plants the transition to flowering is irreversible; once an apex has switched to inflorescence formation it can never go back to leaf formation. A single plant may, however, have inflorescence and leaf formation going on side by side in separate shoots or tillers.

Many species show patterns of inflorescence development which are intermediate between these

Fig. 7.4 Inflorescence developed from axillary bud on a pea plant.

two extremes. The inflorescence of cabbage, for example, carries many small stalkless leaves. These are quite different in appearance from the normal vegetative leaves and are usually described as bracts, implying that they are part of the inflorescence itself. As in cereals and grasses, once inflorescence formation has been initiated the apex cannot switch back to vegetative development.

Let us look a little more closely at this critical event in the growth and development of a shoot, the initiation of flowering. We saw in Section 6.1.1 that the root tip meristem has at its base a group of quiescent cells which divide much more slowly than those of the main body of the meristem. This is true also of the vegetative shoot apex, where the quiescent cells play little or no part in shoot development but seem to be pre-programmed from an early stage to generate reproductive structures. The quiescent cells, however, can burst into activity only once something has activated them.

The initiation of flowering, then, involves the activation of quiescent cells in the apical meristem but exactly how they are activated is far from clear. What is clear is that the apex cannot switch from vegetative to reproductive development unless it is in a suitable physiological state. It has to be *competent* to initiate flowering. This usually means that the shoot must have reached a certain minimum size, in terms of the number of leaves already formed. Clearly this is a device to prevent the initiation of flowering before there is sufficient photosynthetic area to ensure that the whole sequence of events through to seed ripening can be successfully completed.

The transition from purely vegetative to reproductive growth and development is thus timed with respect to the age of the plant and its stage of growth. In some species such as annual meadow grass this is all that appears to be necessary, and apices switch to inflorescence formation automatically as soon as they have formed the minimum number of leaves to become competent to do so. Plants of this species can be found in flower at all times of the year.

But most plants flower at one or perhaps two times of the year characteristic of the species or variety. Apple blossom time in the northern hemisphere is in May, not in August or November. Among grasses, sweet vernal grass always flowers earlier in the year than, say, crested dogstail. Among varieties of perennial ryegrass, Hora habitually flowers about 15 days later than S24 and about 12 days earlier than Perma.

Thus the initiation of flowering is in most species timed not only with respect to plant age but also with respect to the seasons. In this way delicate floral organs are more likely to avoid damage by drought or frost and the plant is programmed to bear fruit and seed within the limits of the growing season. Not only that, but the individuals within a species occupying any area must flower in unison if they are to cross-pollinate successfully. To do this they must be sensitive to environmental cues which signal what time of year it is.

7.2.1   Vernalization

One of these cues, for plants living in climates with a cold season, is a period of low temperature. Biennials such as sugar beet and carrot do not normally flower in the year of sowing. Under the influence of the low temperatures of the ensuing winter, however, the shoot apex switches to inflorescence formation. When warmer conditions return in the spring the inflorescence develops rapidly. The induction of flowering by a low temperature stimulus in this manner is called *vernalization*.

The temperatures which cause vernalization are for most species around 2–8°C. Thus if biennial root crops are sown too early, the young plants are exposed to vernalizing temperatures in the spring and a large proportion will 'bolt' – that is, produce an inflorescence – in their first year. Such plants do not lay down carbohydrate reserves in the root and are useless for their intended purpose.

Winter varieties of wheat, rye and barley differ from spring varieties in several respects, but one of the most important differences lies in their requirement for vernalization. Winter wheat, sown in the autumn, is vernalized in the early winter and those apices which are competent to initiate flowering then do so. Inflorescence development is completed in the following spring and a crop of grain can be harvested in the summer or early autumn. If winter wheat is sown in the spring it will not flower and therefore will not yield grain until its second year.

However, varieties of winter cereals differ in the length of the cold period necessary to induce flowering – that is, in their *vernalization requirement*. A variety with a low vernalization requirement can be sown in early spring, since it will experience a sufficiently long cold period at that time, but a variety with a high vernalization requirement must be sown in autumn to give it the full benefit of winter. Some varieties are 'dual-purpose' varieties,

able to be grown either as winter or as spring cereals. These have no vernalization requirement and are really spring varieties which happen to be hardy enough for autumn sowing.

Some plants cannot be vernalized until they have reached a certain stage of growth, while others, including cereals, can be vernalized at any stage except as the quiescent seed. In either case the initiation of flowering is still dependent on shoots having reached the minimum size characteristic of the species. If vernalization occurs early, apices initiate flowering as minimum shoot size is attained. If it occurs late, all apices which have already attained minimum shoot size switch to inflorescence formation immediately the vernalization requirement is satisfied.

7.2.2   Daylength stimulus

Vernalization tends to synchronize flowering in early spring, but it does not explain how flowering is timed in climates lacking a cold season or in plants which characteristically flower later in the year. How do such plants recognize what time of year it is?

Formerly it was thought that the initiation of flowering was largely a response to temperature but the pioneering experiments of Garner and Allard in the 1920s and later research by others showed convincingly that the main stimulus was the length of the night. At first it was assumed that daylength rather than night length was the quantity that plants measured and responded to, and most plant physiologists still refer to the *daylength stimulus*. For most practical purposes it does not matter whether the plant is responding to the length of the light or the dark period since a 16-hour day in nature automatically means an 8-hour night, a 10-hour day automatically means a 14-hour night and so on.

Some species, including cucumber and tomato, are insensitive to daylength and are known as *day-neutral* plants. They initiate flowering on shoots which have attained the minimum shoot size for competence, at any time when temperature and other conditions permit. Many plants which require vernalization, including carrot and most varieties of winter cereals, are also day-neutral. Probably most plant species, however, require the daylength stimulus for the initiation of flowering.

It should be noted that this stimulus differs from the vernalization stimulus in that it is not effective if applied before minimum shoot size is attained.

Only apices which are competent to initiate flowering can respond to daylength. If both vernalization and a daylength stimulus are necessary, vernalization must come first. Whereas vernalization can prime a plant to flower, daylength can only trip the switch that converts a competent but vegetative apex into a flowering apex.

Apart from flowering, many other processes in plants have been observed to respond to daylength. Such responses are collectively known as *photoperiodism*. They include the onset of bud dormancy in trees, the breaking of bud and seed dormancy in certain species, leaf abscission, the onset and termination of meristematic activity in the vascular cambium of woody plants, tuber formation in the potato and bulb formation in the onion. But most of our understanding of photoperiodism comes from the study of the initiation of flowering in shoot apices.

7.2.3 *Long-day plants*

In most daylength-sensitive plants of temperate climates, the apex can switch from vegetative to reproductive development only when daylength exceeds some critical minimum value. In truth, as we have seen, it is night length rather than daylength that is critical, but for most practical purposes this does not matter, and such plants are known as *long-day plants*. They initiate flowering in spring, when the critical minimum daylength is reached. The shorter the minimum daylength required, the earlier initiation will take place.

Long-day plants include spring varieties of wheat, rye, barley and oats and most temperate herbage grasses. The long-day response can be seen as an adaptation to the temperate climate with a pronounced cold season; plants flower in late spring or summer to ensure completion of seed production before the onset of winter. Many long-day plants, including beets, swedes and turnips, require vernalization before the daylength response can be activated. Commonly the longer the period of vernalization, the shorter is the critical minimum daylength for flowering. In winter oats, short days can substitute for vernalization in priming the plant for initiation of flowering. The plant then needs long days to trip the switch.

In spring cereals, the critical minimum daylength has usually been exceeded by the time the minimum shoot size for competence to initiate flowering is reached. Thus the initiation of flowering is delayed until minimum shoot size (7–9 leaves already

formed at the apex) is attained. The earlier the crop is sown the earlier this stage is reached and the earlier the crop flowers. This in turn permits an earlier harvest of ripened grain, a considerable advantage in areas troubled by poor harvesting weather. In a late-sown crop, development to minimum shoot size proceeds more rapidly, so that a month's difference in sowing date may mean only a few days difference in harvesting date. Yield, however, suffers because of the shorter period of growth.

It may take weeks or even months after flowering initiation for a plant to reach the full flowering stage. The time of year at which a plant appears to flower may not therefore give an accurate impression of the time at which flowering initiation takes place and therefore of its daylength requirement. Thus in a spring cereal when 7–9 leaves have been formed at the apex and the shoot is competent to initiate flowering only about three leaves have emerged and unfolded. By the time five leaves have unfolded, the last of the primordia making up the embryonic inflorescence has been formed at the apex, and flowers and bracts are beginning to differentiate in the primordia already formed. Soon the stem below the inflorescence begins to elongate and swollen leaf nodes become visible. At the stage when the first or second of these nodes is detectable, gametes are beginning to be formed in the developing flowers (Fig. 7.5).

Now the processes of inflorescence formation during the three to five-leaf stage and gamete formation during the early stem elongation stage are very critical ones for the yield of cereal crops. These processes are highly sensitive to a group of herbicides based on auxins, a type of growth regulator which has a major role in the natural control of growth and development in the plant (see Section 8.4.1). The auxin-based herbicides, or phenoxyalkanoic acids, kill a wide range of dicotyledonous weeds without damaging the cereal crop, provided they are not applied during the sensitive growth stages referred to above. These herbicides include well-known materials like MCPA, 2,4-D, mecoprop and 2,4-DB. What effects do these materials have on cereals?

Such herbicides have little or no effect on yield if applied between the five-leaf stage and the first-node-detectable stage, although in some situations stunting of roots, particularly with mecoprop, has been observed. Any damage to the cereal crop is greatly outweighed by the benefits of weed removal. But if these herbicides are applied too

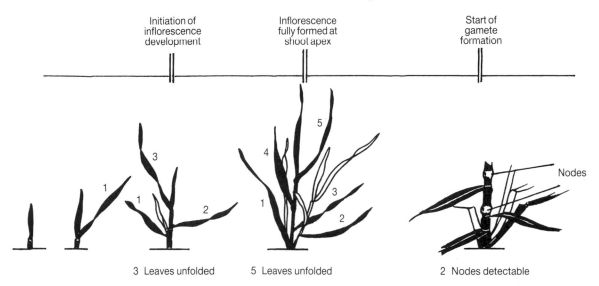

Initiation of inflorescence development

Inflorescence fully formed at shoot apex

Start of gamete formation

Nodes

3 Leaves unfolded 5 Leaves unfolded 2 Nodes detectable

Fig. 7.5 In a spring cereal the stage of growth as measured by leaf unfolding and stem extension indicates the stage of reproductive development at the shoot apex.

late, once gamete formation has begun, they can result in ovules and pollen containing non-functional gametes, which in turn leads to missing grains and therefore reduced yield. If they are applied too early, while flower primordia are still forming at the apex, they typically cause distortion of the inflorescence, often with missing flowers leading to missing grains and again reduced yield (Fig. 7.6).

7.2.4 Short-day plants

Short-day plants can initiate flowering only when daylength is less than a certain critical maximum value. Strictly speaking, they respond to long nights, not to short days. Many species of warm temperate and subtropical climates show this type of response, which can be seen as a device to ensure completion of the life cycle before the onset of the dry season in summer. Most varieties of maize, rice and soya bean are short-day plants. They do not perform well at high latitudes because they are not competent to initiate flowering until at least midsummer, and the initiation of flowering cannot take place until the days shorten to the critical length in late summer. There is insufficient time then for the crop to flower, fruit and ripen seeds before the winter frosts.

In the commercial glasshouse propagation of chrysanthemum, a short-day species, it is frequently desirable to suppress the natural tendency of the plants to flower in winter. To keep them vegetative, supplementary lighting can be used to lengthen the day. It has been found, however, that a very short period of illumination in the middle of the night, breaking the night into two relatively short dark periods, has the same effect (Fig. 7.7). This emphasizes the fact that it is night length, rather than daylength, that provides the cue for the

Fig. 7.6 Distorted barley inflorescences with missing grains, caused by an auxin-based herbicide applied too early, while the shoot apex is still at a sensitive stage of development.

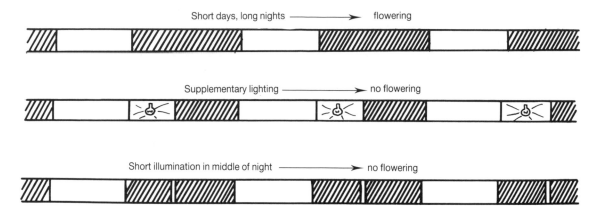

Fig. 7.7 The use of supplementary lighting to prevent flowering in glasshouse chrysanthemums. Short days in winter normally promote flowering, but this can be prevented by artificially lengthening the day or, more cheaply, by providing a short period of illumination in the middle of the night.

photoperiodic response. To induce chrysanthemums to flower in summer, the night must be artificially lengthened by covering the plants. Through a knowledge of the photoperiodic requirements of this species, potted or cut chrysanthemums are now available in flower all year round.

7.2.5 Some complications

The situation in nature is both more complex and more flexible than the simple classification of plants into day-neutral, long-day and short-day groups would suggest. Varieties within species such as barley, maize and soya bean differ in their daylength requirements. Spring barleys are, as we have already indicated, long-day plants, while most winter barleys are day-neutral, switching from vegetative development to flowering as soon as their vernalization requirement has been met. Most maize and soya bean varieties are short-day plants, but some are day-neutral, each shoot initiating flowering as soon as it achieves the minimum number of leaves for ripeness to flower. Similarly, within grass species which occur over a wide range of latitudes, such as some prairie grasses of North America, northerly and southerly types exist with different photoperiodic requirements.

In some species there may be a further photoperiodic requirement for subsequent development. Dwarf beans have no daylength requirement for the initiation of flowering, but development to the fruit stage can take place only in short days. White clover needs short days followed by long days to initiate flowering.

Some plants require only one dark period of the correct length to be induced to flower while others require many. Where a number of dark periods are needed, the response, measured as the percentage of plants induced to flower, may be proportional to the number of dark periods, or it may be an all-or-nothing reaction where after a critical number of dark periods all the plants initiate flowering. How plants measure the length of darkness and use it as a cue for flowering are only partially understood. They are part of the plant's control system and are examined in Chapter 8.

7.3 Structure and function of the flower

7.3.1 Insect-pollinated flowers

We know from the record of fossils in rocks laid down during the Cretaceous period of geological time, around a hundred million years ago, that the rise of the angiosperms to their present prominence took place at the same time as that of the insects. Before the Cretaceous period the only seed plants were gymnosperms – ancestors of today's conifers and cycads – the pollen of which was, and still is, transferred by wind. The advent of flying insects offered a much less haphazard means of pollination and this in turn meant that less copious quantities of pollen needed to be produced.

Insect pollination, however, requires various things which are not necessary in a wind-pollinated plant. First, the plant must have something to offer the insect as an inducement to visit it. In some cases, as, for example, in roses, the pollen itself is what the insect seeks; it provides a highly nutritious food. Some inevitably adheres to the body of the insect and a proportion of this is deposited on the next plant visited by the insect. Where pollen is the inducement, it is clearly essential for the stigmas to be situated close to the pollen-producing anthers, so that there is a good chance of the stigmas picking up some of the pollen accidentally deposited by the foraging insect.

In other species, including brassicas and beans, pollen is supplanted as an inducement to insects by a solution of sugars and other nutrients called nectar which, as we saw in Section 7.1.1, is produced by glands near the base of the flower.

Nectar is particularly attractive and useful to bees, which concentrate it as honey to nourish their larvae. Bees and other nectar-gathering insects tend to have a long snout or proboscis with which they reach down to the nectaries, in the process brushing pollen off the stamens on to their hairy bodies and legs, or depositing pollen from their bodies and legs on to the receptive stigmas. An example of the structure of the male and female floral parts that evolved through such processes is shown in Fig. 7.8.

Fig. 7.8 An example of the structure of (a) male, and (b) female floral parts.

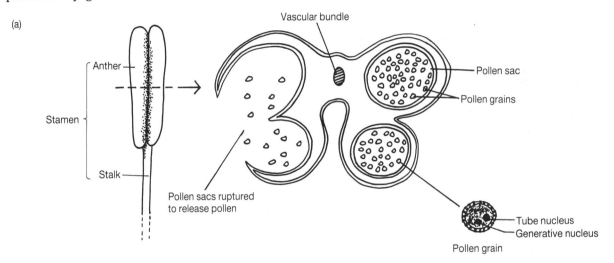

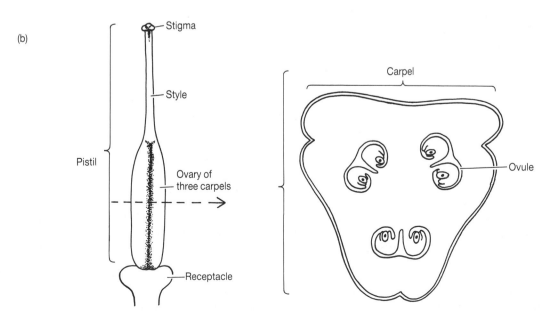

Another requirement of insect pollination is some means of advertisement to the insect that pollen or nectar is available. This is the function of the petals, which are highly modified leaves closely surrounding the stamens. They are conspicuous, usually brightly coloured or white, and often have markings that guide insects to the nectar below. In addition, the petals of many species are strongly scented, and can attract insects from long distances.

Let us look briefly at some variations in the structure of insect-pollinated flowers, concentrating on agriculturally important species. A very common trend is for two or more parts of the flower to become fused together to form a single united structure. This is particularly common among the carpels. In a brassica flower, for example, two carpels are joined side by side to make a single carpel-like organ with an ovary at the base containing several ovules, and a single style and stigma (Fig. 7.9). The term *pistil* is used to refer to the female part of a flower where this consists of a single carpel or, as in the brassica, a single structure made up of two or more carpels fused together.

The ovary of the brassica flower (Fig. 7.10) is set above the receptacle and therefore above the level of attachment of stamens, petals and sepals. For this reason it is described as a *superior* ovary. In many flowers, however, the ovary is partly or wholly embedded in the receptacle, so that the stamens, petals and sepals are attached around or above it. Such an ovary is said to be partly or wholly *inferior*. The flower of apple (Fig. 7.11), for example, has an inferior ovary.

Fig. 7.9 Structure of pistil of brassica flower showing fused carpels.

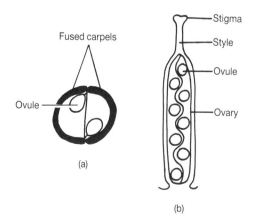

(a)

(b)

Fig. 7.10 Flower of brassica in half-section.

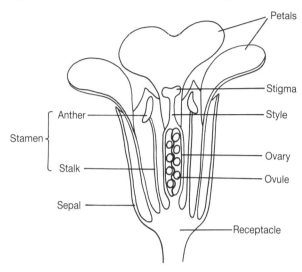

Fig. 7.11 Flower of apple, in half-section.

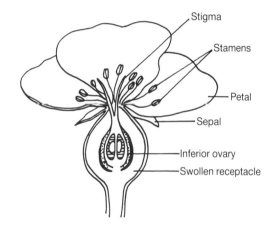

In many species the bases of the petals are joined to form a single skirt-like or tube-like structure called a *corolla*. (The term corolla can also be used to refer collectively to the petals even where these are quite free of one another as in the strawberry, brassica or apple.) In the potato flower (Fig. 7.12) not only are the petals fused together, but the stamens are fused to the corolla so that they appear to be borne on the corolla rather than directly on the receptacle. Just as petals can join together in the corolla, so sepals can join together to form a cup-like or tube-like structure called a *calyx*, around the base of the flower. (Again the term calyx can be used as a collective word for the sepals whether or not they are joined together.)

Fig. 7.12 Flower of potato, in half-section.

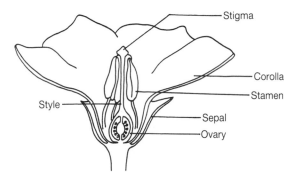

Fig. 7.13 Regular (left) and irregular (right) flowers. A regular flower has several lines of symmetry whilst an irregular flower has only one.

All of the flowers we have looked at so far are fairly unspecialized and attract many different kinds of insect. They all show strong radial symmetry. This arises because all the petals are the same size and shape and are equally spaced around the receptacle. When looked at from above there are many lines about which the flower, in particular the corolla, is symmetrical. Such a flower is said to be *regular*. In Figure 7.13 we contrast this with a more specialized flower which, when looked at from above, has only one line about which the corolla is symmetrical. This bilateral symmetry arises from petals which are of different sizes and shapes and are often unevenly spaced around the receptacle. Such a flower is described as *irregular* and is usually adapted for pollination by a particular group of insects.

A good example of an irregular flower is that of the broad bean (Fig. 7.14). This has a calyx of five united sepals around the base of a bilaterally symmetrical corolla of five petals. The two 'lower' petals are joined together throughout their length to form a keel-like structure. Flanking this are two lateral petals, each with a large dark-coloured spot, and above these is the large and conspicuous 'upper' petal which advertises the flower to distant insects. Held within the keel is a tube formed from the stalks of nine of the ten stamens; the tenth stamen is free of the others and can be pushed aside by a bee reaching in to collect nectar from the base of the tube. Within the tube of stamens is a pistil of one carpel, the ovary of which contains a row of ovules. The style at the top of the ovary is bent upwards just below the stigma.

Fig. 7.14 Flower of broad bean, in half-section.

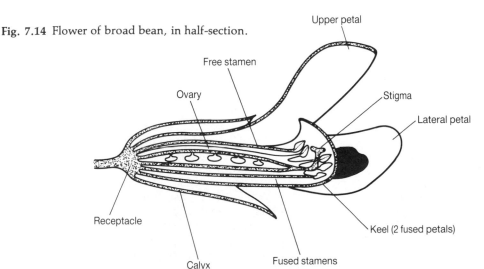

Irregular flowers evolved as a means of restricting the range of insect species attracted to them. This is well illustrated by the bean and other legumes, all of which have the same general flower structure. Only bees and their close relatives can get at the nectar deep within the flower, because these are the only insects with a long enough proboscis to reach the bottom of the stamen tube, and because only heavy insects like bees have sufficient weight to depress the keel which otherwise blocks entry to the stamen tube. Certain species of wild bumble bee commonly make a hole through the base of the calyx and corolla of field bean flowers and get access to the nectar without pollinating. Subsequently honey bees, normally good pollinators, 'steal' the nectar through the same hole. This may lead to reduced pollination and hence reduced yield of beans.

What is the advantage of restricting access to a few species of insect? Because the nectar of beans is available only to bees, for example, bees will tend to concentrate on bean flowers where they do not have to compete with other types of insect for the nectar. They are therefore more likely to bring bean pollen to a bean flower than if they have been indiscriminately visiting dozens of different plant species.

The structure of the bean flower permits one further device to aid pollination. The stamens and pistil within the keel are held there under tension. With the weight of a bee landing on the keel this tension is suddenly released. Pollen is thus forcibly 'flicked' on to the bee's body. At the same time, a membrane over the stigma is ruptured as it hits either the bee or the upper petal of the flower. This makes the stigma receptive and the chances are high that it will immediately pick up some pollen collected involuntarily by the bee in another flower. This effect of a bee on a legume flower is known as *tripping*. Without being tripped the flower cannot be pollinated, thus the chances of self-pollination are reduced. Other devices to promote cross-pollination will be discussed in Section 7.3.3.

We have looked at a few important trends in the evolution of insect-pollinated flowers, but another common trend is for individual flowers to be massed together in an inflorescence, as illustrated in Fig. 7.15. Flowers with relatively small petals can be seen by insects from afar if a sufficient number of these flowers are borne in a cluster. Good examples of this are provided by red clover with its globular head of small irregular flowers or cow parsley with its large flat-topped inflorescence (of the type known as an *umbel*) of individually inconspicuous white flowers (Fig. 7.16). Fig. 6.8 illustrates a similar umbel, that of giant hogweed.

This trend has been carried to its extreme in members of the daisy family. What we think of as a daisy 'flower' is in fact a whole inflorescence (Fig. 7.17). Careful pulling apart of the inflorescence reveals that each of the white 'petals' is the one-sided corolla of a tiny flower. These peripheral flowers are known as *ray-florets*. Similarly, the yellow centre of the inflorescence is made up of large numbers of tiny flowers called *disk-florets*, which have a much reduced corolla. Some other members of the family, such as the dandelion, have only ray-florets, others such as pineapple mayweed have only disk-florets but most, like the daisy, have both.

Fig. 7.15 Some massed or compact inflorescences, showing the umbel of cow parsley (a), the composite, daisy like, inflorescence of mayweed (b) and the inflorescence of clover (c).

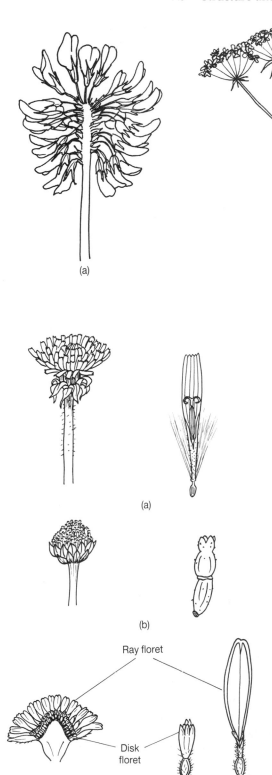

(a)

(b)

Fig. 7.16 (Above) Inflorescence of (a) red clover, and (b) cow parsley. In the cow parsley umbel, each of the stalks of the umbel ends in another, smaller, umbel. This structure is called a compound umbel.

Fig. 7.17 (Left) Types of inflorescence found in the daisy family: (a) dandelion carrying only ray florets; (b) pineapple mayweed carrying only disk florets; (c) daisy carrying both. Reproduced from S Ross-Craig (1961) *Drawings of British Plants, XVI–XVIII.* G Bell & Sons, London, with permission of the publishers.

(a)

(b)

Ray floret

Disk floret

(c)

7.3.2 Wind-pollinated flowers

In the course of evolution many angiosperms have reverted to wind pollination. They retain, however, a flower structure which arose first as a device for insect pollination, but show modifications to make this structure better suited for pollination by wind.

The most immediately obvious difference between a wind-pollinated and an insect-pollinated flower is the absence in the former of a conspicuous corolla. In most wind-pollinated species there is only one set of organs outside the stamens. These organs are sepal-like rather than petal-like and could be called a calyx but more commonly the term *perianth* is used, since it is often not clear whether it is derived from a calyx, or a corolla, or both. (The perianth is a collective term for all the floral organs outside the stamens whether or not these are differentiated into petals and sepals).

The flower of beet (Fig. 7.18) is typical of wind-pollinated flowers. It has a perianth of five sepal-like organs joined together at the base, five stamens

173

Fig. 7.18 Flower of beet, in cross-section.

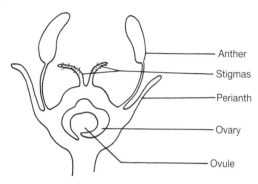

- Anther
- Stigmas
- Perianth
- Ovary
- Ovule

Fig. 7.19 Inflorescence structure in the grass family. Left, panicle of oat; right, spike of wheat. Both panicles and spikes are made up of individual subunits called spikelets.

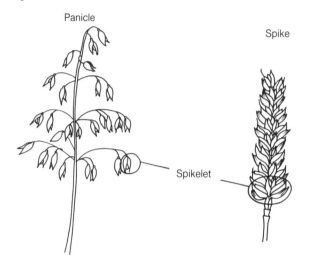

Panicle

Spike

Spikelet

with long stalks to bear the anthers out in the wind, and a pistil with a partly inferior ovary containing a single ovule. The two or three stigmas are long and covered with hairs to help in the trapping of windborne pollen.

Probably the flowers most highly modified in the course of evolution for wind pollination are those of the grasses, including those species now cultivated as cereals. Because of the supreme importance of the grass family in world agriculture, it is desirable to understand something of their flower and inflorescence structure. To reveal the flowers it is necessary to take an inflorescence apart carefully in a series of steps.

The inflorescences or 'ears' of grasses and cereals are of two main types. In wheat, barley and ryegrass the inflorescence is unbranched with a single main axis, or *rachis*, on which a number of unstalked subunits are borne. The whole inflorescence is called a *spike* and the subunits, which are compact little groups of flowers together with accessory structures, are called *spikelets*. In oat, fescue and meadow grass, spikelets are borne on the ends of branched stalks which arise from nodes on the rachis. Such an inflorescence is called a *panicle* (Fig. 7.19).

The spikelets of grasses, whether carried on spikes or in panicles, show a remarkable uniformity of structure throughout the large and very widespread grass family. At the base of every spikelet is a pair of bracts called *glumes*. Sometimes, as in oat, these are large and completely enclose and protect the spikelet during its development; in other species such as barley they are very small and do not appear to serve any useful function. The spike of ryegrasses is unusual in bearing spikelets with only one glume; the protective function of the second glume is taken over by a depres-

sion in the side of the rachis. In all grasses and cereals the glumes and the other parts of the spikelet within them are carried alternately on two sides of a slender stalk, the *rachilla* (Fig. 7.20).

What are the other parts of the spikelet? They are a number of similar structures called *florets*. Note that a grass floret is not analogous to a daisy floret which, as was noted above, is a small flower. A grass floret contains a flower but is not itself a flower. There may be up to twenty florets per spikelet, as in Italian ryegrass, but more commonly there are fewer than ten, for example 3–5 in wheat and meadow grass, 2–3 in oat and only one in barley and timothy.

Each floret consists of a large outer bract, or *lemma*, a smaller inner bract or *palea*, and between them, at last, a single flower (Fig. 7.20). In some species the lemma may carry a bristle or *awn* on its tip (e.g. barley) or its back (e.g. wild oat). The lemma and palea together perform the function normally served by the perianth of wind-pollinated flowers, enclosing the flower as it develops. They remain in position during the development of the fruit or grain and in barley and oats continue to enclose the grain after threshing. When these grains are milled for human consumption, the lemma and palea must be removed.

The true perianth of the grass flower is absent altogether or in some species is reduced to two or occasionally three tiny scales called *lodicules*. The

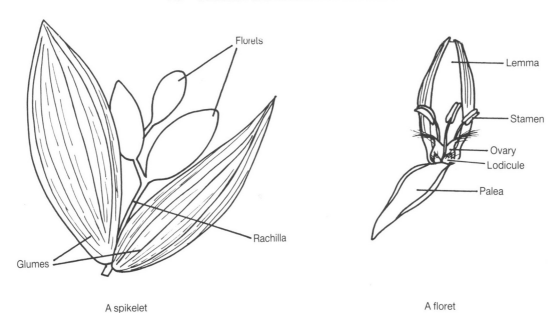

A spikelet   A floret

Fig. 7.20 Structure of a typical grass spikelet (left) and of an individual floret (right). Each spikelet has one or more florets, which in turn contain a single flower.

flower has a pistil consisting of a small round ovary containing a single ovule and surmounted by two long feathery stigmas, well adapted for catching windborne pollen. Around the pistil are three (in rice, six) long-stalked stamens. When the flower is ready for pollination, the lodicules take in water and swell, forcing apart the lemma and palea. The anthers are thrust out on their long stalks to release their pollen to the wind. The stigmas also emerge to catch pollen. Soon afterwards the lodicules lose their water and the lemma and palea close together once more.

The pollen grains of wind-pollinated flowers are produced in far greater quantities and are considerably smaller and lighter than those of insect-pollinated flowers. At certain times of the year the air contains large amounts of pollen, particularly during spells of dry weather when pollen is not washed out of the air by rain. Inhaled pollen causes the allergic reaction known as hay fever in susceptible people. The species primarily responsible for hay fever are obviously those locally frequent and in flower at the time. During spring in Great Britain, for example, the main contributors of pollen to the air are trees such as birch, elm, beech and oak, but during summer grasses are more important. The number of people affected depends not only on the amount of pollen

in the air but on its species of origin. In North America the pollen of ragweeds seems to be particularly allergenic and is responsible for much of the hay fever suffered during the late summer.

7.3.3   Devices to promote self- or cross-pollination

The angiosperm flower, whether it uses insects, wind or some other agency, is thus designed to promote cross-pollination. Some wild species and many agricultural crops are, however, predominantly self-pollinating. In wheat, oats and barley, for example, the anthers release pollen on to the stigmas of the flower before the floret ever opens. Contact between anthers and stigma in the pea flower similarly ensures that self-pollination is the rule and cross-pollination a rare accident. Self-pollination in an agricultural crop has considerable advantages for the farmer, as we shall see in Section 9.3.1.

Many plants, however, including some crop species, have additional devices to discourage or prevent self-pollination and thereby promote cross-pollination, leading to a greater degree of genetic recombination and therefore variability and adaptability in the population. As mentioned in Section 7.3.1, the tripping mechanism of many legume flowers, such as those of bean and lucerne (alfalfa), is such a device, because the first pollen to come in contact with the stigma after its membrane ruptures is likely to be on the body of a bee and it will have been collected by the bee from another flower.

Another device is the existence of two kinds of flower with anthers and stigmas in different positions. This occurs, for example, in *Primula* which has one kind of flower with a short style and anthers set high up on the inside of the corolla tube, and another kind with a long style and anthers set further down. These are known as 'thrum' and 'pin' flowers respectively (Fig. 7.21). Nectar is secreted at the base of the corolla tube. A bee foraging for nectar tends to collect pin eye pollen on its head and thrum eye pollen on its abdomen. It similarly tends to deposit pollen from its head on a thrum eye stigma and that from its abdomen on a pin eye stigma. Thus in either case the pollen deposited on the stigma is likely to have come from a flower of the opposite kind.

Very commonly, both in wind-pollinated and insect-pollinated flowers, the stamens and carpels mature at different times. In most grasses the anthers emerge first from the floret and release their pollen before the stigmas expand. Sweet vernal grass is exceptional in that the stigmas emerge first, followed by the anthers. In either case self-pollination is discouraged but cross-pollination is not inhibited since different flowers mature at different times. Thus, for example, the anthers on flower A may release pollen just as the stigmas on flower B emerge and can catch it.

Fig. 7.21 The two kinds of flower found in *Primula*, which promote cross-pollination.

Self-pollination is prevented rather than merely discouraged in species that have *unisexual* flowers, that is, flowers lacking either stamens or carpels. Such flowers are described as male if they have stamens but no carpels, or female if they have carpels but no stamens. If male and female flowers occur in the same inflorescence or more commonly in separate inflorescences on the same plants, the plant is said to be *monoecious*. If male and female flowers occur on separate male and female plants, the plant is said to be *dioecious*.

Maize has unisexual flowers and is monoecious. The male flowers are in a large panicle called the 'tassel' at the top of the plant, well exposed to air currents for pollen dispersal. The female flowers are in a separate inflorescence, the 'cob', situated in the axil of a leaf midway up the stem. Each of the several hundred female flowers in the cob has a very long filamentous stigma, and these stigmas, or 'silks', emerge in a bunch from the top of the cob (Fig. 7.22).

Separation of the sexes on different inflorescences as in maize means that cross-pollination between flowers on different plants is at least as likely as between flowers on the same plant. In maize, this is further favoured by the tassel maturing slightly earlier than the silks on any one plant. In dioecious plants such as hop, spinach and asparagus, cross-pollination can only occur between flowers on different plants; thus all seeds produced by the females of dioecious species must contain genetic information derived from two parents.

Flowers, in all their diversity, are the means

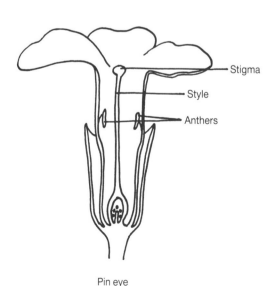

Pin eye

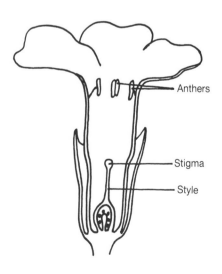

Thrum eye

Fig. 7.22 Plant of maize, showing the male inflorescence, or 'tassel', at the top and the female inflorescence or 'cob', lower down.

whereby sexual reproduction and the recombination of genetic information take place in the angiosperm plant. The degree of recombination is very limited in species which are predominantly self-pollinated, greater in species which cross-pollinate but which do not discourage cross-pollination between flowers on the same plant, greater still in species where 'pin' and 'thrum' flowers, or unisexual flowers, are separated on different inflorescences, even greater where anthers and stigmas mature at different times but where all the flowers on any one plant come to the anther-opening or receptive-stigma stage at the same time, and greatest of all in dioecious species where both self-pollination and cross-pollination between flowers on the same plant are impossible. As we shall see later in this chapter, there may be physiological devices to promote cross-fertilization (that is, fertilization of the ovule by pollen from another plant) superimposed on the various structural devices to promote cross-pollination.

7.3.4   Uses of flowers

Before we leave the subject of flower structure, we should say something about the usefulness of flowers to mankind. The flower, in particular the insect-pollinated flower with its colourful and often scented petals, is a universal symbol of natural beauty and other values and plays an important part in the cultural life of most societies. Cut and potted flowers such as narcissi, tulips, carnations and chrysanthemums are a major commercial crop, and considerable areas of land are given over to the production of seeds, bulbs and rooted plants of flowering ornamentals for our gardens. Flowers are the source of many perfumes and were formerly an important source of dyes, before these could be more cheaply produced by chemical synthesis from petroleum and other organic raw materials. Commercial beekeeping for the production of honey is also entirely dependent on flowers.

In cauliflower the development of the inflorescence is naturally arrested at an early stage and the whole structure becomes grotesquely swollen to form the 'curd' which is eaten as a vegetable. Unlike most vegetative storage organs, the cauliflower curd is rich in protein as well as carbohydrate. In broccoli a number of smaller curds known as 'spears' form in leaf axils.

The structure of the flower is invaluable in the identification and classification of plants. It is much less strongly influenced by the conditions under which the plant is growing, and has tended to evolve more slowly than vegetative features. Family relationships between species can therefore be more readily discerned by an examination of flowers than of leaves, stems or roots.

7.4   From pollination to fertilization

Pollination is an essential prerequisite of fertilization, but the pollen on the stigma is still a long way from the ovules down below. We must now examine how the male gametes, which have been transferred in the pollen, get to the female gametes or egg cells in the ovules.

When a pollen grain lands on the stigma it germinates to produce a slender outgrowth called the *pollen tube*, which grows down into the tissues of the stigma. Pollen grains of one species will not normally germinate on the stigma of another species, except where the two species are very closely related, but how pollen 'recognizes' whether or not it has found the right species is at present a mystery. Pollen grains, with a few exceptions, do not need specific stimuli to prompt them to germinate; most will happily produce pollen tubes when placed in a drop of sugar solution on a glass slide. Inhibitory substances produced by the stigma have been suggested as a possible mechanism to prevent germination, but this would imply a different inhibitor for each species – about quarter of a million in all – a highly improbable state of affairs.

To complicate the issue, many species show *self-incompatibility*, in which pollen from one individual will not germinate, or shows reduced germination or pollen tube vigour, on the stigmas of the same individual. Usually individuals of these species can be arranged in two or more *incompatibility groups* such that the pollen of any member of a group will not grow on the stigmas of any other member of the same group. The group to which an individual belongs is determined genetically; thus close relatives are more likely than distant relatives to be mutually incompatible.

Self-incompatibility occurs in many cross-pollinated crop species, including beet and red clover, and is an obvious adaptation to promote cross-fertilization. Some fruit crops, including apples and pears, show varying degrees of self-incompatibility; since fertilization is a prerequisite of fruit development it is usually essential to grow more than one variety in a single orchard. Trees of the main commercial variety are commonly grown in a lattice arrangement with trees of a pollinator variety distributed at regular intervals (Fig. 7.23). Care must be taken to ensure that the intended pollinator does not belong to the same incompatibility group as the main variety; for example, the apple variety

Kidd's Orange Red cannot be used as a pollinator for Cox's Orange Pippin.

The growth of the pollen tube takes place through the intercellular spaces of the stigma and style (Fig. 7.24). The energy for its growth comes not from the pollen grain, whose energy reserves are very limited, but from the cells of the style. The tube grows unerringly towards the ovary, and once in the ovary towards any ovule that has not yet had its egg cell fertilized. The direction of growth of the pollen tube is thought to be a response to minute concentration gradients of chemicals generated by the ovules.

The pollen grain, at the time it is shed, contains two or sometimes three nuclei. These are the *tube nucleus* (which was formerly thought to control pollen tube growth but is now believed to be unimportant in this respect), and a *generative nucleus*, which may already have divided at this point to form two male gamete nuclei. In all species the generative nucleus has so divided by the time the tip of the pollen tube has reached an ovule.

Fig. 7.24 The growth of a pollen tube down the style to the ovary.

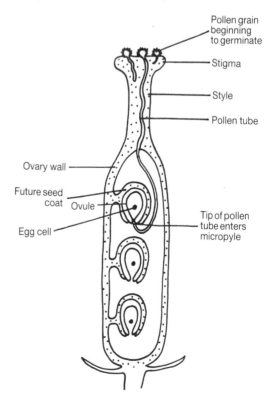

Fig. 7.23 Typical arrangement of trees of the main fruit-bearing (unshaded) and pollinator (shaded) varieties in an apple orchard. Redrawn from *MAFF Bulletin* No. 207 (1972). HMSO, London, with permission of the publishers.

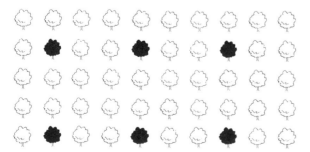

178

As shown in Fig. 7.25, the pollen tube penetrates the wall of the ovule through a tiny pore, the *micropyle*, where it releases the two male gamete nuclei. Here, one of the male gametes fuses with the egg cell to create a single cell from which an embryo will eventually grow. The other male gamete fuses with two other cells in the ovule; the resulting cell gives rise by repeated cell division to the endosperm, which becomes an energy storage tissue helping to nourish the growth of the embryo.

For a good yield of any crop grown for its seeds, such as cereals, beans, peas and oilseed crops, a large percentage of its ovules must be fertilized. This in turn requires effective pollination. In self-pollinated crops like wheat, barley, oats and peas, and in wind-pollinated crops like maize and rye, inadequate pollination is seldom if ever a problem, but in insect-pollinated crops such as field beans, oilseed rape and most fruit crops, yield is often limited by poor pollination. It is common practice to place hives of bees among such crops, giving a yield of honey as a valuable by-product, and this is particularly beneficial where hedgerows and other habitats for wild bees have been removed. Cold or wet weather limits the activity of bees and

prolonged spells of conditions like these can therefore result in reduced yield. Bumble bees will often forage, and therefore pollinate, in conditions that keep honey bees in the hive, thus the introduction of honey bees is not a complete substitute for the maintenance of habitats for wild insects on a farm where insect-pollinated crops are grown. Care must be taken when using agricultural insecticides not to kill pollinating and other beneficial insects along with harmful ones; by choosing the right chemical and applying it at a time when pollinators are not active such damage can often be minimized.

Adequate pollination does not necessarily lead to adequate fertilization, especially in hot or dry conditions. Tomatoes are predominantly self-pollinating, but in glasshouses where the temperature is allowed to rise too high they often fail to form fruit; instead entire flowers are shed by a process of abscission. The grower must therefore keep a careful control of temperature and humidity to ensure fertilization.

Flower abscission happens naturally in most species if the flowers are not pollinated; in hot conditions it may happen following pollination or even following fertilization. The prevention of flower abscission and initiation of fruit development which normally accompanies fertilization of the ovules is referred to as *fruit set*, and it is to the changes following this that we must now turn our attention.

7.5   Seed and fruit development

7.5.1 Seed development

Initially, endosperm development in the seed takes the form of a proliferation of protoplasm with little or no cell wall formation. The resulting liquid endosperm can be seen, for example, if a cereal grain at the 'milk-ripe' stage is crushed with the thumb-nail. Coconut milk is another example of liquid endosperm. Later, cell walls begin to form and the endosperm becomes doughy or floury. Starch is the main substance stored in the endosperm, but cereal grains also contain a considerable amount of protein. In wheat a substantial part of this is the mixture of proteins known as *gluten*, which is responsible for the elasticity of dough made from wheat flour. This elasticity in turn permits the manufacture of a light-textured, risen loaf by enabling retention in the dough of bubbles of carbon dioxide produced by the fermenting action of yeast. In addition, a zone

Fig. 7.25 Entry of a pollen tube into an ovule just prior to fertilization. The first male gamete fuses with the egg cell, the second with two other cells to form the first cell of the endosperm.

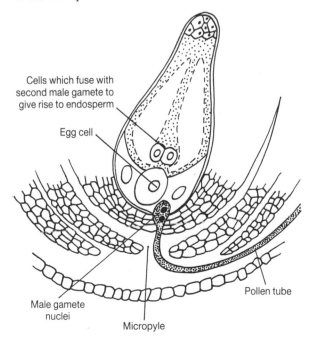

Cells which fuse with
second male gamete to
give rise to endosperm

Egg cell

Male gamete
nuclei

Micropyle

Pollen tube

of endosperm near the seed coat, known as the *aleurone layer*, occurs in cereals (see Fig. 5.2). This is relatively rich in protein and is involved in the production of enzymes needed for the mobilization of carbohydrate reserves during germination of the seed.

Once the endosperm has passed the milky stage, the fertilized egg cell begins to divide. The first division forms two cells, one of which, the *suspensor cell*, acts as an anchorage at the micropyle end of the ovule, thrusting the other cell and the products of its division into the endosperm. As the endosperm nourishes the growing embryo it is gradually consumed.

In some seeds, such as those of peas and beans, the embryo consumes all the endosperm and accumulates a considerable food reserve in its cotyledons. Thus at maturity the seed consists only of a seed coat and an embryo, with no recognizable endosperm. In other seeds such as those of cereals, however, the mature seed retains a large body of endosperm which supplies most of the energy needed for germination and seedling growth.

The development of the seed is, therefore, the normal consequence of fertilization of the ovule. In some species, however, seeds grow directly from the ovules without fertilization. The embryo in such cases is derived from a single cell which is genetically identical to all other cells in the parent plant. Thus no genetic recombination takes place and the offspring do not show the variability expected of the progeny of sexual reproduction. Reproduction by seed which does not involve fusion of gametes and is therefore asexual is called *apomixis*. It is the normal reproduction method of several common wild plants, including dandelion and blackberry. These species probably reproduce sexually very occasionally and thereby maintain the adaptability of the species.

Some apomictic plants, though not requiring the ovules to be fertilized, still need the stimulus of pollination to trigger off seed and fruit development. In such cases it is very difficult to tell whether seed has arisen through apomixis or through normal fusion of gametes. Because of this, apomixis may be a much more common phenomenon than is generally believed.

7.5.2 *Fruit development*

As the ovules develop into seeds, so the ovary containing them swells to become the fruit. The nature of the fruit depends firstly on the nature of the ovary – the number of chambers, the number of ovules per chamber, the arrangement of ovules in the chambers, and so on – and secondly on the way in which the ovary develops after fertilization. The wall of the fruit is called the *pericarp* and is derived from the ovary wall. In the case of an inferior ovary the pericarp is derived partially from the receptacle which surrounds the ovary. The pericarp may be of almost any texture from soft and succulent as in the grape or tomato to hard and stony as in the hazel nut. Many fruits show differentiation of the pericarp into layers – the plum, for example, has a thin outer pericarp forming the skin, a succulent middle pericarp forming the flesh and a stony inner pericarp surrounding the single seed. Some examples of fruits are shown in Fig. 7.26, including two, the apple and the strawberry, in which most of the fleshy material is formed from the receptacle.

Ordinarily fruit development in non-apomictic species is triggered by fertilization of the ovules, and if pollination or fertilization does not occur then flower abscission occurs and no fruit is set. Some plants, however, can form *seedless fruit* in the absence of fertilization – good examples are banana and most varieties of tangerine. With an understanding of the natural hormonal control of fruit development, it is now possible to produce seedless fruits in a wide range of species by the use of growth-regulating chemicals (see Section 8.7).

Fig. 7.26 Some examples of fruits whch develop in different ways from the ovary and, in some cases, the receptacle of the flower.

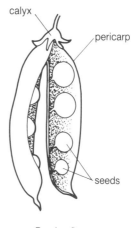

calyx

pericarp

seeds

Pea (pod)

7.5   Seed and fruit development

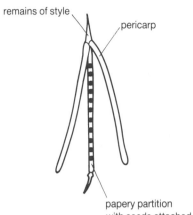

remains of style

pericarp

papery partition
with seeds attached

Rape (capsule)

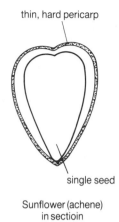

thin, hard pericarp

single seed

Sunflower (achene)
in sectioin

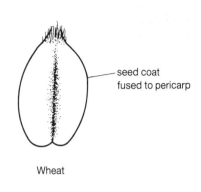

seed coat
fused to pericarp

Wheat

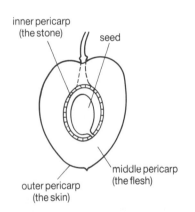

inner pericarp
(the stone)

seed

outer pericarp
(the skin)

middle pericarp
(the flesh)

Plum in section

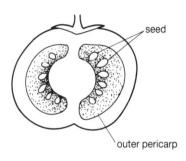

seed

outer pericarp

Tomato in section

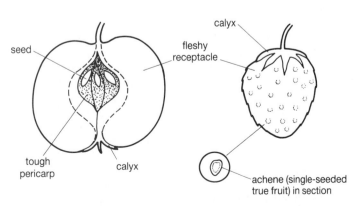

seed

tough
pericarp

calyx

Apple in section

calyx

fleshy
receptacle

achene (single-seeded
true fruit) in section

Strawberry

181

7.5.3 Ripening of seeds and fruits

Once the seed has acquired its full complement of carbohydrates, lipids and proteins for storage in the endosperm or cotyledons and the embryo is well formed, it begins to lose water and dry out. This is the process of *ripening* of the seed. As its moisture content falls, all growth within the seed stops and the embryo becomes quiescent or dormant. The rate at which seeds lose moisture depends on the relative humidity of the atmosphere; in damp or wet weather the process is much slower than in dry or windy weather.

The loss of water by seeds during ripening is not, however, a simple case of evaporation. We saw in Section 5.1.3 that dry seeds imbibe water powerfully because of the strong forces tending to bind water on to protoplasmic colloids; similarly, it must take considerable forces to remove water from seeds. The ripening of seeds is to a great extent an active process, under the plant's control. Seeds of many plants will ultimately ripen even in continuously humid conditions.

In many species drying out of the seed is accompanied by a similar drying of the pericarp of the fruit. The pods of legumes such as peas and beans, for example, become dry and brittle when the seeds are fully ripe. During ripening of cereal grains, which are single-seeded fruits, the seed coat and pericarp both lose water, become lignified and bond tightly together to form a single structure.

The ripening of fleshy fruits such as apples, strawberries and bananas involves rather different processes, some of which are akin to senescence in leaves. The colour of the skin often changes as chlorophyll breaks down. In fruits like bananas which change from green to yellow no new pigments are formed, but in many fruits the green chlorophyll is replaced by orange or red carotene pigments as in tomatoes, purple anthocyanins as in plums, or other types of pigment. At the same time flavour and aroma become more attractive as starch is converted to sugars, sour-tasting substances such as malic and citric acids decline and a wide range of phenolic compounds and volatile esters are synthesized. There is also a marked softening of the fleshy tissues resulting from the enzymic breakdown of the pectin, the material that cements cell walls together. Sometimes the skin develops a waxy surface. The ripening of fruit, like the senescence of leaves, is accompanied by the development of an abscission layer in the stalk, and the fruit is shed if it is not picked first.

In the later stages of ripening, many fruits exhibit a surge in the rate of respiration known as the *climacteric*, followed by a gradual slowing of respiration once again. In some fruits, such as banana and avocado, the climacteric is more pronounced than in others, such as apple and pear. Yet others, including fig, grape and lemon, show no climacteric at all.

Picking fruit hastens the climacteric and ripening in general; in a few fruits such as avocado ripening does not begin until the fruit is picked. Fruit is often picked before the climacteric and is then stored in refrigerated conditions to slow down ripening and prolong storage life. The susceptibility of tropical and subtropical fruits such as banana to chilling injury (Section 4.9.2) restricts the scope for this technique. The climacteric can, however, be delayed by storage in an atmosphere low in oxygen but rich in carbon dioxide, and controlled atmosphere storage systems relying on this have been developed (see Section 11.8.4).

7.5.4 Seed dispersal

The biological functions of the fruit are to protect the developing seeds and to aid in their dispersal. In many species the whole fruit rather than the seed is the dispersal unit and as such may, without serious error, be referred to as the 'seed'. Thus the agricultural 'seeds' of cereals, grasses, sunflowers and lettuces are, strictly, one-seeded fruits, while the 'seed' of sugar beet is a multiple fruit containing one to three true seeds.

Crop seeds must, of course, be harvested before the plant sheds them. During the domestication of crops over the last 10 000 years they have evolved through intentional or unintentional selection by growers to become much less prone to premature shedding. This has come about largely through the natural abscission process being delayed or eliminated. An extreme example is maize, which has become totally dependent on man to break open the ripe cob and separate and sow the seeds. The maize plant could barely survive at all in the wild. Other cereals would have similar though less extreme difficulties. One of the most important differences between cultivated oat and wild oat, that makes the first a very useful crop and the second one of the world's most troublesome weeds, is that cultivated oat retains its grains until they are fully ripe and can be harvested, whereas wild oat forms a very pronounced abscission zone just beneath each grain and sheds its seeds on to the soil before the farmer can harvest them with his crop.

In many wild species the seeds, or fruits acting as seeds, have structural adaptations to facilitate dispersal by natural agencies such as wind, running water and animals (Fig. 7.27). Wind dispersal requires a large surface area relative to the seed's weight. This is provided by long hairs as in rosebay willowherb and spear thistle, or by membranous wings as in birch and sycamore. Seeds of some species float readily and these species, such as giant hogweed, are commonly to be found on the banks of rivers where the seeds have been deposited in times of flood. Other species such as cleavers have seeds with barbed or hooked bristles which adhere to the coats of mammals and the feathers of birds. Many small weed seeds, such as those of chickweed or corn spurrey, are carried in mud on animal and human feet, and on the wheels of farm and other machinery. Some species, including broom and hairy bitter cress, have fruits which split open explosively to propel their seeds a short distance away from the parent plant. Other seeds can survive passage through the digestive tract of animals, and are thus spread in farmyard manure, slurry and human faeces and sewage.

Fleshy fruits evolved as a means of tempting animals to do the work of seed dispersal. Birds and mammals often carry fruits for considerable distances, or they may consume the entire fruit where it grows or falls. In either case a large proportion of the seeds, with their hard, lignified seed coats, are often deposited, undigested, in the animal's droppings. Since soft fruits are designed to attract animals, it is not surprising that they are among the most attractive of plant products to the human palate.

Human activities provide important routes of seed dispersal, especially for weed seeds. Trade is the most significant example of such activities. Seeds may be carried quite accidentally along with a non-agricultural product, or in a plant material associated with it, as in the straw used to pack crockery. More often they are carried in or on an a agricultural product, for example within bulks of grain or in the soil on imported vegetables, the waste from which may be dumped where the weed seeds on it can germinate and grow. The trade in agricultural crop seed such as cereals or grass and pulses, inevitably spreads weed seeds effectively, despite efforts to remove weed seeds and other detritus from the seed before it is sold. Wild oats are an example of an important weed of cereals spread in this manner. An important feature of

Fig. 7.27 Structures on seeds or fruits to aid their dispersal: creeping thistle (a) and sycamore (b) with a parachute of hairs and wings respectively, to aid wind dispersal; cleavers (c) with hooks that attach to animals' fur; giant hogweed (d) with flat wings for flotation to facilitate dispersal by running water; and broom (e) with pods that split open explosively and project the seeds through the air.

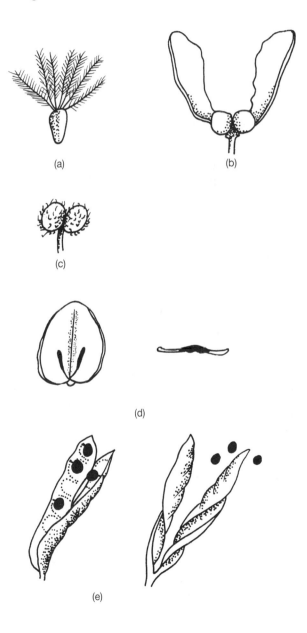

spread through trade is that if often takes place over large distances, and indeed is often intercontinental. In this manner, weeds have been introduced to many new areas of the world, such as North America and Australia where the ecological factors that may have limited their spread in their former habitats are lacking, and they have spread to become serious weed problems. The introduction of ragwort from Great Britain and its rapid spread in Canada, Australia and New Zealand seems to have been an instance of this.

Sometimes, seed of a weed species has been introduced to an area deliberately for use as an ornamental or for other purposes, and the species has subsequently spread and become a weed. Examples include giant hogweed introduced to Scotland as an ornamental, and prickly pear introduced to Australia as a hedging plant.

Summary

1. Flowering, fruiting and seeding are the means by which angiosperm plants carry out sexual reproduction. The flower is a device to ensure adequate genetic recombination, which is essential for the maintenance of variability, and hence adaptability, in a plant population.

Sexual reproduction in the angiosperm is a sequence of events, beginning with the initiation of flowering, continuing through formation of floral primordia at the shoot apex, growth and differentiation of the flowers and inflorescence, flower opening, pollination, fertilization of ovules and development of seeds and fruits, and ending with ripening and seed dispersal. At the same time vegetative organs, particularly leaves, go through a programme of senescence as the nutrients they contain are remobilized to nourish the fruits and seeds.

2. The initiation of flowering involves the activation of previously quiescent cells in the apical meristem and is normally irreversible. Only when a shoot has reached a certain minimum size is its apex competent to initiate flowering. It may then undergo the transition from vegetative to flowering development automatically, or the transition may be dependent on an environmental stimulus.

Many temperate species require a low temperature stimulus, or vernalization, before they can initiate flowering. This is true of most winter cereals and all biennial crops. The effect of vernalization is usually to synchronize flowering in spring.

The initiation of flowering in most plants, including many which require vernalization, is triggered by a daylength stimulus. The majority of cool, temperate species are long-day plants, requiring a certain minimum daylength before flowering can be initiated. Most warm temperate and subtropical species are short-day plants, in which flowering is initiated only if daylength is shorter than some critical value. In fact it is the length of the night, not that of the day that is measured and responded to by plants.

3. The flower first evolved as a device for insect pollination. Insect-pollinated flowers attract insects with their showy and often scented petals and reward them with surplus pollen or with nectar. In visiting successive flowers, insects inevitably deposit pollen collected from previous flowers. Many species have flowers constructed in such a way that self-pollination is unlikely or impossible. In addition, many plants show self-incompatibility, whereby fertilization cannot occur between gametes originating in the same plant.

The flowers of grasses and many other species have in the course of evolution reverted to wind pollination. These do not have showy petals, but produce large quantities of pollen to ensure that some lands on receptive stigmas, which are hairy or feathery and intercept airborne pollen efficiently. Some species, including many agricultural crops, are predominantly self-pollinating.

4. After pollination the two male gametes carried in each pollen grain migrate along the rapidly growing pollen tube to reach an ovule, where one fertilizes the female gamete or egg cell and the other fuses with two other cells to form a single cell from which the endosperm develops. The embryo grows from the fertilized egg cell, nourished by carbohydrate and other food reserves laid down in the endosperm. In many seeds the endosperm is completely consumed and an energy store for germination accumulates in the cotyledons of the embryo itself, but in others much of the endosperm remains to nourish the seedling during germination.

5. As the ovule develops into the seed, which in apomictic plants happens without fertilization, the ovary containing the ovules develops in various ways to become the fruit. The fruit protects the developing seeds and may eventually aid their dispersal.

The final stage of seed and fruit development is ripening. In seeds the main process is loss of water and the onset of quiescence or dormancy. Ripening of fleshy fruits involves changes in colour, flavour and texture and in the later stages is often characterized by a surge in respiration known as the climacteric. Fruits should be picked and stored before this stage is reached.

Seeds and fruits may have various structural adaptations to facilitate dispersal by wind, water and animals. During domestication many crops have lost the ability to shed their seeds, and they can thus be left until fully ripe before harvesting. Human activities, especially trade and the deliberate introduction of plant species to new areas of the world, have been important factors in the spread of weed seeds.

8

Plant Growth and Development:
Environmental Effects and Internal Control

8.1 Rate of growth

Growth is a feature of plants which is so familiar to us that we seldom stop to consider exactly what it means. It has so many different aspects that it is impossible to define precisely. If, however, we wish to say for example that this plant grows faster than that one, or that a plant grows better in one set of conditions than in another, or more critically if we wish to have some measure of *growth rate*, we must clarify which aspects of growth we are talking about.

8.1.1 Growth rate and net assimilation rate

Growth essentially involves the production of new protoplasm and of new cell walls to contain it. An increase in bulk due entirely to the absorption of water, as in seed imbibition, is not true growth, whereas an increase in the amount of *dry matter* (that is, everything other than water) unquestionably is. Indeed one of the commonest meanings of the word growth, at least in an agricultural context, is an increase in *dry weight* of a plant over a period of time. In this sense it is closely related to yield (see Section 11.1.4).

The amount of dry matter present at any time – the plant's dry weight – is a result of the balance between what has been produced by photosynthesis and what has been lost by respiration and other downgrade metabolic activities, consumption by parasites and predators and the death of plant parts. Thus growth rate as measured by increase in dry matter is closely related to *net assimilation rate* which, as we saw in Section 1.9, is the balance between the rates of photosynthesis and of respiration.

But growth may occur without any increase in the amount of dry matter. Since the activities of growth – the laying down of new cell walls, the manufacture of proteins, lipids, nucleic acids and other substances, and the multiplication of cell organelles such as nuclei, chloroplasts and mitochondria – are highly energy-demanding, they must be accompanied by the breakdown of carbohydrate in respiration. However, the amount of new dry matter that can be manufactured in growth is much less than the amount destroyed in respiration because most of the energy released in respiration is lost as heat. Only if photosynthesis is going on faster than respiration does growth appear to involve an increase rather than a decrease in dry weight. If the dry weight of a sugar cane plant could be measured in the evening and again the following morning it would show a fall because of losses in respiration, yet its height would show an increase, indicating that growth had taken place. Germination of seeds similarly involves a loss of dry weight, as stored carbohydrates are broken down in respiration, but the extension of radicles and hypocotyls is unquestionably growth.

In the long term, however, plant growth is utterly dependent on photosynthesis, which makes the carbohydrates to fuel respiration and form carbon skeletons from which other molecules can be built. For this reason a plant which has shown greater net assimilation over its life is a bigger plant, and over a period of days or weeks a plant which averages a greater net assimilation rate is generally a faster growing plant.

The simplest and in many ways the best measure of the growth of a plant is its size. For example, we may measure the length of a root or stem or individual internode, the area of a leaf, or the volume of a storage organ. This method of measuring growth is non-destructive, which means that it can be repeated many times on the same plant or organ over a period.

Growth is often measured by weighing, although, as we have seen, weight is not necessarily correlated with growth, at least in the short term. This is particularly true of fresh weight, since short-term fluctuations in moisture content with weather or soil moisture status can cause substantial changes in fresh weight which have nothing to do with growth. Measurement of dry weight is destructive – it involves killing the plant or organ by drying in an oven – and for repeated measurements it is therefore necessary to sample from a large number of plants. Because of the inevitable variation between plants this introduces a further source of error.

8.1.2   The growth curve

If we measure some unit of plant structure such as the area of a single leaf or the length of a single internode over the whole period of its growth and plot the measurements against time on a graph, we get a curve like that shown in Fig. 8.1. This characteristic S-shaped curve is known as the *growth curve*. The first part of the curve, where the line is getting ever steeper, represents a period of accelerating growth rate. The bigger the organ gets, the faster it grows. This period of accelerating growth is known as the *logarithmic phase* because the logarithm of the size of the organ is increasing at a constant rate.

In the middle of the growth curve there may or may not be a short *linear phase* when growth rate is no longer accelerating but remains constant. This is followed by the *maturing phase* when growth rate slows down and eventually stops. The organ may remain at its maximum size for quite a long time thereafter, indeed for the rest of its life.

Various processes are going on during these phases of growth. In the very early part of the logarithmic phase cell division is the main process, but for most of the logarithmic phase the organ is growing primarily by cell enlargement. During the linear phase cell enlargement becomes progressively confined to a smaller proportion of the bulk of the organ, as growth ceases in the earlier formed parts, and this trend becomes more marked in the maturing phase.

In the growth curve for a whole plant the linear phase is usually longer than in the curve for a single organ. We can readily see why this should be. During the linear phase the plant is adding internodes or leaves one after another at a fairly steady rate. As young organs are growing logarithmically, older ones have reached the maturing phase or have stopped growing. Because of the indeterminate habit of growth this can go on for quite a long time before the plant reaches its maximum size or begins to flower.

How can we compare the growth rates of two plants, when growth rate varies so greatly with the age of the plant? One way might be to compare growth rates during the linear phase, when these are more or less constant, but it is often difficult to determine when a plant is in the linear phase. Plant breeders who wish to compare growth rates in large numbers of plants, in order to select fast-growing individuals, prefer to make their measurements as early as possible because small plants require less space and because the breeders do not have to wait so long to get the information they need. They

Fig. 8.1 The characteristic S-shaped growth curve of plants and plant parts.

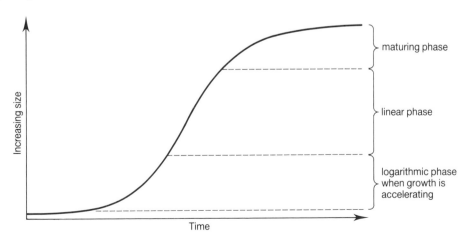

therefore measure the plants during the logarithmic phase of growth.

In the logarithmic phase, as we have seen, the bigger a plant or part of a plant is, the faster it grows. Indeed the rate of growth at any one time is proportional to the size of the plant at that time. In other words, the ratio of growth rate to plant size, whether this be measured as length, area, volume or dry weight, is constant throughout the logarithmic phase. This ratio is known as *relative growth rate* and is a good measure of the vigour of a young plant.

8.1.3 Environmental factors affecting growth rate

Clearly over an extended period of time, any factor influencing net assimilation rate will influence the rate of growth. We saw in Section 1.9 that the major environmental factors affecting net assimilation rate are light intensity, temperature and water availability, which affects carbon dioxide supply to the leaf by determining whether the stomata are open or closed. Any one of these factors may be a limiting factor to growth as well as to net assimilation.

Growth is, however, an even more complex process than photosynthesis or respiration and it is therefore influenced by the environment in more ways than these. Very commonly, for example, the main limiting factor to the growth of agricultural crops is the soil nitrogen supply. Evidence for this is seen in the response to nitrogen fertilizer in terms of increased growth rate. Once the nitrogen supply is rectified some other soil nutrient – most likely phosphorus or potassium – may become the limiting factor.

Other soil conditions may limit growth through adverse effects on water or mineral uptake or by physical impedance to root penetration. These include excessively low or high pH, salinity, waterlogging and compaction. Alternatively, growth may be restricted by the diversion of assimilated carbohydrate and other nutrients to parasitic microbes or animal pests. Weeds reduce crop growth rate by shading or by using soil water or minerals which would otherwise be available to the crop. This is an aspect of *competition* between plants which we shall deal with in Section 11.4. Many other external influences on crop growth could be mentioned, ranging from atmospheric pollution and the side-effects of applied herbicides

to damage by animals' feet, tractor wheels, wind or frost.

Temperature, light and water availability have direct effects on the rate of growth in addition to the indirect effects that they exert through net assimilation rate. These direct effects, though they are worth mentioning here, are not nearly as significant for the growth and yield of crops as the influences of the same factors on net assimilation rate.

Extension growth by cell elongation has been found to be most rapid at temperatures around 20–25°C. The ultimate length attained by plant organs, however, is often greatest at lower temperatures, around 10°C.

In very low light intensity or in complete darkness the extension growth of shoots is extremely rapid. This produces long, spindly stems, hypocotyls or leaf stalks which are yellow, because without light chlorophyll cannot be synthesized. It is an effect known as *etiolation* and is a means whereby a buried plant, or one densely shaded by other plants, can rapidly reach sunlight. Potato tubers sprouted in the dark produce weak, etiolated shoots which are easily knocked off during planting. It is better to sprout potatoes in the light when shorter, more sturdy shoots are produced.

Finally, water stress has a direct effect on growth rate through the development of large negative water potentials in growing cells. These water potentials seem to restrict the enlargement of cells.

8.2 Direction of nutrients in growth and development

Clearly growth and development in the whole plant is in some way programmed. Early in the life of the plant leaves are formed, and only once sufficient photosynthetic capacity has been established does the plant start flowering or filling storage organs such as tubers or bulbs. A balance of growth must be maintained between the various parts and events must happen in the right order. All this requires that the water and nutrients needed for growth and development must be directed from where they are absorbed, manufactured or stored to where they are needed at any one time.

The site from which a mineral or organic nutrient moves in the plant is a *source* whether it has been absorbed, manufactured or stored there. Young roots are sources of water and minerals, actively photosynthesizing leaves are sources of carbohydrates and organic nitrogen compounds such as amino acids, and senescent organs and storage sites within the plant are sources of many types of nutrient. These nutrients are directed along the xylem and phloem translocation systems to be utilized or stored at their destinations, which are known as *sinks*.

The main sinks are developing structures such as young leaves and roots, flowers, fruits and seeds, and storage organs. A single organ may be a source of one nutrient and at the same time a sink for others. Transpiring, photosynthesizing leaves, for example, are a sink for water and a source of carbohydrate. Some important patterns of nutrient movement between sources and sinks are shown in Fig. 8.2. Clearly, the control of movement of nutrients within plants is an important feature of the control of plant development.

8.2.1   Relationships between sources and sinks

The rate of growth of a sink such as a fruit or storage organ and the final size it can attain are determined, to a considerable extent, by the rate at which nutrients are supplied to it. This supply must

Fig. 8.2 Some patterns of flow of water and nutrients between sources and sinks in a schematic plant.

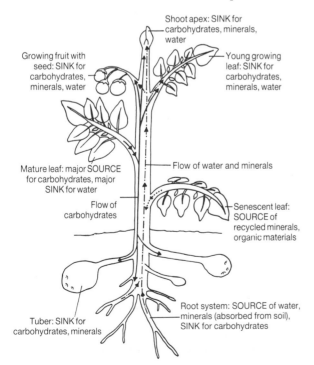

Shoot apex: SINK for carbohydrates, minerals, water

Growing fruit with seed: SINK for carbohydrates, minerals, water

Young growing leaf: SINK for carbohydrates, minerals, water

Mature leaf: major SOURCE for carbohydrates, major SINK for water

Flow of water and minerals

Flow of carbohydrates

Senescent leaf: SOURCE of recycled minerals, organic materials

Tuber: SINK for carbohydrates, minerals

Root system: SOURCE of water, minerals (absorbed from soil), SINK for carbohydrates

be controlled in some way if a balance of parts is to be maintained in the plant. But what controls the rate of supply? Is it the rate at which the sources can produce the nutrients? Is it the rate at which they can be translocated? Or is it the rate at which the sink can utilize them? The answer could be very important in deciding what limits crop yield (because harvestable organs are, by their nature, sinks). This in turn would give us an indication as to how we might go about improving yield.

Such evidence as we have suggests that the limitation is unlikely to lie in the rate of translocation. Traffic jams are apparently not a major problem in nutrient supply within the plant. A weakness in translocation, however, is evident in the relative inability of plants to transfer materials laterally from one bundle of phloem tissue to another because of the restricted number of vascular connections across the stem or root. In sugar beet each leaf appears to supply a particular part of the root system and similarly in sunflower each leaf supplies a particular sector of the seed-head. Monocotyledons, including grasses and cereals, with the base of each leaf completely sheathing the stem have to some extent overcome this problem as each leaf connects with the complete ring of phloem in the stem at the node.

Equally, the evidence we have suggests that sources generally do not limit nutrient supply to sinks, except in the case of nitrogen, phosphorus and potassium which are often below optimum levels for plant growth in the soil. Carbohydrate sources, mainly leaves, are seldom limiting. Tobacco, for example, has been shown to have more photosynthetic capacity than it needs for most of its life. Maize often contains large amounts of unutilized mobilizable carbohydrate in its stems at harvest. It is probably significant that in many crops the rate at which seeds are filled is not markedly influenced by fluctuations in the rate of photosynthesis with changing weather conditions.

It would appear, therefore, that it is the capacity of the sink itself that usually limits the rate at which it can receive nutrients. This is difficult to prove, because changes in the sink can have a pronounced feedback effect on the sources. An increase in sink capacity often induces an increase in the strength of the source. In apple trees, for example, the rate of photosynthesis in leaves near developing fruits may be 50% greater than in more distant leaves.

The direction of nutrients, then, is controlled largely by sink demand, rather than by the output of sources or the ability of the phloem to translocate. What happens when several sinks are competing and the overall demand for nutrients exceeds supply? In clusters of fruit as in strawberry or tomato, for example, what controls the sink strength of each fruit and decides which will grow?

It is certainly not the absolute size of the sink, for as one fruit reaches maximum size its sink strength diminishes and another may start to grow. Relative growth rate is, however, important; the faster a sink is growing, the greater is its drawing power. But it is unclear whether it grows faster because it is receiving more nutrients, or it attracts more nutrients because it is growing faster.

The nearer a source is to a strong sink, the greater is the tendency of the sink to dominate the nutrient supply from that source. Thus, in wheat, barley and oats the products of the uppermost leaf – the one just below the ear – go to swell the grain while the others support the roots, and in crops like pea and soya bean that have axillary inflorescences each inflorescence is nourished by the leaf in whose axil it is borne. Such nutrient flow patterns are not rigid but adapt to circumstances. If, for example, the uppermost leaf of a cereal plant is removed, the next leaf supports the ear. Nonetheless, damage to the uppermost leaves of cereals usually results in poor grain filling, giving considerable yield loss.

In root crops such as sugar beet the nutrient flow pattern is dominated by a single, very strong sink. In these crops all the leaves, even the uppermost, supply the underground storage organ.

8.2.2 Recycling of nutrients

Closely linked to the direction of growth and development through source–sink relationships is the process of nutrient recycling. Of the nutrients flowing into a growing organ such as a tuber or fruit only a portion derives directly from current photosynthesis or mineral uptake by roots, though all must at one time have done so. The rest come from one of two other types of source. First, they may come from previously manufactured or absorbed nutrient reserves held either in long-term storage, as in a tuber or bulb, or in short-term storage in tissues of leaves, stems and roots not specifically modified as storage organs. Second, they may be released from plant structures through the processes of senescence.

It is important to realize that plants recycle and reuse in new organs many materials that were previously part of the structure of earlier formed organs. Without the ability to move these materials

into new organs such as developing seeds, many crop plants could not concentrate sufficient nutrients, particularly minerals, into the harvested part and yield would be lower.

Plants' ability to recycle materials results from the ways in which their structures are maintained. Some structures, especially the cellulose cell walls, remain fixed throughout the life of a plant and the molecules embedded in them cannot be moved. On the other hand molecules, membranes and organelles in the cytoplasm of cells are constantly being broken down and their apparent persistence arises from their constant replacement. Thus the contents of cells are in a continuous state of turnover. By adjusting the balance between the rate of breakdown and the rate of replacement the plant permits these materials to be built up, maintained, or broken down and their constituents mobilized for transport and reuse elsewhere. This gives great flexibility to the development of the plant. Mobilizable materials in a shaded-out leaf or in redundant roots or stems are not lost or inaccessible but can be dismantled for transport to organs where they are needed. As we saw in Section 6.3 this remobilization of nutrients is particularly marked in senescent organs.

The ability of plants to redistribute nutrients within themselves should not be underestimated and has important practical implications. For example, when the grains of wheat and oats have reached only 25% of their final dry weight, uptake of nitrogen and phosphorus from the soil is thought to be almost complete and the remaining 75% of grain filling is nourished by recycled nutrients. Not surprisingly, therefore, wheat leaves approaching senescence have been found to lose 85% of all their nitrogen and 90% of their phosphorus. Similarly, in apple trees it has been shown that 52% of the nitrogen in the foliage is exported before leaf fall.

As was seen in Section 3.4.3, some mineral nutrients are more mobile in the plant than others and are therefore more readily recycled. Phosphorus, for example, is highly mobile, sulphur intermediate, and calcium highly immobile. Developing fruits and seeds are almost entirely dependent for their calcium supply on that ascending in the transpiration stream in the xylem, as little or none can be carried from senescent leaves in the phloem. For this reason calcium is more likely than other minerals to be deficient in organs that develop late in the season when root growth has declined or ceased. Not surprisingly, then, there are a number of disorders of fruit such as bitter pit of apples and blossom-end rot of tomatoes which are caused directly or indirectly by calcium deficiency.

8.3 Internal programming of development

8.3.1 Activation of the genetic blueprint

A key feature of growth and development in plants is specialization of form and function. We see this at the cell level, as cells differentiate to become, say, photosynthetic parenchyma, sieve tubes or xylem vessels, and at the organ level, as the apical meristem of the shoot, for example, gives rise to a stem and to leaves, and later perhaps to flowers. It might be thought that one way in which the plant could achieve this would be by having cells carrying different genetic blueprints. One blueprint might produce a sieve tube, another a xylem vessel; superimposed on this would be genetic information dictating whether the cell was to be part of a leaf or a stem.

For many reasons, this explanation will not do. For example, stems initiate adventitious roots, and certain differentiated cells sometimes lose their specialized function and give rise by renewed division to different types of cell. It also cannot explain how a single meristem at one stage may be generating leaves and at a later stage flowers, or how another meristem may, depending on environmental conditions, give rise to an ordinary aerial shoot or to a horizontally creeping rhizome.

From our knowledge of the way in which cells and their nuclei divide (see Section 9.2.3) we know that all cells in a plant carry exactly the same genetic blueprint. A convincing demonstration of this is in the artificial culture of single cells. Any cell of the secondary phloem of a carrot root, for example, can in culture give rise eventually to a complete flowering plant. Thus although it has a specialized function in the intact plant each cell retains the whole genetic blueprint and has the potential to make an entire plant. This property of cells we call *totipotency*.

How, then, can cells of identical genetic constitution develop in such widely different ways? The answer is that a large proportion of the genetic information in any cell, probably around 80% at any one time, is *repressed* (i.e. it is not activated). Every cell in a tomato plant, for example, carries a gene (the unit of genetic information) for red pigment, but this gene is activated only in the fruit, and only when that fruit is ripening. Its expression is thus controlled in both space and time.

The growth and development of the plant is directed by the successive unmasking of different parts of the genetic blueprint at appropriate times so that fruiting follows flowering, which follows vegetative growth, which follows germination and so on. Usually entire sections of the blueprint which control major pathways of development such as the initiation of flowering or synchronous leaf senescence are unmasked at once.

The selective repression of genetic information thus depends on the position of a cell in the plant, that is, where it is in relation to other cells, and on the stage of development reached by the plant. Let us look a little closer at the importance of position.

8.3.2 The importance of position

A cell must be able to 'recognize' where it is in the plant so that appropriate parts of the blueprint can be activated, but just how it does this is not understood. It is a very intricate business, as is evident, for example, in the regular distribution of vascular bundles in a stem or the differential growth rates of different parts of a leaf primordium producing a final leaf shape which is so characteristic of a plant species that it can be used in identification.

Cells 'recognize' not only where they are in relation to other cells but how they are orientated with respect to the apex and base of the organ of which they are part, that is, which 'way round' they are. This recognition of one end from another is called *polarity*. It is an ability that is acquired very early in the life of the plant, and once it is established it persists and acts quite independently of gravity. Simple experiments with willow have demonstrated this point. If a portion of willow stem is taken and suspended upright in a humid atmosphere that inhibits desiccation and encourages growth, roots grow at the basal end, and shoots at the apex. If, however, similar shoots are suspended upside down, then roots still grow at the basal end, which is now at the top, and shoots at the apical end, which is now at the bottom (Fig. 8.3).

8.3.3 Selection of a development programme

Another feature of plant development should be taken into account at this point. If single cells are artificially cultured, as mentioned above, they exhibit totipotency and can ultimately give rise to a whole plant. In nature the fertilized egg cell is

Fig. 8.3 Demonstration of polarity in a section of willow stem. Sections are suspended by threads in moist air. One stem section is suspended the right way up (a), and one upside down (b). Even in (b) the shoots develop at the apical end and the roots at the basal end.

(b)

(a)

similarly totipotent. But if a whole leaf or flower primordium is put in a culture medium, it carries on the development programme that it has already begun and produces only a leaf or a flower, not a whole plant.

It seems that once a pathway of specialized development has been embarked upon, it cannot easily be reversed. Once an organ has been initiated by an embryo or a meristem, the subsequent stages of development follow inexorably. This successive commitment in a growing plant of different parts to specific pathways of development is known as *canalization*. It is especially marked in organs of

determinate growth such as leaves and flowers but is less marked in organs of indeterminate growth like stems and roots which retain the capacity to initiate new organs.

A number of basic questions relating to the control of growth and development can thus be asked. How does a cell 'recognize' its position in a meristem or in a plant so that it can follow the appropriate developmental pathway? What imposes polarity on cells, organs and whole plants? How is a particular development programme selected? Why is it that, once the programme has been selected, it can seldom be reversed?

We cannot yet answer any of these questions with confidence. We do know, however, that certain chemical substances produced in the plant act as messengers, influencing the way cells, organs or whole plants develop. These messengers are the plant *hormones*. At very low concentrations in plant tissues they affect growth and development, and generally have their influence in a different part of the plant from where they are manufactured. They must therefore be translocated from their site of synthesis to their site of action.

Some of these hormones have been isolated. We know their chemical structure and many of their properties. We can even make synthetic substances of a similar chemical nature, which act as hormones, often more potent in their effects than natural hormones. These are known as *growth regulators* and are finding many uses in agriculture and horticulture. But we are still very vague about how hormones and growth regulators work. We can list the main groups of naturally occurring hormones and their observed effects but we lack a cohesive overall theory to explain all these effects.

8.4 Natural plant hormones

We recognize five categories of natural plant hormones but others, as yet unidentified, may exist.

8.4.1 Auxins

It has been known for a very long time that when plants receive light from only one direction they tend to grow towards the light (*phototropism*). The coleoptiles of canary-grass seedlings were used by Darwin in 1880 to investigate this effect and coleoptiles have figured prominently ever since in studies of the control of growth and development. Darwin showed that coleoptiles curved towards a

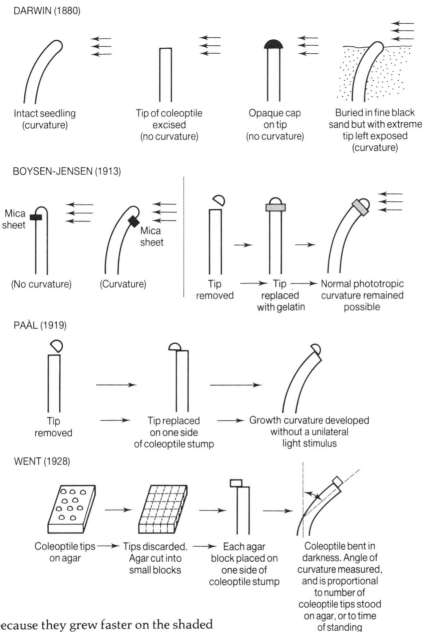

DARWIN (1880)

Intact seedling
(curvature)

Tip of coleoptile
excised
(no curvature)

Opaque cap
on tip
(no curvature)

Buried in fine black
sand but with extreme
tip left exposed
(curvature)

BOYSEN-JENSEN (1913)

Mica
sheet

(No curvature)

Mica
sheet

(Curvature)

Tip
removed

Tip
replaced
with gelatin

Normal phototropic
curvature remained
possible

PAÀL (1919)

Tip
removed

Tip replaced
on one side
of coleoptile stump

Growth curvature developed
without a unilateral
light stimulus

WENT (1928)

Coleoptile tips
on agar

Tips discarded.
Agar cut into
small blocks

Each agar
block placed on
one side of
coleoptile stump

Coleoptile bent in
darkness. Angle of
curvature measured,
and is proportional
to number of
coleoptile tips stood
on agar, or to time
of standing

Fig. 8.4 Some important steps in the discovery of auxins by various researchers. All the experiments were done using grass seedling coleoptiles. Triple arrows indicate the direction of light incident on the coleoptiles. (From T A Hill (1980) *Endogenous Plant Growth Substances*, 2nd ed. Studies in Biology no. 40. Edward Arnold, London.)

light source because they grew faster on the shaded side than on the illuminated side. Extension growth occurs not at the tip but a short distance behind the tip, yet Darwin found that if the tip were cut off or completely shaded with an opaque cap the coleoptile did not curve towards the light, though it continued to grow (Fig. 8.4). Apparently the light stimulus is sensed in the coleoptile tip but the effect is on growth lower down.

A series of experiments by workers in the early part of the twentieth century revealed even more interesting results. If the tip is cut off and then replaced, even with a small block of gelatin separating it from the rest of the coleoptile, the normal curvature towards the light still occurs. Furthermore, if the tip is removed and replaced on one side of the coleoptile stump, the coleoptile curves away from that side, without illumination. The obvious conclusion was that something, almost certainly a chemical, must be diffusing down the coleoptile from the tip and having an influence on growth further down.

That a chemical was indeed responsible was proved by Went in a famous series of experiments in which excised coleoptile tips were placed on blocks of agar jelly and the blocks then placed on one side of the coleoptile stumps in total darkness. The coleoptiles curved exactly as if the tips had been used instead of the agar blocks. The chemical had diffused from the tips into the agar jelly. It was called *auxin*, from a Greek word meaning to grow.

The auxin in Went's experiments was later found to be a substance called indolylacetic acid, or *IAA*, and we now know that this is just one of a number of naturally occurring hormones with similar properties, all of which are classified as auxins.

One of the most important properties of auxins is that they influence the rate of cell division and cell enlargement. At very low concentrations they stimulate, but at slightly higher concentrations they inhibit these processes. To complicate matters further, cells of roots are much more sensitive to auxins than are those of shoots. Roots show their maximum stimulatory response to IAA at the incredibly low concentration of 0.02 parts per billion (around 10^{-10}M), whilst shoots are not affected at this level. At a concentration one million times greater, around 20 parts per million (10^{-4}M), stems show maximum stimulation, but cell division and enlargement in roots is strongly inhibited. Buds are intermediate between roots and stems in sensitivity to auxins (Fig. 8.5).

Another key property of auxins is that they direct phloem translocation. An organ that has a high auxin content becomes a sink for carbohydrates and other nutrients. To what extent this inward migration of nutrients is responsible for the stimulation of shoot growth by auxins, and to what extent the stimulation is a direct influence of the auxins, is not known. The effect is, however, seen very clearly during seed and fruit development. Once an ovule has been fertilized, it very quickly begins to synthesize auxin. Some of this auxin moves into the ovary wall, so that both the ovule

Fig. 8.5 The effects of application of increasing concentrations of IAA on the growth of different plant organs.

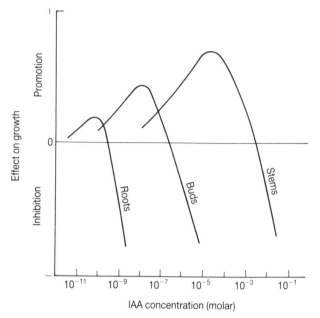

and ovary wall become nutrient sinks, rapidly growing into seed and fruit respectively. If the ovule is not fertilized the fruit normally does not develop, except in the case of seedless fruits like banana which produce auxin without being fertilized.

In the vegetative plant, auxins are produced in the shoot apex and the young leaves around it, and it is to this region that much phloem translocation is directed. A remarkable property of auxins is that their movement in the plant is highly unidirectional. In shoots they move only from the tip towards the base and in roots only towards the root tip. This has nothing to do with gravity, for if the plant is turned upside down the auxin continues to move in the same way. It is not known whether this is a contributory cause of polarity in plants or merely an effect of it.

As auxins are produced at the shoot tip they are continuously broken down in all parts of the plant. This creates a gradient of auxin concentration, high at the shoot tip and vanishingly low at the root tip. Auxin transport takes place from cell to cell in most living tissues, not only in the phloem.

8.4.2   Gibberellins

Another group of natural plant hormones, discovered after the auxins, are the *gibberellins*, the best known of which is *gibberellic acid*. These seem to have little influence on cell division but can powerfully stimulate cell elongation, especially in stem internodes. They are also involved in the breaking of dormancy of buds and seeds.

Gibberellins appear to be synthesized in young leaves and to some extent in roots, and they are rapidly distributed in both phloem and xylem to all parts of the plant. Their transport is not unidirectional like that of auxins.

8.4.3   Cytokinins

A third group of hormones, the *cytokinins*, includes a wide range of chemically related materials, for example *zeatin* and *IPA* (isopentenyladenine). They affect division rather than elongation of cells, and inhibit senescence. Like auxins, they help to create sinks for the direction of phloem translocation, but in this case primarily in the root system.

Cytokinins are produced mainly in the root tip and are carried to the shoot in the xylem. There tends thus to be a gradient of cytokinin concentration opposite to the auxin gradient. Transport of cytokinins is not, however, markedly unidirectional.

8.4.4   Ethylene

The gas *ethylene* (C_2H_4) has many effects on plant growth and development, often mimicking the effects of auxins. It has been suggested, indeed, that auxins may exert their influence by generating ethylene in plant tissues, for one of the observed effects of auxins is to stimulate ethylene production.

Ethylene has, however, effects which are unique and have no connection with auxin action. In particular it stimulates abscission of flowers, fruits and leaves and in some species it helps to overcome seed dormancy. Ethylene is produced by fleshy fruits, where its effects include the stimulation of the various processes associated with ripening. As the fruit ripens, it produces more and more ethylene, tending to accelerate the process. This is an example of *autocatalysis*, the stimulation of a process by a substance produced in the course of that selfsame process. The effect must be guarded against in the bulk storage of fruit. A single ripe apple in a bulk store can set off the ripening of all the unripe apples, unless fresh air is constantly blown through the store to prevent ethylene from accumulating.

Ethylene moves in plants by gaseous diffusion through intercellular spaces, and also in solution.

8.4.5   Abscisic acid

Abscisic acid, often abbreviated to ABA, is a natural hormone which counteracts the growth-promoting properties of gibberellins and is for this reason sometimes described as a growth inhibitor. It is one of probably several such materials occurring in plants. It promotes senescence and seed and bud dormancy, and has also been found to play a role in the opening and closing of stomata. It is made in leaves and moves to buds and other organs but little is known of its pathway of movement.

8.4.6   How hormones work

Some would question whether plant hormones are truly analogous to those which direct growth and development in animals, and whether they should properly be called hormones. Correctly speaking, hormones act as messengers and are characterized by two features. First, they are produced in one part of an organism and, like messengers, move to another part to induce changes in it, normally in its growth and development. Second, the changes they induce are through alterations in the concentrations of and balance between different hormones in the affected organ. Since some plant 'hormones' seem to act principally on the organs in which they are produced, they do not always act as messengers, and on this count could be disqualified as hormones.

Another area of doubt centres on the possible mode of action of hormones. It is believed that all hormones act directly or indirectly by controlling the selective repression of parts of the genetic blueprint and by influencing the permeability of membranes in the cell, but how they accomplish this is still very poorly understood. Nevertheless, enough is known for some intriguing questions to be posed about their mode of operation. For instance, only five main types of plant hormone are known at present, yet there are many thousands of genes in a plant to be regulated in a quite specific manner. Genes which deal with a particular metabolic pathway may be organized into an integrated sequence, but these sequences must still be directed in specific ways and activated at the correct time. Gibberellic acid stimulates hydrolysis of starch in barley grains, thereby accelerating germination, but this is one of very few known examples where a hormone has a specific effect on metabolism which might explain its effect on growth and development. The main picture that emerges is that each group of hormones influences many processes, and often one process is influenced by a number of hormones. Stem elongation, for instance, is influenced by all five groups in one way or another.

How can such a system deliver precise signals to cells and produce specific responses? There are several views as to how this could be achieved. In part the answer could be that cells respond to the balance between concentrations of different hormones, and many examples of such interactions are known. We shall look at some examples in Section 8.5. But interactions are not the whole story. The response of plants, organs or cells to a hormone often changes as they grow and develop. A particular response may occur only after a particular stage in the life cycle has been reached. It is as if the plant, organ or cell has to be in the right condition before it can respond to the hormonal signal. In other words, it has to be *competent* to respond. We saw, for example, that a shoot apex could not initiate flowering in response to daylength until it was competent to do so; undoubtedly this reflects the fact that the apical meristem becomes competent to respond to a hormonal signal to initiate flowering only when the shoot has reached a certain stage of development. The specificity of response of organs to hormones may thus be largely a property of the organ itself, and not of the hormone, which may merely provide a signal for the organ to follow one of a few clearly defined possible courses of development.

It may not even always be the same hormone that provides the signal. Thus in growing wheat coleoptiles, the early stages of elongation have been found to be gibberellin-dominated, the middle stages cytokinin-dominated and the final stages auxin-dominated, though all three groups of hormones are always present to a greater or lesser degree. It is the responsiveness of the cells as much as the hormonal signals that changes in the course of plant development.

It is widely accepted that a number of factors, including hormonal balances and the competence of cells and organs, must be correct before a change in development is triggered. According to one

viewpoint, the missing factor may be different in different organs and at different stages of development. The plant shows a response whenever the missing factor is supplied, either naturally or by an experimenter.

A more recent idea now gaining ground is that a whole series of factors, including the above ones, all act on one central controlling mechanism in growth and development. Accumulating evidence suggests that this may well be the concentration and distribution of calcium in the protoplasm.

An open mind could be particularly useful here, for it is not unlikely that the final picture of the control of growth and development in plants will contain parts of all the above mechanisms.

8.5 Hormone interactions

Let us now illustrate the importance of interactions between different types of hormone in some key processes of growth and development: stem elongation, apical dominance, root initiation, senescence and abscission, and dormancy.

8.5.1 Stem elongation

That auxin is essential for internode elongation in stems can be demonstrated by cutting off the shoot apex and applying a paste with or without IAA in it to the cut surface (Fig. 8.6). If the experiment is conducted in darkness, to avoid the complicating effects of light acting through the phytochrome and related systems (see Section 8.6.2), internode extension is strongly reduced in the absence of applied IAA. With IAA, however, the internodes grow normally. The auxin applied in the paste simply replaces that produced naturally by the shoot apex.

In plants which in the vegetative phase have a rosette habit, including biennial root crops such as sugar beet and turnip, applied gibberellins override the natural control mechanisms preventing internode extension and induce bolting, that is, the rapid elongation of the stem that normally occurs prior to flowering. Dwarf varieties of certain crops, including maize and pea, which have short internodes and are agriculturally useful because of their greater standing ability and ease of harvest than tall varieties, appear to have lost part of their ability to synthesize specific gibberellins. If sprayed with gibberellin at the appropriate stage of development, stem elongation is promoted and they are converted into tall plants. This is usually accompanied

Fig. 8.6 Effect of removal of the shoot apex on internode elongation in stems. When IAA is applied to the cut surface normal internode growth takes place, but without IAA growth is inhibited.

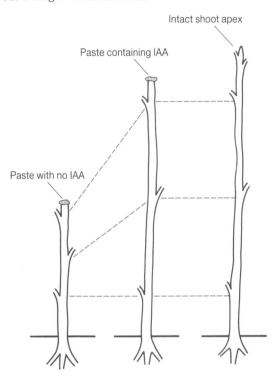

Intact shoot apex

Paste containing IAA

Paste with no IAA

by an increase in IAA levels in the stem, indicating, it is thought, an interaction between auxins and gibberellins in the control of stem elongation.

8.5.2 Apical dominance

We saw in Section 6.2.2 that the shoot apex exerts an influence on axillary buds below it, preventing their growth. We can now see that this influence is the polar movement of auxin. Because of the natural concentration gradient of auxin in the stem, the upper axillary buds are subjected to a higher auxin concentration than those lower down. The auxin level near the shoot apex is such that stem growth is stimulated but bud growth strongly inhibited. Further down, the auxin level is lower and bud growth is only weakly or not at all inhibited. Thus apical dominance becomes weaker and weaker the further the shoot apex is removed from the axillary buds.

Removal of the apex by pruning cuts off the auxin supply to axillary buds, which then grow to become branch shoots. The key role of auxin in maintaining apical dominance can be demon-

strated by cutting off the apex and applying IAA to the cut surface. This prevents the axillary buds from growing as in the intact plant.

Until recently it was believed that the control of apical dominance was as simple as this. We now know that ethylene, produced in cells as a result of auxin action, prevents bud growth and we cannot be sure if apical dominance is a direct auxin effect or an ethylene effect triggered by an auxin signal. It has also been found that cytokinins can release axillary buds from apical dominance. A condition seen in many trees known as *witches' brooms*, characterized by proliferation of branches (Fig. 8.7), results from attack by certain pathogens which produce cytokinin-like substances, disrupting the plant's natural control over branching from axillary buds. It may be, therefore, that the natural cytokinin concentration gradient from shoot base to tip complements the auxin gradient in the opposite direction in maintaining apical dominance and allowing lower buds to develop as branches.

Fig. 8.7 Witches' brooms on laburnum, a proliferation of branches due to a pathogen which causes the plant to lose control over the sprouting of axillary buds.

8.5.3 Root initiation

While a high cytokinin/auxin ratio seems to favour the development of branch shoots from axillary buds, the reverse seems to be true of the development of branch roots. The high cytokinin concentration and very low auxin concentration near the root tip inhibits lateral root formation, but further back, where the cytokinin/auxin ratio is lower, branch roots can be initiated.

The formation of adventitious roots by stems is similarly stimulated by auxin and inhibited by cytokinin. Indeed one of the earliest effects of auxin on plant growth to be discovered and put to commercial use, as we shall see in Section 8.7.1, is its assistance of rooting in stem cuttings. In the artificial culture of callus tissue resulting from disorganized cell proliferation, a high cytokinin/auxin ratio in the culture medium has been found to favour shoot development, and a low cytokinin/auxin ratio to favour root development. If an appropriate balance is maintained between cytokinins and auxins, both shoots and roots form from the callus.

8.5.4 Senescence and abscission

Cytokinins, auxins and in some cases gibberellins delay or inhibit senescence and abscission of leaves, flowers and fruits, while abscisic acid promotes them. Again the balance between stimulatory and inhibitory hormones is the key to whether senescence and abscission occur or not.

In the normal course of events a flower, the ovules of which have not been fertilized, forms an abscission layer and drops off. Ovule fertilization, however, results in auxin production which prevents abscission from taking place, thereby allowing fruit set. Sometimes abscission occurs prematurely in immature fruits on a tree; it has been found that this generally coincides with low auxin levels in the fruit and can be corrected by application of a synthetic auxin. It may be that the effect of auxin is an indirect one, creating a translocation sink which brings cytokinins from the root, thereby preventing senescence and abscission.

Ethylene does not influence senescence but strongly promotes abscission. In leaves and fruits it appears that ethylene is formed in the later stages of senescence or ripening and this is largely responsible for abscission. It is interesting to note that while early application of auxin can inhibit abscission by delaying senescence, late application can stimulate abscission by causing the formation of ethylene.

8.5.5 Dormancy

Abscisic acid promotes dormancy in buds and seeds, whereas gibberellins and ethylene break dormancy. Yet again it is the balance of different types of hormone, in this case primarily abscisic acid and gibberellins, that determines whether a bud or seed is dormant or not. The close correlation in deciduous trees between leaf senescence and the onset of bud dormancy is readily explained since both processes are stimulated by increasing levels of abscisic acid relative to other hormones.

8.6 Environmental cues for development

8.6.1 Autonomous versus environmentally directed development

Many of the patterns of development we have described are scarcely or not at all affected by external conditions, provided these are broadly suitable for growth. Once a leaf, flower or root is initiated, for example, its subsequent pattern of development proceeds with indifference to happenings in the environment around it. Branching patterns as controlled by apical dominance, sequential leaf senescence, and the setting and ripening of fruits and seeds – all these are programmed genetically and are not dependent on environmental signals. These developmental patterns are described as *autonomous* – literally, self-determined.

We have some sketchy ideas as to how autonomous development might work. At the molecular level, some of the large molecules in plant cells such as enzymes and nucleic acids have exact shapes which allow them to fit together automatically in a process of self-assembly. On a larger scale, the cells of a plant may pack together automatically in certain ways because of their shape. In addition the spacing of structures such as stomata on a leaf, or leaf or flower primordia on an apical meristem, or ovules in an ovary, is probably controlled by competition between them for nutrients, perhaps in combination with the effects of inhibitors diffusing out of the structures already formed.

Many aspects of development, however, take place in response to environmental stimuli. The plant must be able to 'sense' these stimuli, and to use the sensory information to direct development. As in autonomous development, hormones play a major role in control.

The most important stimuli acting as environmental cues for development are light, temperature and gravity. Plants use these cues first and foremost to recognize what time of year it is, so that development can be entrained to the changing seasons, and also to direct growth and development in relation to the position in which the plant finds itself and to make various more short-term adjustments.

act by releasing natural hormones in plant tissues, for example *ethephon* which releases ethylene.

Other growth regulators are substances of similar but not identical chemical structure to natural hormones. These synthetic structural *analogues* of natural hormones have similar properties to them and can be classified in hormone groups, for example naphthylacetic acid or *NAA* is an auxin and a structural analogue of IAA, while *benzyladenine* is a cytokinin and a structural analogue of IPA. Some compounds have hormone-like chemical structure but do not show hormone activity. These may block the action of natural hormones by occupying their binding sites on other molecules and may thereby modify the internal control of growth and development. They include *antiauxins*, which impede auxin activity, and *antigibberellins*, which impede gibberellin activity.

However, the majority of substances used commercially as growth regulators have little or no chemical similarity to natural hormones. These may be divided into two categories: substances which interfere with hormone manufacture or transport in the plant, and substances whose action is unrelated to the functioning of natural hormones.

We shall now review briefly some of the most important groups of synthetic growth regulators, and some of their uses in agriculture and horticulture. It should be appreciated that although some techniques such as the use of 2,4-D as a weed killer and IBA as a growth regulator have been in use for decades, the technology of growth regulation is still advancing.

8.7.1   Auxins

IAA itself is seldom used as a growth regulator. It is rapidly broken down in the plant to restore the normal IAA level; we have already seen how a natural concentration gradient is maintained from shoot tip to root tip by continuous degradation of IAA balanced by continuous production at the shoot tip.

Structural analogues such as NAA and *IBA* (indolylbutyric acid) are not broken down so readily and therefore have more pronounced and more persistent action. One of the earliest uses of these materials was for the induction of rooting in stem cuttings. Commercial *rooting powders* contain NAA or IBA dispersed in talc.

We saw that apical dominance is enforced by auxin produced by the shoot apex. Pruning releases

axillary buds from apical dominance, and in fruit trees this sometimes results in excessive proliferation of branches below a cut. To prevent this, NAA-based *pruning paints* are frequently applied to the pruning wounds.

Auxins have various commercial uses in fruit production related to the functions of natural auxin in fruit set and ripening. In some cases gibberellins can be used instead of auxins. Poor fruit set in tomatoes, cucumbers and many other crops can be corrected by auxin or gibberellin treatment, as can problems of premature abscission of fruit. Auxins and gibberellins are widely used in the production of seedless fruits from unpollinated or unfertilized flowers.

Far and away the most important application of synthetic auxins has, however, been as selective herbicides, and although they are not, strictly speaking, growth regulators when applied in this way it is worth looking briefly at their herbicidal properties.

IAA has the chemical structure shown in Fig. 8.9. It consists of a benzene ring linked through a pyrrole group to acetic acid. If the pyrrole link is replaced by an oxygen atom, we get *phenoxyacetic acid*, a substance with only weak auxin activity. The same is true of structural analogues such as phenoxypropionic and phenoxybutyric acids. If, however, we substitute chlorine atoms or methyl groups for some of the hydrogen atoms in the benzene ring we obtain products with very pronounced auxin activity. They include 2,4-D, 2,4,5-T, MCPA, MCPB, mecoprop and several others, and the structures of some of them are shown in Fig. 8.9.

Some of these are used as auxin growth regulators, especially in fruit crops, but applied at higher dose rates they are toxic to plants and can be used as herbicides. How they act is not well understood but the upsetting of natural auxin gradients and the balance of auxins with other hormones must play a part. Before killing susceptible plants they produce severe distortion of growth, often characterized by a twisting of stems and leaf stalks and a proliferation of adventitious roots. In woody plants they characteristically cause defoliation, probably as a result of ethylene formation, which is a common effect of auxins in plant tissues.

A major feature of the auxin herbicides, which resulted in a revolution in cereal husbandry, is that at doses of approximately $1 \, \text{kg ha}^{-1}$ they are virtually harmless to the cereal crop (if they are applied between the five-leaf and first node detecta-

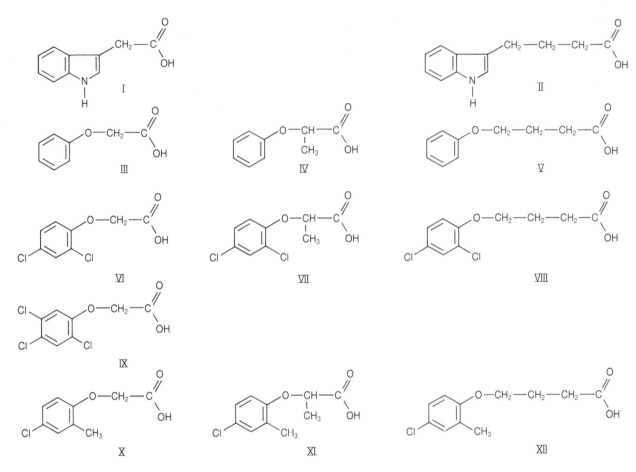

Fig. 8.9 Structures of IAA, IBA and some synthetic phenoxyalkanoic acids. I and II are auxins used commercially as plant growth regulators. III, IV and V have only very weak auxin activity but are the parent molecules of a series of very active auxins, VI to XII, used commercially as selective herbicides. I, indol-3-ylacetic acid (IAA). II, 4-indol-3-ylbutyric acid (IBA). III, phenoxyacetic acid. IV, 2-phenoxypropionic acid. V, 4-phenoxy-butyric acid. VI, (2,4-dichlorophenoxy)acetic acid (2,4-D). VII, 2-(2,4-dichlorophenoxy)propionic acid (dichlorprop). VIII, 4-(2,4-dichlorophenoxy)butyric acid (2,4-DB). IX, (2,4,5-triclorophenoxy)acetic acid (2,4,5-T). X, (4-chloro-2-methylphenoxy)acetic acid (MCPA). XI, 2-(4-chloro-2-methylphenoxy)propionic acid (mecoprop). XII, 4-(4-chloro-2-methylphenoxy)butyric acid (MCPB).

ble stages of growth) and highly injurious to a wide range of dicotyledonous weeds. The mechanism of this selectivity is poorly understood, but it appears to be related to differential retention on the leaves of crop and weeds and differential entry into the plant, rather than differential toxicity once the herbicide is in the plant. Herbage grasses are also highly tolerant of auxin herbicides, which can therefore be used for control of dicotyledonous weeds in grassland as well as in cereals.

The range of weed species killed – that is, the *weed spectrum* – depends on the particular herbicide. For example, MCPA and 2,4-D have fairly similar target weed spectra, while mecoprop is more effective against some weeds including chickweed but less effective against others such as hemp-nettle. The herbicide 2,4,5-T is particularly effective against woody perennials such as gorse. Phenoxybutyric acids like MCPB and 2,4-DB have a narrower weed spectrum. These materials are not in themselves toxic to plants but certain plants convert them to the corresponding phenoxyacetic acids (MCPA and 2,4-D), which are toxic. Plants which cannot convert them are resistant, while plants which can convert them are susceptible. Among resistant species are most legumes; with the introduction of the phenoxybutyric acids the auxin herbicides extended their range of uses into such crops as clover, lucerne and peas and could now be used in cereals undersown with legumes.

Though still in widespread use, the auxin herbicides are of declining commercial importance, having been partially superseded by other classes of herbicides that are active at even lower dose rates and have other advantages.

8.7.2 Gibberellins

There are few synthetic structural analogues of gibberellic acid, and most commercial applications of gibberellin involve gibberellic acid itself. As we have seen, it has a range of uses in the promotion of fruit development, where it has similar effects to auxins.

An important use of gibberellic acid is in the breaking of dormancy. Fruit trees in parts of the world where winters are mild, such as California, may not receive sufficient of a chill to overcome the dormancy of their buds. Spraying with gibberellic acid breaks the bud dormancy. Similarly in rhubarb the dormancy of underground buds can be broken before they have experienced their full winter chill, thus enabling early 'forcing' in warmth indoors. Gibberellic acid is also used by maltsters to break the dormancy of newly harvested barley and by horticulturists to stimulate the germination of many kinds of seed.

The shoot elongation effects of gibberellic acid have not been exploited to the same extent. It is, however, sometimes used to increase the length of leaf stalks in celery, particularly when the crop has been grown in less than perfect conditions.

8.7.3 Cytokinins

So far, synthetic cytokinins have found little application in practical agriculture and horticulture. Two of them, however, benzyladenine and *kinetin*, have applications in the artificial culture of plant tissues, which is becoming an important technique for rapid propagation (see Chapter 10). Among other effects, they counteract the inhibitory effects of auxin on bud growth. Cytokinins, by virtue of their inhibition of senescence, are also used to prolong the life of cut flowers.

8.7.4 Ethylene-generating agents

Ethephon, the most widely used ethylene-generating agent, stimulates the germination of some 'difficult' horticultural seeds which cannot be induced to germinate by gibberellic acid. Ethephon is also used on seeds, the germination of which is being stimulated by gibberellic acid, to counteract the undesirable stem elongation effects of gibberellic acid on the seedlings. Ethylene has been found to initiate flowering in pineapple, one of the few species that can be induced to flower by simple hormone treatment. Application of ethephon at an appropriate stage synchronizes flowering in the pineapple crop, enabling the entire crop to be harvested in a single operation.

8.7.5 Substances interfering with hormone manufacture or transport

The best known of several hormone transport inhibitors is *TIBA* (triiodobenzoic acid), which blocks the movement of auxins and gibberellins. It thereby releases axillary buds from apical dominance, creating a more bushy growth habit, which can be beneficial in some crops such as soya bean.

Of substances affecting hormone manufacture the best example is furnished by *chlormequat*, probably the most important single growth regulator in commercial use at the present time. Its action is to block the synthesis of gibberellin. Gibberellin promotes shoot growth, and abscisic acid inhibits it by antagonizing the action of gibberellin. Now both hormones are made in the plant from the same starting material. Thus when gibberellin synthesis is blocked by chlormequat more of the starting material is made into abscisic acid. The decrease in gibberellin and increase in abscisic acid results in inhibition of shoot elongation. The inhibitory effects of chlormequat can be reversed by treatment with gibberellic acid.

The main use of chlormequat is in cereals, where it is used as a stem-shortening agent to reduce the risk of lodging. It is applied just before stem elongation begins, and is much more effective in wheat than in barley or oats. This is because it is broken down more rapidly in barley and oats and a single application is insufficient to maintain the effect throughout the period of stem elongation. In barley a related substance, *mepiquat chloride*, does not suffer this disadvantage and is widely used.

Sometimes an increase in grain yield has been obtained following the use of chlormequat just before stem elongation, even in the absence of lodging. This has been explained as arising from the more favourable partition of assimilated dry matter between grain and straw, and also from the greater root/shoot ratio giving better drought tolerance. It has been found that chlormequat applied to barley at a much earlier stage of growth can also give increased yield. It appears that chlormequat exerts a transient effect whereby apical dominance is weakened so that more tillers form and later there are more grain-bearing ears per plant. The weakening of apical dominance may be a direct effect of chlormequat, or it may result from reduced growth of the main shoot, freeing nutrients for the development of extra tillers.

8.7.6 Other growth retardants

Another well-known growth regulator, the effect of which is to retard growth, is *maleic hydrazide*. It acts as a chemical pruning agent by stopping cell division in the shoot apex, which results in cessation of shoot elongation. This, unlike the effect of chlormequat, is not reversible with gibberellic acid. At the same time it frees axillary buds from apical dominance.

Its main uses are as a replacement for the pruning of hedges and street trees and the suppression of growth of rough grass in situations where mowing is impracticable. Another growth retardant, *amidochlor*, is more suitable for sports or ornamental turf because it causes less discoloration. *Paclobutrazol* is increasingly being used as a chemical pruning agent for trees in streets and under power lines.

Maleic hydrazide is also used as a pre-harvest treatment for onions, potatoes and other crops to inhibit sprouting of the tubers or bulbs while in store. Other sprouting inhibitors, applied post-harvest, include chlorpropham and tecnazene.

The herbicide *glyphosate* is used as a growth retardant in sugar cane to stop extension growth and thereby permit a greater content of sugar to accumulate in the cane prior to harvest.

8.7.7 Desiccants

Chemicals applied to crops before harvest to desiccate green material and thereby facilitate harvesting and limit the spread of disease are perhaps not truly growth regulators, but it is convenient to mention them here. All desiccants are also active as herbicides, and it is usually important that they are not translocated into the harvested part of the crop. Examples of desiccants used in the potato crop are *diquat* and *metoxuron*.

Summary

1. Growth is an irreversible increase in size, usually but not invariably accompanied by an increase in the amount of dry matter present. Early in the life of an organ or whole plant, growth is logarithmic, the rate of increase in size being proportional to the size already attained. Later there is a linear phase when the rate of increase in size is constant, followed by a maturing phase when growth slows down and stops.

Growth rate over a prolonged period is closely related to net assimilation rate and is therefore strongly influenced by light, temperature and water availability, which affects the opening and closing of stomata. Growth is, however, even more complex than net assimilation and is influenced by many other environmental factors such as mineral nutrition, the physical impedance of the soil and the presence of weeds, pests and diseases.

2. Growth of plant parts is directed through the translocation of organic and inorganic nutrients from sources to sinks. Harvestable organs such as seeds and tubers are sinks; it appears that their growth is limited not primarily by source supply or by the efficiency of translocation but by their capacity to attract nutrients. In the later stages of growth their demand for nutrients may be met largely by recycling from other organs, especially senescent leaves.

3. Every living cell in a plant has the genetic potential to make an entire plant, but genes are selectively repressed in different parts of the plant and at different stages in its growth. Cells not only 'recognize' where they are in the plant and develop accordingly but also show polarity, the ability to tell one end of the organ from the other. Once a particular development programme has been selected by a cell or meristem, this programme can seldom be reversed. This canalization of development, the recognition of position in the plant, polarity, and the selective repression of genes are controlled to some extent by hormones, substances which affect growth at minute concentrations in plant tissues and which generally have their influence some distance from where they are made.

4. There are five main groups of naturally occurring plant hormones. Auxins, including IAA, influence cell division and enlargement, and create sinks for phloem translocation, for example in fertilized ovules. They are manufactured in the shoot apex and move in one direction only, towards the roots. Gibberellins stimulate cell elongation, especially in stem internodes, acting in conjunction with auxins. Their transport is not unidirectional. Cytokinins influence cell division and inhibit senescence, and they create sinks for phloem translocation, primarily in the root. They are formed in the root tip and move towards the shoot but their transport is not as markedly unidirectional as that of auxins. Ethylene stimulates the ripening of fleshy fruits, abscission of fruits and leaves and in some seeds the breaking of dormancy. Abscisic acid counteracts the growth-promoting properties of gibberellins and promotes senescence and seed and bud dormancy. How so few hormones can have so many different effects is not well understood, but it appears that it is the competence of cells and organs to respond to hormonal signals, rather than the hormones themselves, that varies in different parts of the plant and at different stages of growth.

5. Most growth responses involve interactions between hormones. Apical dominance, for example, depends on the balance between auxins and cytokinins, low auxin or high cytokinin levels allowing axillary buds to grow into branch shoots. Initiation of branch or adventitious roots, on the other hand, is stimulated by high auxin and low cytokinin concentrations. Senescence is promoted by abscisic acid and inhibited by cytokinins, while dormancy is promoted by abscisic acid and inhibited by gibberellins. The balance of stimulatory and inhibitory hormones in each case determines the course of development.

6. Many aspects of plant growth and development are autonomous, proceeding entirely under internal control. Others are sensitive to environmental cues such as light, temperature and gravity. This enables development to be entrained to the seasons. Low temperature, for example, breaks the dormancy of many seeds and buds and vernalizes shoot apices, priming them to flower. Photoperiod, measured in some way by phytochrome, provides stimuli for flowering, the breaking and induction of dormancy, and synchronous leaf senescence. Both low temperature and photoperiod are sensed in such a way that the plant is not misled by aberrant cues.

Environmental cues also allow seeds and other parts of plants to recognize where they are, for example whether they are buried in the soil or are densely shaded by other vegetation. The main stimuli are light intensity, perceived by phytochrome, and atmospheric carbon dioxide concentration around buried seeds. Etiolation is a common response to low light intensity.

Plants respond to gravity and the direction from which light comes by orientating themselves with respect to these stimuli. Roots and shoots are geotropic, growing down and up respectively in response to gravity, while shoots are phototropic, growing towards light. Nastic responses to environmental stimuli are not directional and often they are not true growth responses but result from changes in turgidity of special cells. Plants also show internal rhythms on a 24 h cycle, so that diurnal movements, for example of stomata, continue when the plants are placed in constant darkness.

7. Growth regulators are synthetic substances which show hormone-like activity when applied at low concentrations to plants. At higher doses many of them are toxic to plants and can be used as herbicides. Auxins such as NAA and IBA are used in rooting powders to induce formation of adventitious roots, and in pruning paints to prevent excessive proliferation of branches following the removal of apical dominance by pruning. Auxins and gibberellins are used to promote fruit set, including the production of seedless fruit. Gibberellins are used to break dormancy, for example of fruit trees in climates where they do not receive a chilling stimulus in winter.

Some growth regulators interfere with the manufacture or transport of natural hormones in the plant. Chlormequat inhibits gibberellin synthesis and leads to reduced extension growth, for example in wheat, where it can increase yield by preventing lodging at high rates of fertilizer application. TIBA blocks the movement of auxins and gibberellins and can release axillary buds from apical dominance. Growth retardants such as maleic hydrazide, paclobutrazol and the herbicide glyphosate find a variety of commercial uses, for example as a replacement for the pruning of trees or the mowing of grass and in limiting extension growth in sugar cane. A number of chemicals are used as desiccants to facilitate harvesting and to limit the spread of disease.

9

Plant Variation, Inheritance and Breeding

9.1 Variation in plants

Progress in plant agriculture depends to a great extent on an ability to select the husbandry measures and inherited plant characters which give the best crops, and it is important to understand how this is done.

If the plants in any crop are examined, it will be found that no two are identical in appearance. There is always a degree of variability within a crop, and this variability is much greater if the entire species is considered. Some pea plants, for example, have white flowers whilst others have red flowers. There are in this case no intermediate types with various shades of pink flowers. Variation in flower colour in peas is therefore said to be *discontinuous*; characters showing discontinuous variation are almost always inherited, that is to say, the variation is genetic in origin.

Most of the variation in any plant population is not discontinuous like this. As an illustration, the total leaf areas of forty-five individual swede plants from a field were measured seven weeks after the crop was sown. The leaf areas ranged from 187 to 1216 cm^2 and, as shown in Fig. 9.1, did not fall into just a few clearly delimited categories. Instead there was a whole spread from the smallest to the largest, with most of the plants falling near the middle of the range. This is an example of *continuous* variation.

Almost every variable attribute of agricultural importance that it is desirable to encourage in a crop, such as number of seeds produced, protein content, relative growth rate or disease resistance, varies continuously from plant to plant. These characteristics are sometimes controlled by the genetic inheritance of the plant and sometimes by environmental influences, but usually it is the combined effects of inheritance and environment that lead to continuous variation.

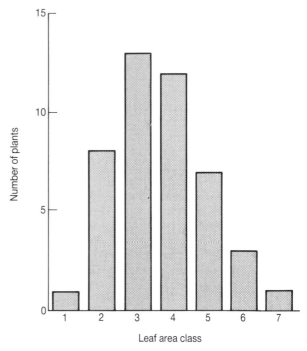

Fig. 9.1 Frequency distribution of forty-five swede plants in seven leaf area classes (1: 0–200 cm^2; 2: 200–400 cm^2; 3: 400–600 cm^2; 4: 600–800 cm^2; 5: 800–1000 cm^2; 6: 1000–1200 cm^2; 7: 1200–1400 cm^2). This is a typical example of continuous variation in a plant populaiton.

9.1.1 Variation and sampling

Whatever the cause of variation, there is a problem of measurement. Because of the range of behaviour, examining a single plant cannot provide a measure of the crop as a whole. To illustrate this point, when a grower wishes to assess whether a crop is ready for harvesting he looks at the stage of maturity reached not by one plant but by many in different parts of the field. Similarly, the weight of a single grain of wheat or barley is no guide to the average grain size, an important attribute of any consignment of grain, since it affects its quality, whether for feeding, milling, malting or seed.

To determine the true average grain size it is necessary to *sample* from the bulk of grain being tested. Any sample of one thousand grains taken from one place in the bulk will contain a range of sizes and the mean weight of these grains will probably be a truer estimate of the mean weight of all the grains in the consignment than would be the weight of any one grain. In agriculture, the total weight of one thousand grains, the *thousand-grain weight*, rather than the mean weight of the grains, is commonly used as a measure of grain size and quality.

However, the estimate obtained from a single sample of one thousand grains may still be inadequate, for there are numerous causes of variability in grain bulks and not all the variability has been taken into account. For example, in any mass of grain, whether in sacks or in bulk, the smaller grains tend to settle towards the bottom. Thus a sample from the top will usually show a greater thousand-grain weight than one from near the bottom. It is necessary therefore to take several samples from different parts of a sack or store, as illustrated in Table 9.1, to obtain a realistic estimate of thousand-grain weight.

Variability occurs not only in sacks of grain and in other stored materials, but also in soils and crops in every agricultural field, and this necessitates the use of sampling procedures whenever any factor such as soil pH, crop yield or date of flowering is measured. Note, however, that no matter how many samples are taken, the mean value obtained is merely an estimate (unless the entire population is measured, which is seldom practicable or desirable). The more samples that are taken, the more precise the estimate becomes. Fortunately it is seldom necessary to know the exact value of any mean. It is, however, frequently useful to have some measure of the precision of an estimate, that is, how near the exact value it can be relied upon to be, so that it can confidently be stated that the true mean lies inside a certain narrow range.

Such a measure of precision or reliability is provided by the *standard error of the mean*. Frequently, in reports of scientific investigations, means are presented along with their standard errors. It is not the place here to explain how a standard error is calculated, but the interested reader will find an explanation in any elementary textbook of statistics. What is more important for the present purpose is to understand what information, in essence, is provided by the standard error.

For example, in a weed control experiment the mean density of chickweed plants in a particular barley field was 30 m^{-2}, with a standard error of 2.5 m^{-2}. This might be presented as 30 ± 2.5 plants m^{-2}. In another field the density was 30 ± 1.5 plants m^{-2}. The standard error this time is smaller, indicating that the mean value presented for this field (30) is more likely to be close to the true mean. Just how close it is likely to be can be determined from published tables, but, as a general rule of thumb, it is 95% probable that the difference between the true mean and the estimate is less than twice the standard error. Thus a value of 30 ± 2.5 plants m^{-2} implies that the true mean probably lies somewhere between 25 and 35, while a value of 30 ± 1.5 plants m^{-2} indicates that the true mean probably lies in the range 27–33. Note, however, that there is still a 5% (or 1 in 20) chance that the true mean lies outside this range.

Why was there a more reliable measure of the mean weed density in the second field than in the first? It may have been because more samples were taken in the second field, leading to a more accurate estimate and thus a lower standard error. Alternatively, equal numbers of samples may have been taken in both fields but variation in chickweed density may have been less in the second field than in the first. Note therefore that the standard error of the mean indicates the precision of the estimate of the mean, but it does not necessarily indicate how variable the population is.

To measure the degree of variability, another statistic known as *variance* is used. Again the concern here is not with the method of calculation but with the kind of information that a variance

Table 9.1 The thousand-grain weights of ten samples drawn from a single bulk of grain in store; all values adjusted to 16% moisture content

Sample no.	1	2	3	4	5	6	7	8	9	10	Mean
Weight(g)	38.0	43.5	35.4	42.7	43.8	40.2	37.4	45.9	41.1	33.4	40.1

Table 9.2 Lengths of 20 randomly selected heads of wheat and timothy. Note that the mean length is similar but the variance of the data is much greater for timothy than for wheat.

Sample no.	Length (mm)	
	Wheat	Timothy
1	68	95
2	65	70
3	80	87
4	75	39
5	79	48
6	80	110
7	73	65
8	65	58
9	70	48
10	80	54
11	66	99
12	75	105
13	72	89
14	85	53
15	70	108
16	90	81
17	70	60
18	92	53
19	73	58
20	70	63
mean	74.90	72.15
variance	60.62	507.29

figure conveys. Table 9.2 gives data showing the lengths of ears of a sample of twenty wheat plants from a single crop, and the lengths of twenty ears of the herbage grass timothy, again from a single crop. Both samples have a similar mean, but the individual ears of timothy show a much greater spread of values than those of wheat. Timothy is thus more variable in ear length than wheat, and this difference in variability can be quantified by comparing the variance of the two sets of data.

9.1.2 Variation and agricultural experimentation

The variability of plants has a major influence on the design of experiments to find improvements in agriculture. Consider, for example, an experiment with two potted bean plants, A and B, in which a possible growth retardant chemical is applied to A, leaving B untreated. It is hoped that a comparison will show if the treatment produces a dwarfing effect on A. If A grows less tall than B, can it be concluded that its growth has been reduced by the chemical?

The answer is no, for there are alternative explanations. Some bean plants inherit a genetic make-up that causes them to grow taller than others. Also, many environmental factors influence the growth of plants, including water supply, mineral nutrition, light intensity, temperature and disease. Unbeknown to the experimenter, one or more of these influences, including hereditary effects, may vary sufficiently between the plants to cause the observed difference in height. These factors, which have nothing to do with the experimental treatment, thus produce *background variation* in many plant characters. There are a number of ways in which this background variation can be dealt with in an experiment.

Suppose a large number of bean plants are divided randomly into two lots, which are planted out into two plots in the field, and one plot is then treated with the supposed growth retardant. If the mean height of plants in the treated plot is less than that of plants in the untreated plot, does this now prove a dwarfing effect of the chemical? Again, it does not. The difference is now less likely to be due to hereditary differences between plants, since samples of the population, not just two individuals, are being compared, and plants with a genetic tendency towards being smaller should be present in approximately equal numbers in both plots. But there remains the difficulty that no account has been taken of the possible variation in environmental influences between plots. The treated plot might just be receiving less water or poorer mineral nutrition than the untreated, causing the plants in it to grow less. The same would apply if we compared two whole fields of beans, one treated and one untreated. In part, the error of comparing just two plants has been repeated.

If several plots receive the treatment and an equal number are left untreated, and the plots mixed randomly in the field, then this would get over the difficulty to a considerable extent. It is much less likely that environmental factors could influence several randomly distributed treated plots more than the untreated ones in such a way as to cause a consistent difference in height. If such a difference is observed, we can fairly conclude that it is probably caused by the treatment, not by background variation.

This sort of experiment is called a *trial*, and much of modern agricultural practice derives from field experiments with this structure, comparing two or more varieties, treatments or products. Plots receiving the same treatment are called *replicates*. Normally the plots are grouped into blocks, each block containing one replicate of each treatment (Fig. 9.2). A similar logic underlies the design of trials with animals and also laboratory or glasshouse experiments.

Even in a replicated experiment the possibility remains that the difference in height measured between treated and untreated beans could be due to background variation, not to the treatment. Obviously, if the measured difference in mean height is much larger than the difference being caused by background variation (for the determination of which, see below), then it is much less likely to be a chance effect of background variation and more likely to be a real effect of the treatment.

What is needed is a set of mathematical techniques for comparing the background variation with the variation apparently due to the treatment, to measure the likelihood, that is the probability, that

the observed differences are really treatment effects. Background variation can be quantified by measuring the variation among untreated replicates, as this cannot be due to treatment, and this measure is known as the *error variance*. This can then be compared with a measure of variation between plots receiving different treatments, namely the *treatment variance*.

A number of statistical techniques, foremost among them the *analysis of variance*, have been developed to aid in the interpretation of results of experiments in this manner, to help decide whether new crop varieties or treatments have a beneficial effect on characters such as crop height, disease resistance or yield. The use of these techniques strongly influences the design of field trials and laboratory experiments. Analysis of variance indicates how probable it is that any difference – in the earlier example between the mean height of treated and untreated bean plants – is a real effect. If the probability is greater than, say, 95% that the difference does result from treatment, the difference is said to be *statistically significant* at that probability level. There always remains some probability, however, in this case 1 in 20, that the difference in height has nothing to do with the treatment but arises by chance due to background variation.

Fig. 9.2 A simple field trial consisting of twelve plots, arranged in three replicate blocks of four treatments. The treatments are independently randomized among the four plots of each block. The trial is surrounded by untreated areas as in the foreground of the photograph.

It is essential that claims made for new products, machines or techniques are based on properly conducted trials, and claims not based on these should be treated with scepticism. No one should be misled by an apparent yield increase following the use of some new product in a single unreplicated field demonstration, especially if the increase is small.

9.1.3 Variation and crop improvement

The plant-to-plant variation that exists in all crop situations and in wild plant populations arises, as was seen above, from two main sources. Variation resulting from differences in soil fertility, the depth at which the seed was sown, the proximity of weed plants, the chance infection by pathogenic fungi, and so on, can be termed *environmental variation*. Variation resulting from inborn differences between the plants themselves is known as *genetic variation*.

Variation arising from purely environmental causes cannot be transmitted from one generation to the next, i.e. it is non-heritable. Genetic variation is, however, by its nature heritable. A plant which is genetically tall will tend to produce tall offspring. If, however, its tallness is entirely due to its having found a particularly favourable environment, its offspring will probably be no taller than average.

As there are two main sources of variation in plants, two basic approaches to the improvement of crop production are available. First, a uniformly favourable environment for crop growth can be provided by good soil preparation, drainage or irrigation as necessary, fertilizer application, control of weeds, pests and diseases, and so on. This approach, which may be called *husbandry improvement*, is largely up to the farmer himself, though research and development by agricultural institutes and other agencies may point the way.

The second approach is *crop improvement*, that is, the development of crop plants which are genetically better suited to the purposes for which they are grown, giving higher yields of better quality produce and often being easier to grow and harvest. In the developed world crop improvement has largely been taken out of the hands of farmers themselves and is now carried out by state or commercial plant breeders. It should be appreciated, however, that our present-day crops are the products of hundreds, in some cases thousands, of years of both intentional and unintentional improvement by farmers and that in most crops modern

breeding so far is just the icing on the cake.

Whether carried out by a peasant farmer or by a large, modern plant-breeding institute, the essential feature of crop improvement is the *selection* of the most agriculturally desirable plants from the range of variation present. As has been seen, much of the variation that exists is environmental in origin, and selection from purely environmental variation gives no crop improvement. However, where the variation is at least partly genetic in origin, selection of the better individuals will lead to an improvement in the following generation.

By saving seed only from these more desirable specimens and repeating the process over many generations a distinct 'breed' of the crop is brought into existence. Where this is accomplished by farmers themselves the 'breed' is known as a *land race* and is characteristic of the local area in which it has arisen. Where the new 'breed' is the outcome of a scientifically conducted breeding programme it is known as a *cultivar* or more commonly in layman's terms a *variety*.

The question is often posed, which has been more important in increasing crop yields, husbandry improvement or crop improvement? In one study in New York State, fifty-year old varieties of winter wheat were compared with modern varieties in trials over a ten-year period. On average, the modern varieties outyielded the old varieties by about 50%. During the previous fifty years average farm yields of winter wheat in New York State had increased by approximately 100%. It was concluded that crop improvement was responsible for about half the increase in farm yields and that husbandry improvement was responsible for the other half.

It is, however, impossible to separate entirely the effects of genetic and environmental improvement on crop yield. For example, barley is higher-yielding on most soils when given nitrogen fertilizer at the rate of 150 kg ha^{-1} than at 120 kg ha^{-1}, but only if the additional weight of grain can be supported by the straw. Otherwise, lodging frequently occurs and the grain fails to ripen or cannot be picked up by the combine harvester. Newer varieties of barley tend to be shorter and stronger-strawed and can therefore accept higher nitrogen levels without lodging. In this case it is the interplay of crop and husbandry improvement, rather than either on its own, that has brought about the increased yield.

It is generally true, not just in the case of barley, that modern plant breeding has not so much increased the *potential* yield of crops (that is, the yield obtainable in ideal conditions), as improved the *reliability* of yield. Thus, in the absence of lodging an old variety may perform just as well as a new one, but in those years when the ripening crop is battered by wind and laden with rain the new variety is more likely to remain standing and deliver its full yield to the combine harvester. Much of the emphasis in the breeding of many crops is on disease resistance. Again, in the absence of disease an old, susceptible variety is at no disadvantage but in those years or situations when conditions favour the disease a new, more resistant variety will perform much better.

9.2   The mechanisms of inheritance

Crop improvement is possible only where the characteristics desired are at least partially heritable. Thus it is necessary to understand how the inheritance of characteristics is accomplished during plant reproduction.

Consider the plant's seed. However small, it contains all the information necessary to guide the growth and development of the complete plant, however large. Somehow, through the seed, the parents transmit accurately all the characteristics with which they endow their offspring. Furthermore, this vast amount of information must be packaged within single cells since the full range of characteristics of the male parent is contained in the pollen grain and of the female parent in the egg cell. It was seen in Section 8.3.1 how individual cells of a mature plant, though specialized in their development and function, are genetically totipotent, that is, each single cell contains all the information necessary to organize a complete plant if suitable conditions are provided.

Thus there are four fundamental questions to be dealt with in this section. How is the information coded? How is the code transmitted with great accuracy to all the cells of a plant? How is it transmitted from generation to generation? And how is it decoded to be expressed as flower structure, leaf shape, seed size or the hundreds of other features of the plant?

If seeds taken from a single plant are sown, the resulting crop raises a fifth question. The progeny, except in certain exclusively self-pollinated crop species, are all different. It can be shown that these differences are genetic, not merely environmental. Why is it, then, that the progeny, though they resemble each other and their parents, are not identical to them?

Finally, how does all this genetic variation come about in the first place? This, too, is an important question for, as has been seen, without this variation new varieties cannot be produced, nor in nature could evolution take place, since this is driven by a natural selection of the fittest of the different variants to survive and reproduce.

9.2.1   Nature of the genetic code

If cells are examined during cell division, when division of the hereditary material must be taking place, some clues to the nature of the genetic code can be found. When cells are about to divide, some very distinctive rope-like bodies appear in the nucleus (Fig. 9.3). These bodies are called *chromosomes* and they hold the key to many of the questions raised above.

It has now been clearly established that most of the genetic material of higher plants, as in animals, resides in the chromosomes of the nucleus. There is a little outside the nucleus, mainly in chloroplasts and mitochondria, but this need not be of concern at present. On analysis, chromosomes have been found to contain approximately equal amounts of the material called *deoxyribonucleic acid* (*DNA*) and a type of protein known as *histone*. In addition, a further substance, *ribonucleic acid*

Fig. 9.3 Cord-like chromosomes appearing in the nucleus of a cell about to divide in the root tip of *Lilium regale*. Reproduced from J McLeish and B Snoad (1972) *Looking at Chromosomes*, 2nd ed. Macmillan, New York, NY, with permission of the publisher.

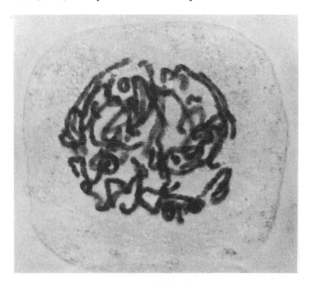

(*RNA*) is present in small amounts. Though differing only slightly in chemical structure from DNA, RNA has a quite distinct biological role.

DNA is based on a sugar called deoxyribose. The basic building block of DNA is a molecule of deoxyribose with a phosphate group and a molecule of any one of four chemicals called *bases* attached to the sugar. Such a combination of sugar, phosphate and base is called a *nucleotide* (Fig. 9.4). The four bases found in DNA are *adenine*, *cytosine*, *guanine* and *thymine*. RNA differs from DNA in having an extra oxygen atom on the sugar molecule (making it ribose instead of deoxyribose) and in having another base, *uracil*, in place of thymine.

The nucleotide building blocks are linked together by their phosphate groups to give long, chain-like molecules (Fig. 9.5). In DNA, two such chains are linked together along their length by bonds that form between pairs of bases, the whole double strand being twisted on itself as shown in Fig. 9.6. The structure of DNA is rather like a ladder in which the two sugar-phosphate chains form the side rails and the paired bases the rungs, with the entire structure given a twist. This structure, first elucidated by Watson and Crick in 1953, is known as the *double helix*, and the pairing of bases is, as will be seen, a highly significant feature of it.

A chromosome contains many such double strands of DNA, much folded back on themselves and coated in histone. It is now known that DNA is the basis of heredity. With apparently little variety in its structure, it must carry a great deal of detailed information. A mechanism must exist to decode this information to produce the enzymes, carbohydrates, membranes and other components of the living cell. Furthermore, each DNA molecule must reproduce itself with great precision and reliability every time a cell divides.

All materials of the cell are synthesized from simpler raw materials with the aid of the proteins known as enzymes. Indeed, enzymes play a central role in all cellular processes. Enzymes, like all proteins, are long chains built up of amino acids, of which over twenty different kinds exist in the plant cell. The properties and therefore the biological function of an enzyme depends on the exact order of amino acids in the chain. Could it be that the order of bases in DNA or RNA in some way dictates the order of amino acids in proteins and therefore which enzymes are formed? If so, it can now begin to be seen how information in the form

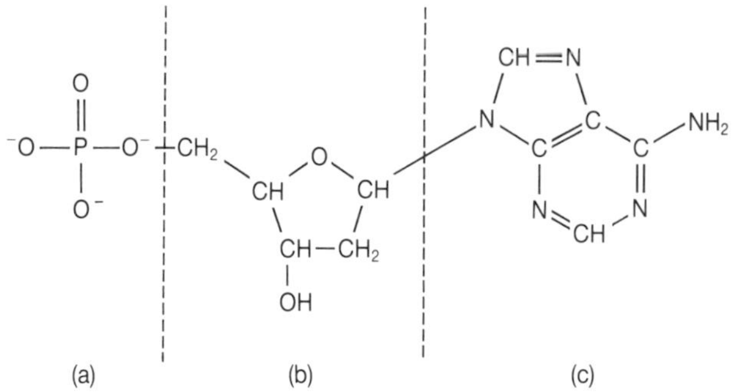

Fig. 9.4 Structure of the nucleotide deoxyadenosine-5′-phosphate. A nucleotide consists of a phosphate group (a), a sugar, in this case deoxyribose (b), and a base, in this case adenine (c).

Fig. 9.5 The structure of a small part of a DNA chain, consisting of a backbone of alternating sugar and phosphate units, with a sequence of bases attached to the sugar units.

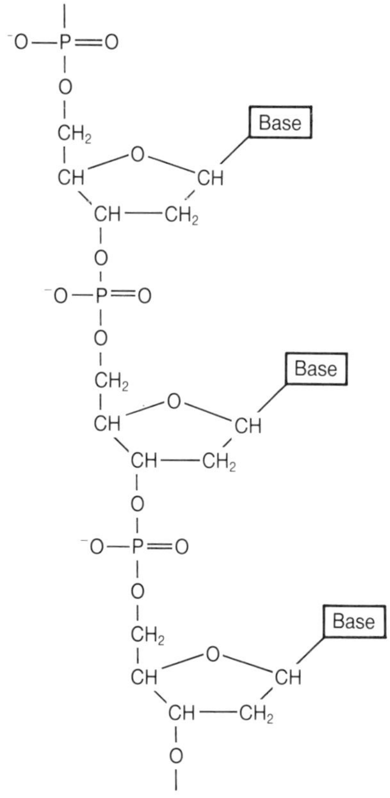

of a sequence of bases could specify how a cell can or cannot function.

The proof that this is indeed the case came in a series of experiments in which strands of artificial

RNA were allowed to interact with various amino acids. An RNA chain consisting solely of uracil-bearing nucleotides, if mixed with a blend of various amino acids in solution, constructs a protein chain consisting entirely of the single amino acid phenylalanine. It does not interact with any of the other amino acids present. Evidently the base uracil on RNA codes in some way for phenylalanine.

However, since there are only four different bases on any RNA molecule and there are more than twenty different amino acids to be coded for, a single base cannot code for a single amino acid. From further experiments it was established that three neighbouring bases code for a single amino acid. The sequence uracil-uracil-uracil, for example, codes for one molecule of phenylalanine

(Fig. 9.7). Such a trio of bases is called a *codon*.

There are sixty-four possible permutations of three bases from a stock of four, and therefore sixty-four different codons. Of these, three code for no amino acid and each of the others codes for one, and only one, amino acid. Some amino acids are coded for by two or more different codons.

Still more significantly, if RNA chains are assembled with codons in a particular order then the resulting protein exactly reflects this order in its order of amino acids. Quite clearly, the code is not only read, but read in precise sequence. In the living cell the order of amino acids in a protein as dictated by the RNA base sequence gives rise to an enzyme of exact structure and therefore of specific function. The process by which a series of codons

Fig. 9.6 The structure of the DNA molecule, showing two long chains of nucleotides (a) joined by bonds between the bases, and the two chains twisted into a double helix (b). (● = phosphate group, ○ = deoxyribose unit, A = adenine, G = guanine, C = cytosine, T = thymine)

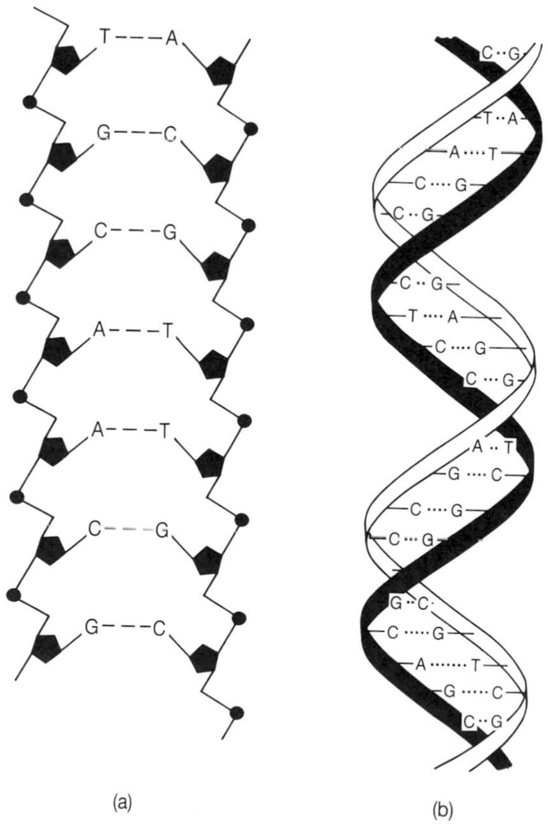

Fig. 9.7 An artificial RNA molecule carrying only uracil bases (U) codes for a protein chain consisting entirely of phenylalanine (PHE) units. Each codon of three U bases provides the information to build one PHE molecule into the protein chain.

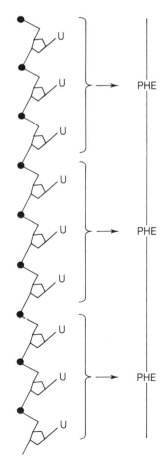

(a)   (b)

on RNA constructs a protein by causing specific amino acids to link together in a sequence reflecting exactly the sequence of codons is called *translation*.

9.2.2 How the genetic code is expressed

It is DNA, not RNA, that carries the 'master copy' of the code for any protein in a plant cell. But DNA is largely restricted to the nucleus, and protein assembly from amino acids is known to take place mainly in the cytoplasm outside the nucleus. It is the role of RNA to copy the code from the DNA, carry it out into the cytoplasm and there interact with amino acids to form protein chains.

There are several types of RNA in the cell, only one of which fulfils this task. It is called, appropriately, *messenger RNA (mRNA)*, and is assembled in the nucleus from free nucleotide molecules following the base sequence dictated by DNA. As shown diagrammatically in Fig. 9.8, mRNA then

Fig. 9.8 Translation of the DNA code into an amino acid sequence in a protein chain. (1) The code (sequence of bases) is copied on to messenger RNA (mRNA). (2) The mRNA moves from the nucleus into the cytoplasm. (3) Ribosomes attach to the mRNA molecules and read the base sequence, one codon at a time. (4) For each codon the ribosome assembles the appropriate amino acid, drawn from solution in the cytoplasm, into a growing protein chain.

migrates to the cytoplasm where it associates with minute bodies called *ribosomes*. With the aid of other types of RNA known as *transfer RNA (tRNA)* and *ribosomal RNA (rRNA)*, the ribosomes 'read' the sequence of bases on the mRNA, one codon at a time, and bring together amino acids from solution in the exact sequence specified (Fig. 9.9).

A section of DNA in which the bases collectively code for a single complete protein is called a *gene*. Each chromosome carries a large number of genes. In peas, one gene codes for an enzyme that is responsible for the production of red pigment from a colourless substance in the petals of the flower. Plants that possess this gene have red flowers; plants that lack it have white flowers – or rather, as will be seen later, white-flowered plants do not entirely lack the gene, but have one whose base sequence is in some way defective and therefore codes for a protein which does not have the enzyme activity needed to make the red pigment.

Because what is seen is the red pigment, not the enzyme involved in its manufacture, it is spoken of as the gene for red flower colour. Similarly genes for short internodes, wrinkled seeds or resistance to wilt disease are referred to, though in each case the gene is really producing a protein which in some way functions to give the character in question.

But how can the mere specification of proteins

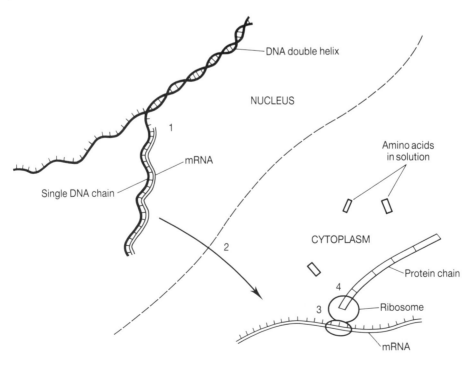

give rise to the complex and highly organized structure and function of cells and whole plants? The answer is by no means fully known, but a major part of it lies in the intricate three-dimensional shapes that protein molecules assume. Apart from

the strong covalent bonds linking the amino acids in sequence in a protein chain, attractions exist between certain amino acids in different parts of the chain and these attractions cause the molecule to fold into a precise shape. In turn, these foldings can induce further convolutions of the molecule which are held in place by a variety of chemical bonds (Fig. 9.10). Little of this complicated process, often called the 'higher order' organization of the molecule, is random. The positions of non-covalent bonds in a protein molecule are determined by the sequence of amino acids and it seems that the range of shapes which a particular protein chain can assume may be very limited or even restricted to only one.

Fig. 9.9 Translation of codons. A = adenine, C = cytosine, G = guanine, T = thymine, U = uracil. ALA = alanine, ILE = isoleucine, MET = methionine, SER = serine.

Fig. 9.10 Three-dimensional structure of a protein molecule. The shape of the molecule is maintained by non-covalent bonds which form between amino acids in different parts of the protein chain.

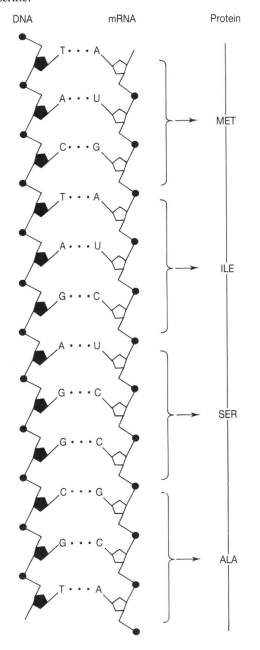

It is this three-dimensional shape of an enzyme that decides which, if any, chemical reactions it can catalyse and gives enzymes their remarkable specificity. As far as is known, the assumption of shape and the resulting acquisition of a specific function is spontaneous, given the right amino acid sequence. All that DNA has to do is specify that sequence.

This matter of self-assembly can be carried a stage further, for proteins and other large molecules can spontaneously bond with one another to form aggregates of particular shape and composition. Thus complexes of enzymes able to carry out a whole series of related actions can be formed, and their products may similarly aggregate spontaneously to form structural complexes in the cell. This is the basis of subcellular architecture and is a further stage in the expression of the genetic code.

Structures such as cell membranes are very probably assembled by this means and their ability to repair themselves rapidly if the cell is damaged is due to their basic components being held in an advanced state of readiness in the cytoplasm. Self-assembly by spontaneously and precisely aggregating molecules is almost certainly an important factor in the construction of cell walls and subcellular organelles such as ribosomes, chloroplasts and mitochondria.

9.2.3 Transmission of genes during cell division

The form in which the genetic code is held in the cell has been briefly described, as has the manner in which it is decoded to guide the growth and development of the plant. The problems of how the code is passed on in its entirety whenever a new cell comes into being by division of a pre-existing cell, and how it is transmitted through the gametes in sexual reproduction will now be considered. All cells of a plant normally have an identical genetic constitution, but during sexual reproduction variability enters the picture, giving rise to progeny which are not genetically identical with either parent or with each other.

On the face of it, these two activities of precise reproduction of a genetic code and deliberately induced variation in it are incompatible, but plants overcome this by having two distinct methods of cell division. First, the method that achieves simple reproduction of the genetic code will be described.

It is obvious that if the whole complement of genes was simply partitioned when a cell divides, each daughter cell would receive a reduced complement. It is known, however, that every cell has a full complement. To ensure this, cell division has to be a precisely controlled affair in which a complete set of genes is replicated.

In any active meristem, cells may be observed at all stages of division. After completing one division, each daughter cell may divide again, but not before a period of apparent inactivity sometimes called the 'resting stage'. In fact, as will be seen later, there is no 'rest' during this period, but it is a time of great metabolic activity in the nucleus.

During the so-called 'resting stage' the chromosomes are not visible as separate bodies, indeed the DNA and protein of the nucleus form a rather featureless granular mass. When a cell is about to divide, however, the chromosomes contract and thicken, becoming visible as rod-shaped or cord-like structures of differing lengths (Fig. 9.11). The number and appearance of the chromosomes are identical in all normal dividing cells of a plant and are characteristic of the species to which it belongs. For example, barley and pea have fourteen chromosomes in each dividing nucleus, onion has sixteen and maize has twenty.

Each chromosome consists of two parallel threads or *chromatids* lying close together. Each chromatid carries a copy of every gene on that chromosome; thus the two chromatids carry identical sets of genes. Just before a cell divides, a system of slender fibres known as the *spindle* forms across it, from points of attachment at opposite ends of the cell. Some of the fibres attach to the chromatids. The fibres contract, pulling one of each pair of chromatids to either end of the cell. Each original chromosome is therefore split in two, and the two daughter nuclei each have exactly half the DNA complement of the parent nucleus. They have, however, a full set of genes, since each chromatid is an exact copy of the other one to which it was originally attached and which is now in the other nucleus.

The process of nuclear division just described is called *mitosis*, and occurs in all cell division except where sexual gametes are being formed. Mitotic cell division is complete when the contents of the two daughter nuclei once again become featureless and a new cell wall forms between them to isolate them in two separate cells.

This, however, cannot be the full story as it leaves the new cells with only half the full DNA complement, and it is known that cells divide repeatedly over long periods of time. This could not be sustained unless new DNA is synthesized.

In fact synthesis of DNA does take place, during the so-called 'resting stage' between one division and the next. Each DNA double strand splits down the middle to form single chains of nucleotides. Each of these then reassembles a new double strand by attachment of individual nucleotides to the now vacant sites on the bases of the single strand, as shown in Fig. 9.12.

This reassembly is not random, for a nucleotide containing the base adenine can attach only to one containing thymine and vice versa, and similarly one bearing cytosine can attach only to one bearing guanine and vice versa. (In the copying of DNA on to messenger RNA, uracil rather than thymine is the base that attaches to adenine, but otherwise the process is very similar.)

(a)

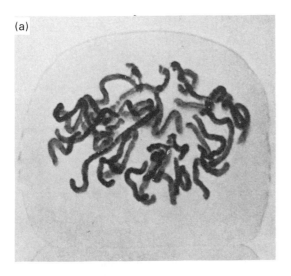

(a) In the early stages of mitosis the chromosomes contract, thicken, and become visible.

(c)

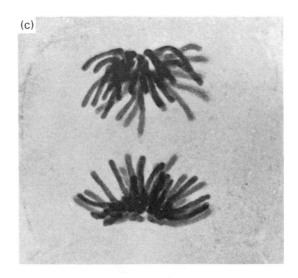

(c) The spindles then contract separating the chromatids of each chromosome and pulling them towards the opposite ends of the cell, where they form the two daughter nuclei (d).

(b)

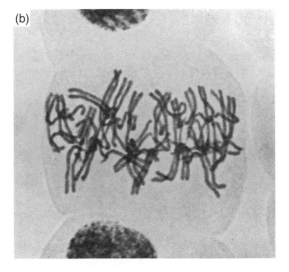

(b) As mitosis progresses, each chromosome can be seen to consist of two parallel threads (chromatids) lying together. Slender fibres (spindles), which are not visible in this photograph, are attached from the waist-like area of the chromosome, where the chromatids are joined, to each end of the cell. At this stage the chromosomes become aligned along the centre of the cell.

(d)

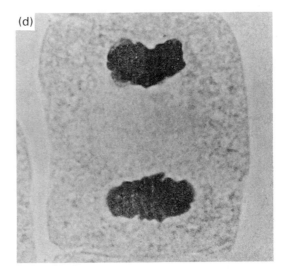

Fig. 9.11 The stages of mitosis in root tip cells of *Lilium regale*. Reproduced from J McLeish and B Snoad (1972) *Looking at Chromosomes*, 2nd ed. Macmillan, New York, NY.

Thus the order of bases on the newly assembled DNA is determined precisely by the order of bases on the parent molecule. The entire process of mitosis and DNA assembly therefore provides for the exact replication of the genetic code and its passage from cell to cell.

Fig. 9.12 Synthesis of DNA by assembly of nucleotides in the sequence dictated by pre-existing DNA molecules, leading to a doubling of the DNA content of the plant cell nucleus.

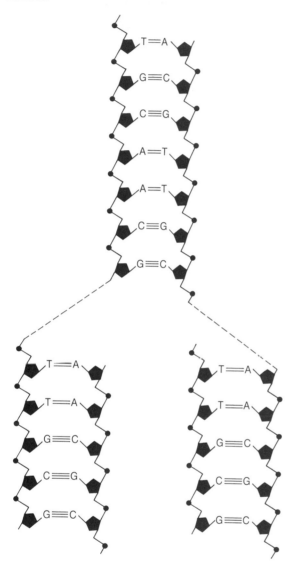

9.2.4 Transmission of genes to gametes in sexual reproduction

The question of variability in the sexual progeny of a single plant in the face of exact replication of the DNA code can now be considered. The explanation lies in how the special sex cells, or gametes, are generated by cell division from normal vegetative cells in the flower.

Each plant is the product of two gametes, a female egg cell and a male gamete contained in a pollen grain, becoming united in the process of fertilization. The fertilized egg cell, and the entire plant that grows from it by repeated mitotic cell division, thus has two sets of chromosomes, one inherited from the male and one from the female parent. (In self-pollination the male and female parent are one and the same.) Since each set of chromosomes has a complete set of genes, the plant has two genes for each character.

The chromosomes inherited from the two parents are normally indistinguishable from one another when examined under the microscope, and form a series of pairs. The two members of a pair are said to be *homologous*. Because there are two of each kind of chromosome in the cells of a plant, these cells are described as *diploid*. Barley, for example, has a diploid chromosome number of fourteen, made up of seven homologous chromosome pairs.

Homologous chromosomes carry genes affecting the same characters. For instance, at the same location on both chromosomes of a pair there will be a gene affecting, say, flower colour. The two genes are not, however, necessarily identical. Thus the gene on one chromosome may confer red flowers and its counterpart on the other chromosome white flowers.

In the formation of male and female gametes a special kind of cell division takes place in which the number of chromosomes is halved. This process is known as *meiosis* (Fig. 9.13). One of its most significant features distinguishing it from mitosis is that when the chromosomes become visible in the early stages, the two homologous chromosomes of each pair lie side by side. Each chromosome, as in mitosis, consists at this stage of two parallel chromatids, but when the spindle forms it pulls apart not the two chromatids of each chromosome, but the two whole chromosomes of each pair.

Soon after this division is completed a second division of each of the two resulting nuclei takes place, this time with the chromatids of each chromosome moving to opposite ends of the cell as in mitosis. Thus meiosis of a single diploid cell gives rise to four cells, each with only one complete set of chromosomes. Such cells are said to be *haploid* and give rise to the gametes by mitosis. The gametes are therefore also haploid.

When the whole chromosomes are separated during the first division of meiosis, it is a matter of pure chance which one of each pair goes to a partic-

(a)

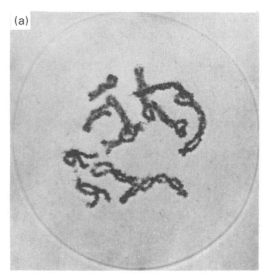

Fig. 9.13 The stages of meiosis in root tip cells of *Lilium regale*. Reproduced from J McLeish and B Snoad (1972) *Looking at Chromosomes*, 2nd ed. Macmillan, New York, NY.

(b)

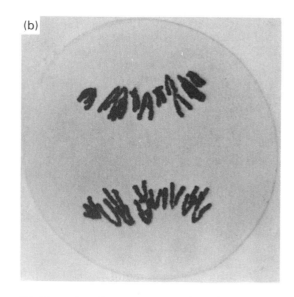

(a) Early stages of meiosis showing pairs of homologous chromosomes lying side by side, each pair joined at one or more points.

(d)

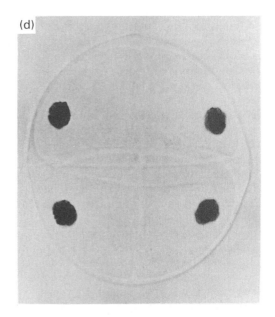

(b) After the chromosome pairs align in the centre of the cell, the spindles pull whole chromosomes, not chromatids, towards the opposite ends of the cell, where they form daughter nuclei (c) and a new cell wall begins to form between them.

(c)

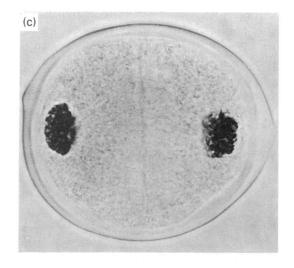

(d) Soon after this, a second cell division takes place, this time with the chromatids of each chromosome moving to opposite ends of the cell as in mitosis, to give a set of four cells each with only one set of chromosomes.

ular end of the cell. Thus each daughter nucleus contains some chromosomes which were originally inherited from the male parent and some from the female parent. For example, if the dividing diploid cell carries on a particular pair of homologous chromosomes one gene for red flowers and one gene for white flowers, it is a matter of chance which one will end up in any particular haploid nucleus. Overall, however, equal numbers of haploid cells will be produced containing 'red' and 'white' genes.

On another pair of chromosomes there may be one gene for hairy leaves and one for hairless leaves. Again, following meiosis equal numbers of haploid cells will be produced containing 'hairy' and 'hairless' genes. But whether a particular cell carries the 'hairy' or 'hairless' gene is completely independent of whether it carries the 'red' or 'white' gene on another chromosome. This is the principle of *independent assortment* and is one of the processes leading to a reshuffling of genes, which is known as *recombination* and is a major cause of the genetic variability seen in the progeny of sexual reproduction.

If independent assortment of chromosomes were the sole cause of recombination, it would follow that genes located on the same chromosome would always stick together. Yet it can be shown that some recombination does occur between genes on the same chromosome. The reason for this can be found in what happens in the early stages of meiosis when homologous chromosomes are lying side by side. The two chromosomes of each pair make very close contact all along their length. Frequently two chromatids, one belonging to each chromosome, break at one or more points and rejoin crosswise as shown in Fig. 9.14. This is known as *crossing over*; it enables homologous chromosomes to exchange genes and is therefore a further source of recombination.

Fig. 9.14 Crossing over between chromatids of homologous chromosomes, showing two of the chromatids with a mixture of genetic material from both parent chromosomes after crossing over.

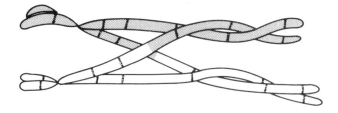

When haploid gametes unite in the process of fertilization, the full diploid condition is restored. The gametes are, as shown above, extremely variable in their genetic constitution because of the recombination that has taken place during meiosis. The random mating of male and female gametes results in still more mixing of genes. Thus it is not surprising that variability of offspring is the hallmark of sexual reproduction.

Figure 9.15 summarizes the main events in mitosis and meiosis and the chief differences between these superficially similar processes.

9.2.5 The origin of genetic variation

The last of the questions posed at the beginning of this section was, how does all this genetic variability arise in the first place? Why, for instance, do some versions of a gene code for red flowers while other versions code for white flowers?

It was seen that the DNA base sequence in a gene codes for an amino acid sequence in a protein that may be an enzyme. Mistakes occasionally occur in the replication of DNA, in the copying of the DNA code on to messenger RNA, or in the translation of the RNA code into a protein. Of these, by far the most serious are mistakes in replication of DNA, since they alter the 'master copy' which will subsequently be replicated and translated. Such mistakes are called *mutations*. They include the omission of one or more bases from the sequence, the insertion of extra bases, or the substitution of one base for another.

A DNA mutation results in a faulty RNA base sequence which either makes the 'wrong' protein or is so confusing that it cannot be translated into an amino acid sequence at all. Even if the 'wrong' protein differs from the 'right' one in only a single amino acid in a critical location in the molecule, its enzyme activity may be seriously impaired or destroyed.

Since most enzymes in plants perform an essential or at least a useful function, it follows that most gene mutations are deleterious. Very occasionally, however, a mutation occurs which is beneficial in that it creates a new protein conformation with improved enzyme activity or even with some novel activity which is useful to the plant. Plants possessing such a mutated gene will be at a selective advantage and will increase in the population. Mutations like this are the basis of all genetic variability and therefore of evolution.

Mutations are induced by the background of

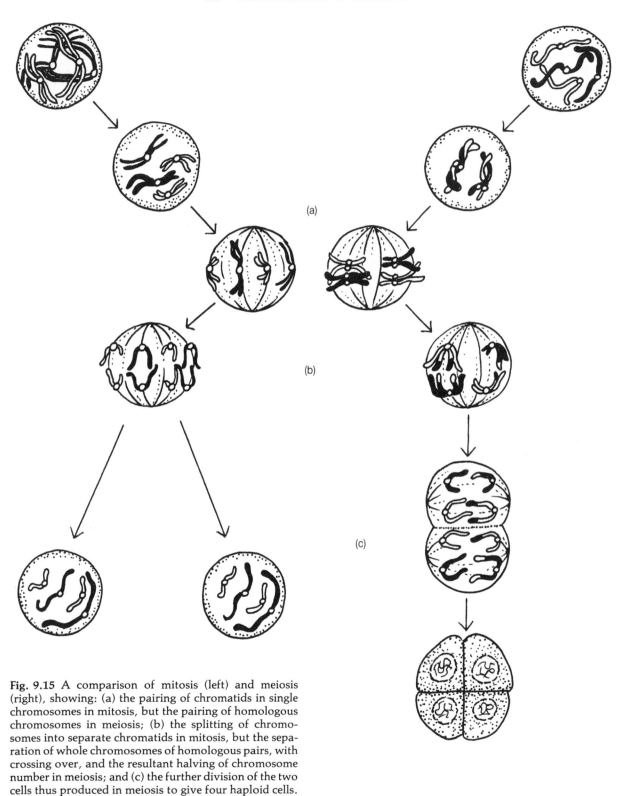

Fig. 9.15 A comparison of mitosis (left) and meiosis (right), showing: (a) the pairing of chromatids in single chromosomes in mitosis, but the pairing of homologous chromosomes in meiosis; (b) the splitting of chromosomes into separate chromatids in mitosis, but the separation of whole chromosomes of homologous pairs, with crossing over, and the resultant halving of chromosome number in meiosis; and (c) the further division of the two cells thus produced in meiosis to give four haploid cells.

natural radioactivity and are a fairly common occurrence in nature, but only because of the vast numbers of individual plants and the thousands of genes held in each of the millions of cells of every one of these plants. A plant breeder looking for a mutation in a particular gene in, say, barley would have to examine perhaps a hundred million plants to find a single occurrence of the mutation. Clearly this is totally impracticable. However, the frequency of mutations can be greatly increased by artificially treating plants with short-wave radiation, including X-rays, or with chemicals that have mutagenic activity.

Breeders sometimes use techniques like these to generate increased genetic variability for selection. Mutagens and X-rays, however, besides being hazardous to use, are highly undiscriminating, and large numbers of undesirable mutations are induced in addition to the occasional desirable one. An example of the successful use of mutations is the induction of the 'erectoides' gene in barley. This mutated gene supplements another gene naturally present in barley and is responsible for extreme shortening of the straw. It has been incorporated into a number of commercial varieties, including the popular Scottish malting variety Golden Promise.

Methods have recently been developed for artificially adding genes which are not part of the normal complement of genes found in a given plant species. These methods are collectively known as *transformation technology* and are outlined in Section 9.6.

9.3 Patterns of inheritance in plants

We have given an outline of the mechanisms of inheritance in plants; we can now consider how these give rise to the patterns of inheritance seen when plants reproduce.

In every diploid cell there are, as has been noted, two genes for every character. For instance, at position A on one chromosome there may be a gene affecting leaf hairiness. On the homologous chromosome position A will again be occupied by a gene for leaf hairiness, but this gene may not be identical to the other one in its DNA base sequence. Basically, it is the same gene but in a different form. We say that these are two *alleles* of the gene for leaf hairiness.

If the two genes at position A on the two chromosomes are identical, that is, the same allele of the gene, the plant is said to be *homozygous* for

that gene. If two different alleles are present, affecting leaf hairiness in different ways, the plant is said to be *heterozygous* for that gene.

Now suppose one allele, *H*, confers hairy leaves and the other, *h*, confers hairless leaves. The genetic constitution of a plant with respect to a character, in this case leaf hairiness, is its *genotype*. The appearance of the character in question as expressed in the growing plant, that is, whether the leaves are hairy or hairless, is the *phenotype*. A plant with the genotype *HH* is homozygous for *H* and is hairy in phenotype. Similarly a plant with the genotype *hh* is homozygous for *h* and is hairless in phenotype. But what of the heterozygous plant with genotype *Hh* – is it hairy, hairless or somehow intermediate in phenotype?

Usually one allele of a gene has a dominating influence over the other in deciding the phenotype of a heterozygous plant. In this case it may be allele *H*. Thus genotype *Hh* gives rise to hairy leaves, just like genotype *HH*. The only plants with hairless leaves are those with genotype *hh*. The *H* allele is said to be *dominant*, because it masks the influence of the *h* allele, which is said to be *recessive*.

These interactions between alleles of genes have important consequences for the inheritance of characters, but, as we shall see, these consequences vary according to whether the progeny result from cross- or self-pollination.

It was seen in Section 7.3.3 that in some species self-pollination is the rule and that in others cross-pollination is much more frequent. In some species such as red clover self-pollination cannot lead to fertilization because of self-incompatibility. Species which are totally or predominantly cross-pollinated are said to be *outbreeding*. In these species genetic recombination permits much greater mixing of genes than is possible in *inbreeding* species which are predominantly self-pollinated. A normally outbreeding plant may, however, be forced to inbreed if it is a member of a very small population in which it can exchange genes only with close relatives whose genotypes are similar to its own. This has important consequences, which will be examined later.

9.3.1 Inheritance of one character in an inbreeding species

The patterns of inheritance in an inbreeding species such as wheat, barley, oats or peas will be discussed first. For simplicity, the discussion will be confined initially to a single character, the expression of which is controlled by a single gene pair.

If the plant is homozygous for the gene in question, meiosis in the stamens and ovules will give rise to gametes all of which carry the same allele of the gene (Fig. 9.16). Thus fertilization following self-pollination can result only in progeny with the same homozygous genotype as the parent plant. All progeny will be genetically identical with each other and with the parent in respect of that gene. An inbreeding plant which is homozygous for a particular gene is therefore said to be *true-breeding* for that gene.

Meiosis in a heterozygous individual, for instance a hairy-leaved plant with genotype Hh, results in the formation of two types of gamete, in this case H and h. As Fig. 9.16 shows, gametes carrying the dominant (H) and recessive (h) alleles are produced in equal numbers. If the plant is self-pollinated, the chances of two identical gametes (H and H, or h and h) coming together are exactly the same as the chances of two different gametes (H and h, or h and H) coming together. Thus 50% of the progeny will be heterozygous. Of the homozygous progeny half will show the dominant phenotype, in this case hairy leaves, and half the recessive phenotype, in this case hairless leaves.

All the heterozygous progeny will show the dominant phenotype, thus overall 75% of the progeny will have hairy leaves and 25% will have hairless leaves. This simple 3:1 ratio is a *Mendelian ratio*, so-called because of its discovery by Mendel,

an Austrian monk who, by virtue of his experimental and quantitative approach to the study of plant inheritance, became the father of the science of genetics.

In his classical early studies, Mendel observed two strains of peas which he found were true-breeding for flower colour. One strain had red flowers and the other white flowers. It is now recognized that these strains must have been homozygous. Red-flowered plants had genotype RR and white-flowered plants had genotype rr. Although the peas were naturally self-pollinating, Mendel artificially cross-pollinated some flowers by removing their own stamens and then placing pollen from the other strain on their stigmas. He thus *hybridized* the two strains.

He harvested the resulting seed and planted it out to grow the 'first filial' or F_1 *generation*. All the F_1 plants bore red flowers. It is now known that every single one of these plants was heterozygous for flower colour, having an R allele contributed by one parent and an r allele by the other parent (Fig. 9.17). When Mendel allowed his F_1 plants to self-pollinate, the resulting F_2 *generation* showed the characteristic Mendelian 3:1 phenotype ratio of red to white flowers.

If the F_2 generation is again self-pollinated, it can be anticipated from Fig. 9.17 that the homozygous

Fig. 9.17 Cross-pollination of red (genotype RR) and white (genotype rr) flowered peas to give red (genotype Rr) flowered F_1 progeny, followed by self-pollination of the F_1 progeny to give red (genotypes RR or Rr) and white (genotype rr) flowered F_2 progeny in a 3:1 ratio.

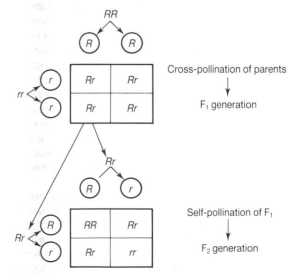

Fig. 9.16 Inheritance of a gene H for hairy leaves in a self-pollinating plant. If the parent plant is homozygous (a), gametes (represented by circles) are all genetically identical and all progeny (represented by boxes) have the same genotype as the parent. If the parent plant is heterozygous (b), two kinds of gametes are produced which, when they come together at random, produce 50% heterozygous (Hh) and 50% homozygous (HH or hh) progeny. The phenotype of a plant with genotype HH or Hh is hairy, that of a plant with genotype hh hairless. Thus self-pollination of a heterozygous parent gives progeny 75% of which are hairy and 25% hairless.

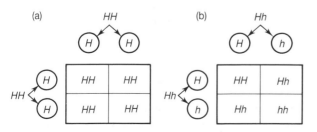

plants will breed true but the heterozygous ones – still 50% of the progeny – will continue to produce offspring showing dominant and recessive characteristics in the 3:1 ratio. The production of different phenotypes by self-pollination of a heterozygous individual is known as *segregation*. With repeated self-pollination the production of heterozygous plants in the progeny halves with each generation – 25% in the F_3, 12.5% in the F_4, and so on. By the F_{10} the population is almost entirely homozygous, and little further segregation takes place.

Any modern variety of an inbreeding crop such as peas or barley represents a selection from the progeny of a single F_2 plant, and may indeed be the selected progeny of a single F_5 or F_6 plant. Such a variety is known as an *inbred line*. Not only are all the plants in an inbred line homozygous and therefore true-breeding, but they are all homozygous for the same allele of any one gene, and are thus very uniform. This is because plants homozygous for the undesired alleles have been removed by selection.

Uniformity and true-breeding behaviour have great advantages for the farmer. The field performance, maturity date, yield and quality of produce of an inbred line are predictable within narrow limits, such variation as exists being almost entirely environmental in origin. Furthermore, uniformity in ripening time makes for ease of harvesting and uniformity of size facilitates grading and marketing. On the other hand, it will be seen later that uniformity has drawbacks when the crop is attacked by disease.

9.3.2 Inheritance of more than one character in an inbreeding species

So far the inheritance of only one character has been considered. When two or more pairs of genes are considered together the situation is a little more complex.

As an example, suppose that in a true-breeding strain of wheat two genes *A* and *B* are present as the dominant alleles. As the strain is true-breeding both genes must be in the homozygous condition, all plants having the genotype *AA* in respect of gene *A* and *BB* in respect of gene *B*. This genotype can be abbreviated to *AABB*. Gametes produced by meiosis in this strain are all of one genetic constitution, namely *AB*.

Similarly, in another strain homozygous for the recessive alleles *a* and *b*, that is with the genotype

aabb, all gametes will be *ab*. Hybridization of the two strains will bring together *AB* and *ab* gametes and result in an F_1 generation with the genotype *AaBb*, that is, heterozygous for both genes, as illustrated in Fig. 9.18.

What happens when the F_1 plants are allowed to self-pollinate to give rise to an F_2 generation? To answer this, consider first the types of gametes that the heterozygous F_1 plants can produce. Meiosis in a heterozygous plant results, as has been seen, in the separation of the dominant and recessive alleles. Any individual gamete will contain either *A* or *a* and it will also contain either *B* or *b*. If *A* (or *a*) and *B* (or *b*) are located on different chromosomes they will assort independently at meiosis; that is to say, whether a particular gamete carries *A* or *a* has no effect on whether it carries *B* or *b*. *A* is just as likely to occur along with *b* as with *B*; likewise *a*

Fig. 9.18 Cross-pollination of two strains differing in two genes, followed by self-pollination of the F_1 progeny to give an F_2 generation.

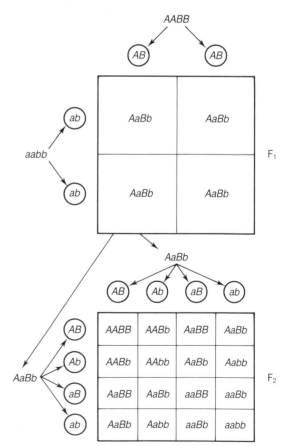

may occur together with either B or b. Thus four types of gametes – AB, Ab, aB, ab – are produced in equal numbers.

When any one of four types of female gamete can be fertilized by any one of four types of male gamete, the outcome can be illustrated in a 4×4 matrix, as shown in Fig. 9.18. It can be seen from the matrix that the resulting F_2 plants will be of nine possible genotypes. However, there are only four possible phenotypes, because plants of genotypes AA and Aa are identical in appearance, as are plants of genotypes BB and Bb. The four phenotypes may be written $[AB]$, $[Ab]$, $[aB]$, and $[ab]$. As Table 9.3 shows, these are produced in a 9:3:3:1 ratio. Two of the phenotypes, $[AB]$ and $[ab]$, are the same as the parental strains which were originally hybridized, but two, $[Ab]$ and $[aB]$, represent new combinations of characters and are described as *recombinant* phenotypes.

However, if the two genes A and B are on the same chromosome pair they will not assort entirely independently at meiosis. Instead A will tend to stay with B and a with b. There is, as has been mentioned, some recombination between genes on the same chromosome because of crossing over, so that the recombinant phenotypes $[Ab]$ and $[aB]$ will appear in the F_2, but with reduced frequency. Genes on the same chromosome showing a reduced frequency of recombination are said to be *linked*.

Whether or not the genes are linked, the number

Table 9.3 Phenotypes of F_2 progeny of a cross between strains differing in two genes, as illustrated in Fig. 9.16.

Genotype	Phenotype			
AABB	[AB]			
AABb	[AB]			
AaBB	[AB]			
AaBb	[AB]			
AABb	[AB]			
AAbb		[Ab]		
AaBb	[AB]			
Aabb		[Ab]		
AaBB	[AB]			
AaBb	[AB]			
aaBB			[aB]	
aaBb			[aB]	
AaBb	[AB]			
Aabb		[Ab]		
aaBb			[aB]	
aabb				[ab]
Total	9	3	3	1

of possible phenotypes which can result from hybridization of two true-breeding strains differing in two genes is four. If there are three gene differences, the number of possible phenotypes is eight ($[ABC]$, $[ABc]$, $[AbC]$, $[Abc]$, $[aBC]$, $[aBc]$, $[abC]$, $[abc]$) and with four gene differences the number rises to 16. It can be easily shown that where there are n gene differences, the number of possible phenotypes in the F_2 generation is 2^n. Hybridization of two distinct inbred lines which might differ in 20 or more genes is thus a potent means of generating the genetic variability that the plant breeder needs for the selection of new and desirable combinations of characteristics.

9.3.3   Inheritance in an outbreeding species

Let us now consider the inheritance of characters in an outbreeding species such as white clover or sugar beet. If the alleles A and a of a gene are equally abundant in a breeding population, then the chance that A and a gametes will fuse is the same as that identical gametes (A and A, or a and a) will fuse. There is thus a high degree of heterozygosity in the population, and this does not dwindle with each successive generation as in inbreeding species. Fresh recombination of genes takes place with each generation, so that varieties of outbreeding crops tend to be much less uniform or predictable than varieties of inbreeding crops.

Selection in an inbreeding species, as has been seen, is able to produce inbred lines which show great uniformity and predictability because of their high level of homozygosity. Selection in an outbreeding species cannot achieve this because heterozygous genotypes cannot be eliminated. Selection can, however, achieve some reduction in variability by the elimination of extreme types and the removal of certain undesirable traits controlled by dominant alleles of genes. Characteristics controlled by recessive alleles cannot be so easily removed because many heterozygous individuals, not distinguishable in phenotype from individuals homozygous for the dominant allele, conceal the unwanted allele and pass it on to their progeny. Some of these will inherit the recessive allele from both parents and express it in their phenotype.

Selection in an outbreeding species need not be preceded by deliberate hybridization, but the breeder should start with as variable a population as possible to maximize the opportunities for recombination. Characters in which environmental variation is great and genetic variation relatively small are particularly difficult to select for in an outbreeding crop. How, for example, is it possible to tell the difference between a plant which is genetically short and one which is genetically long but is stunted for some environmental reason? Obviously it is useful to know, so that only those plants which are genetically short may be selected.

In practice, a procedure known as *progeny testing* is used. This procedure is based on the fact that genetically short plants will produce more short progeny than environmentally stunted plants. Plants are selected on the basis of characteristics possessed by their progeny, not on the basis of their own characteristics.

9.3.4 Quantitative inheritance

Most of the attributes of a crop in which breeders are interested, such as disease resistance, protein content or stem length, are controlled not by single gene pairs but by whole groups of genes working together. Such an assemblage of genes operating on a single observable character is called a *polygene*. Polygenes have several important consequences for crop improvement.

Suppose the resistance of a crop to a particular disease is conferred by dominant alleles R_1, R_2 and R_3 of three genes operating as a polygene. Recessive alleles r_1, r_2 and r_3 lead to susceptibility, but only in the homozygous condition. A highly resistant plant carries dominant alleles of all three genes. Any of these genes may be homozygous (e.g. R_1R_1) or heterozygous (e.g. R_1r_1) but this does not affect the degree of resistance. The resistant plant thus has various posible genotypes, such as $R_1R_1R_2r_2R_3r_3$ or $R_1r_1R_2R_2R_3R_3$. A highly susceptible plant has the recessive alleles of all three genes and must be homozygous for all of them. Its genotype is $r_1r_1r_2r_2r_3r_3$.

If the three genes influence disease resistance equally it might be possible to identify four classes of plant with respect to resistance, depending on whether none, one, two or all three of the genes are present as the dominant allele. If, as is more likely, the genes have unequal effects, so that a plant carrying dominant alleles R_1 and R_2 has a different degree of resistance from a plant carrying R_1 and R_3 or R_2 and R_3, there would be eight resistance classes (Fig. 9.19). In practice, however, it would be very difficult to identify these because differences between classes would be small and would be masked by environmental variation. Instead plants would show a continuous spectrum of disease resistance. Identification of classes would be even more difficult if more than three genes were involved.

This is the explanation for inherited continuous variation. The character in question, in this case disease resistance but many other important characters behave similarly, is said to show *quantitative inheritance*. Selection by plant breeders for such characters is not as easy as in the case of Mendelian inheritance, where the phenotypes are readily distinguishable. For example, it is not simply a case of keeping the short and eliminating the tall, or keeping the resistant and eliminating the susceptible individuals. Some minimum standard of acceptability has to be set, and plants which do not come up to this standard are rejected. This involves large numbers of time-consuming measurements and assessments. Even then many genetically substandard plants may be inadvertently selected because of environmental variation.

9.3.5 Inbreeding depression

It was seen earlier that outbreeding species are characterized by a high level of heterozygosity, maintained by regular cross-pollination. If such a species is forced to inbreed, the level of heterozygosity falls rapidly because any plants homozygous for a particular gene produce only homozygous progeny, while heterozygous plants produce homozygous and heterozygous progeny in equal numbers.

It is an almost universal feature of increasing homozygosity in an outbreeder that it brings with it a decline in vigour, characterized, for example, by poor seed germination, slow growth rate and reduced disease resistance. This decline in vigour is known as *inbreeding depression* and in crops usually results in reduced yield.

It has been suggested that inbreeding depression arises directly from loss of heterozygosity and that this is because individuals which are heterozygous for certain key genes possess some advantage over individuals which are homozygous for either of the alleles of any of these genes. No plausible mechanism can, however, be proposed to explain why a heterozygous genotype should perform better than either of the two homozygous genotypes.

A better explanation of inbreeding depression lies in the fact that many recessive genes are highly deleterious or even lethal in individuals homozygous for those genes. In heterozygous individuals they are masked by their dominant alleles and confer no disadvantage. Inbreeding increases the proportion of homozygous plants and therefore the damage or mortality rate due to deleterious recessives.

More insidious, however, is the deleterious gene which is a member of a polygene. To explain this, consider again the example of polygene $R_1R_2R_3$ which controls disease resistance (Fig. 9.19). Suppose that in an outbreeding population the recessive and dominant alleles of each gene are equally abundant. Then for each gene the dominant allele

Fig. 9.19 Sets of plants in a population carrying genes R_1, R_2 and R_3 for disease resistance. There are eight possible categories of plant (1–8), including plants carrying none of the three genes (category 1) and plants carrying all three (category 8).

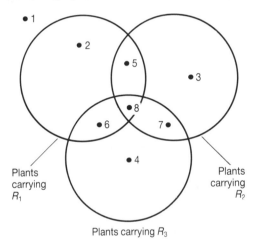

Plants carrying R_1

Plants carrying R_2

Plants carrying R_3

for resistance (e.g. R_1) will occur in 75% of the population either as genotype R_1R_1 or as R_1r_1. The dominant allele will be missing in the 25% of the population with genotype r_1r_1. Taking all three genes together, the proportion of highly resistant plants in the population, that is, those possessing dominant alleles for R_1, R_2 and R_3, will be 75% of 75% of 75%, or 42%. The proportion of plants not at all resistant to the disease, having genotype $r_1r_1r_2r_2r_3r_3$, will be 25% of 25% of 25%, or just 2%.

With random cross-pollination the proportion of highly susceptible plants remains at this low level with each generation. (In fact, the proportion may even fall because of natural selection operating against susceptible individuals.) Inbreeding, however, produces an increase in the proportion of homozygous plants. After a few generations of inbreeding, about 50% of the population will be homozygous for the recessive allele of any one of the three genes and 50% will be homozygous for the dominant allele.

Thus with continued inbreeding, the proportion of highly resistant plants falls to 50% of 50% of 50%, or 12.5%, while the proportion of highly susceptible ones rises from 2% to 12.5% (Fig. 9.20). Homozygosity therefore greatly lowers the average level of disease resistance in the population. Similar

Fig. 9.20 Proportions of a plant population carrying the dominant alleles of genes R_1, R_2 and R_3 for disease resistance, after several generations of (a) outbreeding, or (b) inbreeding. Note how outbreeding leads to a greater proportion of the population carrying two or three genes as the dominant alleles, and therefore a higher level of disease resistance in the population.

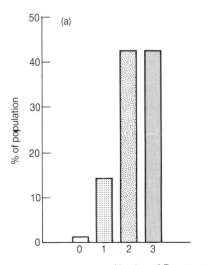

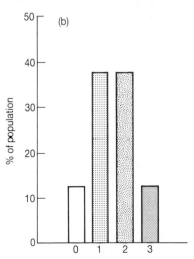

Number of R genes present as dominant allele

reductions will also occur in other aspects of plant vigour.

The plant breeder must therefore be careful when performing selection in an outbreeding crop not to remove so many individuals that the remaining ones are forced either to self-pollinate or to cross-pollinate with close relatives. As a general rule, twenty-five plants in a population are the absolute minimum to avoid inbreeding depression setting in. Varieties of outbreeding crops such as sugar beet or field beans tend to be compromises between the need to minimize genetic variability so as to give a uniform, predictable crop, and the need to preserve some genetic variability in order to prevent inbreeding depression.

There are two reasons why naturally inbreeding species, or species which in the course of domestication as crops became inbreeding, do not appear to show inbreeding depression. Firstly, natural or artificial selection over many generations has tended to purge the species of deleterious recessive genes, leaving only individuals which are homozygous for the dominant alleles of these genes. Secondly, it may not be absolutely true that these species do not show inbreeding depression. There is evidence for this in the fact that when inbreeding species are made heterozygous by hybridization of inbred lines, the progeny show increased vigour and yield. This interesting and useful phenomenon is called *hybrid vigour* and is playing an increasingly important role in the breeding of new high-yielding varieties of certain crops.

9.3.6 Hybrid vigour

Hybrid vigour is the converse of inbreeding depression. Just as inbreeding depression accompanies homozygosity, hybrid vigour accompanies heterozygosity. Once again it is not heterozygosity itself that causes hybrid vigour. The cause lies in the reduced number of deleterious alleles of polygene systems which are present in the homozygous recessive condition.

Suppose the number of grains in a cereal ear is controlled by a polygene of five members $ABCDE$, and that there are two inbred lines of similar ear size, one with genotype $AABBCCddee$ and the other with genotype $aabbCCDDEE$. Both have three of the five genes in the dominant form. When the two inbred lines are hybridized, all the F_1 plants inherit the genotype $AaBbCCDdEe$. They thus possess dominant alleles of all five genes; the fact that the plants are heterozygous for four of these

does not affect the degree to which the genes affect ear size. The hybrid will have more grains per ear than either of its parents and this, together with other aspects of hybrid vigour, will almost certainly lead to higher yield. In one experiment with barley, for example, the F_1 generation of a hybrid of two inbred lines was 25% higher yielding than either of the parental lines. New varieties produced by conventional breeding methods seldom outyield existing varieties by more than 2–3%.

Another agriculturally useful feature of F_1 hybrids is their great uniformity, arising from the fact that they all have the same genotype. Note that this can be true only if the parental lines are entirely homozygous.

Hybrid vigour and uniformity are of such obvious value that it is not surprising to find a great deal of research effort going into the breeding of F_1 *hybrid varieties*, which consist entirely of the F_1 progeny of a cross between two distinct inbred lines. However, the difficulties of producing such varieties on a commercial scale have, up to the present time, limited the impact of this technique to only a few crops. Most of these crops are outbreeders rather than inbreeders and the main advantage of the F_1 hybrid varieties has lain in uniformity rather than hybrid vigour.

To understand some of the problems of commercial F_1 hybrid production we need to consider an outline of the breeding methods involved. First of all, how can cross-pollination be enforced, and how can it be restricted to cross-pollination between the two parental inbred lines, rather than between plants of the same inbred line?

One inbred line is chosen to be the male parent. No seed will be saved from it, but its pollen is allowed, or encouraged, to reach the stigmas of the other inbred line, which has been chosen to be the female parent. At the same time steps are taken to ensure that the female parent, from which seed will be collected, produces no pollen of its own. It therefore cannot self-pollinate and all its seed must result from cross-pollination by the male parent.

There are two principal ways in which it can be ensured that the female parent produces no pollen. The first is to remove all its stamens before they mature. This is not possible on a large scale except where the plant has unisexual flowers in separate male and female inflorescences, as in maize. Here it is simply a case of cutting off the tops of the plants, a process that can be easily mechanized. Much research has gone into the use of growth regulators and selective gametocides to inhibit pollen formation; various chemicals are now commercially available or in development.

The second way to enforce cross-pollination is to use a female parent possessing genes that prevent stamens being produced or cause stamens to be aborted. This is called *male sterility* and has been found in several species. However, where the F_1 hybrid is to be grown for its seed, as in cereals, it is essential that it should not inherit the male sterility, otherwise no seed could be produced. Thus the male parent must carry a restorer gene which, when passed on to the F_1 progeny, counteracts the male sterility gene inherited from the female parent.

Another problem in the production of F_1 hybrid varieties is the enforcement of inbreeding in both male and female parents. This is particularly difficult in naturally outbreeding species. Fortunately maize, though a natural outbreeder, is not self-incompatible, and inbreeding can be enforced simply by isolating the plants from all external sources of pollen. The inbreeding of a female parent carrying a male sterility gene is clearly impossible, since such a plant produces no pollen. This difficulty is overcome by having a set of maintainer plants which carry the restorer gene; apart from this gene they are genetically identical to the female parent plants. Exactly how this is achieved is a complex matter which is beyond our present scope. The fundamental principles of F_1 hybrid seed production are summarized in Fig. 9.21.

A single crossing of two inbred lines to give F_1 hybrids often produces seed in only small quantities, which is thus too expensive for commercial use. This has been overcome by crossing four inbred lines over two years, a technique called the double cross, which produces much larger quantities of F_1 hybrid seed. Commercial hybrid maize seed production is based on the double-cross technique.

By far the greatest contribution of F_1 hybrid varieties to agricultural production has been in maize. F_1 hybrid varieties are also commercially available in tomatoes, Brussels sprouts, onions and, in the United States of America, sugar beet and sorghum. Hybrid wheats are of increasing commercial importance now that many of the technical difficulties of producing large quantities of seed economically have been overcome.

One major difference between F_1 hybrids and conventional varieties is that the seed of an F_1 hybrid variety cannot be saved for a succeeding generation. This is because the F_2 will consist of a

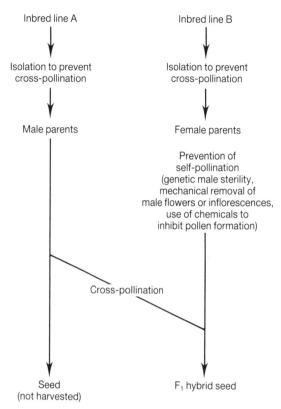

Fig. 9.21 An outline of F_1 hybrid seed production.

wide range of segregating individuals, some with good agricultural characteristics but many useless. The yield will be well below that of the F_1 and the enormous variability of the F_2 will be quite unacceptable.

9.4   Selection in crop breeding

The backbone of crop improvement is not, however, the development of specialized breeding techniques such as those that have given rise to F_1 hybrid varieties, but the process of selection from a pool of genetic variation. In modern plant breeding stations selection is a carefully programmed, rigorously scientific procedure but in essence it is the same process that has gone on over the thousands of years that man has been growing crops. Before considering some of the practicalities of present-day selection, it is worthwhile putting them into their historical perspective.

9.4.1 Crop evolution

Long before man began cultivating plants, most of the species now known as crops were already in existence and, like all species, were slowly evolving in response to natural selection. Through this natural evolution they became well adapted to the climate, soil and other features of the environment of the region where they occurred. Wild ancestors of many crops such as barley and potatoes are still to be found in their native lands and form an extremely important reservoir of genetic variability on which plant breeders can draw now and in the future. Techniques are available to transfer single genes or small groups of genes from wild species to their cultivated relatives, and these techniques have proved particularly successful in enhancing disease resistance in certain crops.

Wild plants were first taken into cultivation about 10 000 years ago in the 'fertile crescent' of the eastern Mediterranean and Mesopotamia. The essential difference between a wild plant which is gathered for food and a crop which is harvested, is the saving of seed from one year's harvest to sow for the following year's crop. Whenever seed is saved, artificial selection is practised, whether the grower intends it or not.

In the very early stages, selection was probably entirely unintentional. The more productive plants would probably supply more seed than those less productive, thus yields would tend to increase over the centuries without any conscious effort on the part of the growers. Later, seed from the best plants was consciously selected to provide next year's crop.

This selection eventually gave rise to crops which differed quite markedly from their wild relatives. When a crop reaches this stage of its evolution it may be said to be *domesticated*. Often the genetic changes accompanying domestication are such that the species is no longer able to survive in the wild. For example, domesticated cereals, peas and beans do not shed their seeds at maturity as their wild relatives do, but retain them even after they are fully ripe. This trait is essential if heavy losses are to be avoided during harvesting but it makes the domesticated plant unfit to survive in the wild for more than a few generations. Only if the seed is harvested and deliberately sown can such a plant survive.

Wheat, peas and flax were all domesticated in the Middle East by 8000 years ago, and within the next thousand years barley also was domesticated in the same region. Meanwhile, in Peru and Mexico runner beans had been domesticated by 8000 years ago and maize by 7000 years ago. Other crops domesticated at various times up to 2000 years ago include rye, lucerne (alfalfa), turnip and radish in the Middle East, oats in Europe, rice in south-east Asia, soya beans in China and potatoes in the South American Andes. Later domestications up to 250 years ago include swede, hop and raspberry, all in Europe. Crops which have been even more recently domesticated include strawberry and sugar beet. Grasses and clovers grown for forage have been subject to deliberate selection only for the past hundred years or so and are still barely distinguishable from their wild progenitors. They can hardly be said to be domesticated.

Once domesticated, crops were taken to new lands by migrating people and there they continued to evolve, becoming adapted to local conditions. In this way distinctive *land races* came into being. Until the mid-nineteenth century agriculture throughout the world was based almost entirely on these land races, but their disappearance *en masse* from the modern agricultural scene has been recognized as a major cause of loss of potentially valuable genes.

Even in crops which are almost entirely inbreeding, land races contain a considerable amount of genetic variation. A land race of barley, for example, may be thought of as a mixture of inbred lines, with a small amount of heterozygosity maintained by occasional cross-pollination between different lines. The essential contribution of modern scientific plant breeding has been to select from land races particular lines that represent the best that these land races have to offer. With hybridization, breeders have also been able to combine in a single variety good features derived from more than one land race.

Modern varieties are far more uniform than the land races they have replaced. This has great advantages, as has been seen, for ease of growing and harvesting and for predictability of field performance and quality of produce.

There are, however, drawbacks which become particularly evident when the crop is attacked by disease. Just as natural plant populations show genetic variation, so do fungi and the other agents of crop disease. When a particular strain of a fungus infects a crop, the plants which are more susceptible to that strain will suffer greater loss or damage than the more resistant ones. If the crop has very little genetic variability, all plants are likely to be equally susceptible or resistant, but it is unlikely that they will be resistant to all strains of the fungus. Thus if an infective strain does strike, all plants may be attacked and losses will be very heavy. Some interest is being shown in the growing of mixtures of varieties to combat this problem. Efforts are also being made to produce varieties homozygous for characters where this is desirable, such as height, but heterozygous for other characters including some forms of disease resistance.

9.4.2   Genotype-environment interaction

Some of the difficulties facing the plant breeder in trying to select the best plants from a range of variation will now be considered. One or two of these difficulties have already been alluded to. For example, much of the variation in a population of plants is environmental and thus not heritable. This is particularly true of the very characters in which plant breeders are most interested, such as plant height, yield and chemical composition. As was seen earlier, progeny testing helps in the task of selection where this is a problem. Another difficulty mentioned previously arises from quantitative inheritance, which means that arbitrary standards have to be set and large numbers of laborious measurements have to be performed.

A third problem in selection is that the best genotype for one agricultural situation may not be the best for another. Consider, for example, two barley varieties A and B. At low levels of nitrogen fertilizer application A outyields B but at higher levels the reverse is true (Fig. 9.22). This arises because A has greater tolerance of low fertility, but B shows greater capacity for response to increased fertility. This is an instance of *genotype-environment interaction*, a situation where the relative performance of two or more genotypes differs in differing environments.

The Scotch yellow turnip varieties Findlay and The Wallace offer another example. Findlay was bred by selection from The Wallace of individuals showing particularly good resistance to club-root disease. In the presence of the organism which causes club-root, Findlay outyields The Wallace by virtue of its greater resistance to the disease. Where the organism is absent, or present only at a low level, The Wallace outyields Findlay. Findlay's disease resistance is of no advantage in this situation and some (probably unrelated) deficiency in Findlay is shown up.

Because of genotype-environment interaction

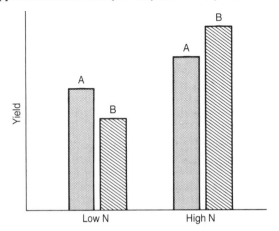

Fig. 9.22 An example of genotype-environment interaction. At low nitrogen fertilizer application rates variety A yields more than variety B. At high nitrogen application rates variety B outyields variety A.

breeders must be careful to perform selection in a range of environments. There is otherwise a danger that they may select types which perform well only under the particular conditions in which they are working.

9.4.3   Simultaneous selection of characters

Yet another difficulty for the breeder arises through the necessity to select simultaneously for several characters. Suppose that for each character the breeder wishes to reject 90% of plants and retain only the best 10%. If he is selecting for two characters, he will require 100 plants in order, finally, to have a single one. For three characters, he will need 1000 plants, and for four, 10 000 plants. This is the largest number that would be manageable in practice.

Commonly, however, the number of characters examined in a selection programme is in the range 10–20. The astronomical number of plants that would be required makes it impossible to reject 90% of plants on the basis of each character. Perhaps only 50% of plants can be rejected for each character. Consequently many plants which are far from ideal have to be retained. It is likely, therefore, that there will, for the foreseeable future, remain scope for improvement in crop varieties.

Very often, two or more of the characters in which the breeder is interested tend to be associated with one another. For example, straw strength in barley (important in the prevention of lodging) is generally accompanied by increased strength in the ear (helping to prevent pre-harvest shedding of grain). Similarly, in swedes high dry matter content, and therefore feeding value, is correlated with frost hardiness.

These are examples of positive correlation, where the task of the plant breeder is simplified, since he needs to select only for one of the characters and the other will automatically follow. Many correlations, however, are negative. For example, high grain yield is associated in wheat with low protein content. In this case selection for one character tends to favour rejection of the best plants in respect of the other character. This makes the breeder's task more difficult.

Correlation of characters arises from two causes. The first is linkage, where the genes controlling the two characters are located close together on the same chromosome. As was seen earlier, recombination is possible between linked genes, so that correlation resulting from linkage can, if desired, be broken.

The breeder cannot, however, break a correlation arising from the other cause, which is where a single gene or set of genes, acting at some fundamental biochemical level, affects a number of measurable characters. This is known as *pleiotropy*. An example is yield and dry matter content in potatoes; because the potato plant can only photosynthesize a limited amount of carbohydrate to put into its tubers, the more and the larger tubers it produces, the less carbohydrate and the more water will be contained in a given weight of tubers. It will always be true, therefore, that the highest-yielding potato varieties will tend to have the wettest tubers, while the highest quality, high dry matter tubers will tend to come from relatively low-yielding varieties.

9.5 Polyploidy

Up to this point all vegetative plant cells have been considered to be diploid, having two homologous sets of chromosomes in the nucleus. In some species, however, it is common for plants to have more than two sets of chromosomes in each nucleus. Such plants are said to be *polyploid*. If the normal haploid chromosome number for the species is n, a plant with $2n$ chromosomes in each vegetative nucleus is, as we know, *diploid*. One with $3n$ is *triploid*, one with $4n$ *tetraploid*, one with $5n$ *pentaploid*, one with $6n$ *hexaploid* and so on. Where there is an odd number of chromosome

sets, as in triploids and pentaploids, reproductive fertility is usually seriously impaired. In many cases the plant is completely sterile. This is because during meiosis the normal pairing of chromosomes cannot properly take place since one set will have no homologous set to pair with, and gametes fail to form.

Triploids occasionally arise naturally when for some reason an ovule does not undergo meiosis but instead produces a diploid egg cell. Fertilization of this egg cell with a haploid male gamete brings into being a triploid individual. Triploids can also result from the natural or artificial crossing of a diploid plant, which produces haploid gametes, with a tetraploid which produces diploid gametes.

Triploid plants often show vigorous vegetative growth and in some species produce excellent seedless fruit. All cultivated bananas, for example, are triploid, and must be propagated vegetatively as they bear no seed. Some sugar beet varieties are triploid and these tend to be higher yielding than diploid varieties. They produce little or no seed; each generation of a triploid variety has to be made afresh by the hybridization of a diploid and a tetraploid parent, using methods similar to those employed for the production of F_1 hybrid seed.

A simple doubling of chromosome number, where all four sets of chromosomes are entirely homologous, gives rise to an *autotetraploid*. This sometimes happens naturally in a diploid plant by failure of chromosomes to separate during mitosis. A single daughter nucleus is formed with four sets of chromosomes; any cells subsequently produced by normal mitosis from this nucleus will also have tetraploid nuclei. The chemical colchicine can be used to generate autotetraploids artificially; it acts by disrupting spindle formation during mitosis.

Because autotetraploids have four homologous sets of chromosomes instead of two, pairing of chromosomes during meiosis is frequently upset by 'competition' among homologous chromosomes. As a result, autotetraploids generally show reduced reproductive fertility compared with diploids, although they are not usually completely sterile as is common in triploids. The large nuclei of autotetraploids are frequently associated with large cells, giving rise to larger, more succulent plants than the corresponding diploids. Autotetraploid varieties of ryegrasses and red clover are, in general, more palatable and digestible than diploid varieties, but their increased succulence leads to greater susceptibility to frost and disease.

There is another type of tetraploid, in which two

sets of chromosomes are derived from one parent and two from another, rather distantly related, parent. This is an *allotetraploid*. In many cases the two original parents belong to different species and the allotetraploid is an *interspecific hybrid*.

Normally interspecific hybrids which are diploid are of low reproductive fertility, because the chromosomes are not entirely homologous and fail to pair properly in meiosis. If chromosome doubling takes place in a diploid interspecific hybrid, whether naturally or following colchicine treatment, an allotetraploid is produced. Allotetraploids are usually fertile, often no less so than diploids. Each chromosome has only one partner that it can properly pair with in meiosis, so that gamete formation is not inhibited as in autotetraploids.

Some allotetraploids are grown as crops; an example is oilseed rape, which is fertile enough to be grown for its seed. It belongs to the same allotetraploid species as the swede, and this species arose naturally as an interspecific hybrid between kale and turnip. Radicole is an example of an artificial allotetraploid produced by crossing kale with radish. It is a forage crop combining some of the characteristics of its rather distantly related parents.

Sometimes an allotetraploid crosses with a diploid and a further chromosome doubling takes place to give rise to an *allohexaploid*. Such a plant has six sets of chromosomes, two derived from each of three separate parents. Allohexaploids, like allotetraploids, are usually fully fertile. Two outstanding examples are oats and bread wheat. Both have diploid and allotetraploid relatives which are or have in the past been cultivated as crops. In general the allohexaploid species are higher yielding and produce higher-quality grain for the intended purposes than either diploids or allotetraploids.

Polyploidy has not, however, produced any improvement in barley, all commercial varieties of which are diploid. The plant breeder finds it very difficult to predict the likely performance of a polyploid – there appear to be no general rules applicable to all species.

9.6   Transformation technology

Until now, crop improvement has been limited by the range of genes occurring naturally in a crop species, that is, the natural *genome* of the species.

Some minor exceptions to this have been mentioned, for example the artificial creation of interspecific hybrids such as radicole (Section 9.5) and the use of chemical mutagens or short-wave radiation to produce mutated genes (Section 9.2.5). These techniques have not yet had broad impact on agriculture. Now, however, a rapidly emerging series of techniques promises to release crop improvement from the limitation imposed by the natural genome of a crop. These techniques could, within a few decades, make as much difference to crops as the entire history of domestication and improvement to date. Collectively known as *transformation technology*, the new techniques are also sometimes referred to as 'recombinant DNA technology' or, more figuratively, as 'genetic engineering'.

9.6.1   What transformation technology can do for plant agriculture

It will be recalled that the DNA present in a plant, animal or microbial cell provides instructions for the assembly of proteins. A particular segment of DNA, that is, a gene, carries the information needed to make any number of identical protein molecules. If the protein in question functions as an enzyme, then that DNA segment determines whether or not a particular metabolic reaction can occur, and if it does occur, how efficiently.

Transformation technology enables DNA from almost any source to be inserted into the genome of a plant, animal or microbial species, thereby providing the instructions for making enzymes or other proteins that are totally foreign to that species in nature. The possibilities this opens up for plant agriculture are virtually limitless, as can be seen from a few examples.

The earliest examples of transformation technology to have an impact on plant agriculture have arisen from the insertion of genes not into crop plants themselves but into bacteria associated with those plants. It is known that frost damage to a number of crops, including potatoes, often results from the presence of bacterial cells on the leaf surface. These cells contain certain proteins which act as nucleation centres for the formation of ice crystals. Genetically transformed bacteria, which produce a protein that does not encourage the formation of ice crystals, can be sprayed on to frost-sensitive foliage, where they rapidly multiply and displace the natural bacterial population. As a result, if frost occurs the leaves of the crop are much less likely to suffer injury from the formation of ice.

A second example, of potentially much wider application, relates to a bacterial species called *Bacillus thuringensis*, or B.t. for short. B.t. produces a natural insecticide, which is toxic to a broad range of insect pests of agricultural importance. Segments of DNA which provide the information needed for production of this insecticide have been identified in the B.t. genome, isolated and inserted into the genome of certain bacterial species which naturally occur in soil and which tend to associate with the roots of crop species. These bacteria with the B.t. gene inserted can be applied to the crop seed prior to sowing, and as they multiply around the growing plant they protect the roots of the crop from attack by certain insect pests. The insecticide they produce is located precisely and exclusively where it needs to be in order to give this protection. In this case transformation technology has provided a highly efficient and environmentally safe delivery system for a natural insecticide.

Another arena in which microbial transformation promises to benefit plant agriculture is in the development of more efficient strains of nitrogen-fixing *Rhizobium* bacteria for inoculation of the seeds of leguminous crops such as soya bean. Research is also going on into genetically engineered microbes that could fix nitrogen in association with non-leguminous crops such as maize and wheat, greatly reducing their requirement for applied nitrogen fertilizer.

Concern has been expressed in many quarters that the release of genetically engineered microbes into the environment may have some undesirable and irreversible effect, which at present cannot be foreseen. While this concern can often be dispelled by thorough explanation of the facts, it is true that once any microbe is released into a situation where it can naturally multiply and spread it is impossible to bring it back into 'captivity'. For this reason it is likely that the greatest impact of transformation technology on plant agriculture will be through the genetic engineering of crop plants themselves rather than of microbes.

Among the probable applications of transformation technology to crop improvement, changes in the chemical composition of the harvested produce rank highly. Demands of consumers and processors are steadily becoming more and more stringent (see Chapter 12) in relation to nutritional quality and flavour; examples of products that would command a high price are tomatoes that keep their flavour after chilled storage and transport, or soya beans which provide oil with a lower content of saturated and higher content of unsaturated fatty acids than present varieties (Section 12.1.6). While some of these changes are possible through conventional plant breeding, they can probably be achieved much faster through the use of transformation technology. In principle, a plant can be engineered to yield almost any natural product, including those far beyond the scope of conventional breeding. On the horizon is the transformation of oilseed rape to produce raw materials for biodegradable plastics; some visionaries foresee sperm oil being harvested not from whales but from the cornfields of Iowa.

Transformation technology may also provide the means to adapt crops to harsh environmental conditions, particularly those causing severe yield losses through water stress (see Sections 2.5 and 2.6). One approach is to identify, isolate and insert into crops a gene for an altered ribulose diphosphate carboxylase enzyme in which the side reaction that gives rise to photorespiration in C_3 plants (Section 1.6.5) is reduced or eliminated. Without this side reaction, the enzyme would be capable of fixing carbon dioxide more efficiently. This in turn would permit photosynthesis to take place with the stomata closed for a greater part of the day, thereby maintaining to some extent the productivity of the crop under the water stress conditions that cause stomatal closure. It should be emphasized that no gene for such an altered ribulose diphosphate carboxylase has yet been identified and that it is by no means certain that the elimination of photorespiration is possible or even desirable in C_3 plants.

Much of the effort in the application of transformation technology to crops is going into pest and disease resistance. DNA segments that code for natural bactericides, fungicides, nematicides and insecticides or for enzyme systems that produce them can be inserted into plant genomes just as they can into microbial genomes. Thus a plant can be engineered to possess a very effective chemical defence mechanism against attack by specific pests or pathogens. It is predicted that genes will to a considerable extent replace applied pesticides as a means of protecting crops from pests and diseases. They will also provide a means of protection against certain pathogens, including viruses, for which no chemical control method now exists.

What transformation technology will probably

not be able to do, at least in the foreseeable future, is provide an alternative to the use of herbicides for weed control. While a pest- or disease-controlling chemical can be delivered to its target via the crop plant, the target for a weed control compound is at some distance from the crop plant. However, while genes inserted into crops for the production of herbicides are unlikely to provide satisfactory weed control, genes providing tolerance of applied herbicides are expected to lead to new ways of controlling weeds selectively in crops.

Already a number of crop species have been engineered for tolerance of various herbicides not normally selective in those crops. For example, tomato, cotton, soya bean and oilseed rape plants have been produced which show high levels of tolerance of glyphosate, a herbicide which is normally extremely injurious to all these species. Because of the broad spectrum of annual and perennial weeds which can be controlled with glyphosate, and the high degree of environmental safety of this compound, new alternatives are being opened up for weed control in a range of crops where at present it cannot be used. Other herbicides for which genetically engineered crop tolerance has been reported include imidazolinones, sulfonylureas, bromoxynil and glufosinate. The tolerance mechanism varies. In some cases the inserted gene codes for an enzyme that rapidly degrades the herbicide to an inactive derivative. In others, it is the enzyme at the biochemical site of action of the herbicide that is altered. A herbicide normally acts by inhibiting one or more enzymes in an essential metabolic pathway; if the enzyme is modified in such a way that its essential function is unimpaired but it no longer interacts with the herbicide, then the plant possessing the modified enzyme will be tolerant of the herbicide. Thus insertion of a gene for such a modified enzyme will confer herbicide tolerance. Another way to achieve tolerance is to insert a gene that 'over-produces' the relevant enzyme, so that even in the presence of the herbicide enough enzyme molecules are still active to enable the plant to function normally.

9.6.2   How DNA is inserted into plant cells

It is beyond the scope of this book to describe in detail the methods by which particular DNA segments are identified, isolated from the genome in which they naturally occur, and introduced from one organism to another. In order to complete the picture, however, some basic techniques should be outlined.

One important technique on which transformation technology relies is *gene sequencing*, in which the sequence of bases (adenine, cytosine, guanine and thymine) on a strand of DNA can be accurately mapped. Another is the use of certain enzymes to cut a DNA strand between any base and the next one in the sequence. With these powerful techniques, and the ability to relate a particular DNA segment to a particular protein, it is possible to isolate any desired piece of DNA. To isolate a complete functional gene, it is necessary not only to include that part of the base sequence that codes for a protein, but also some additional sequences that are involved in regulating the expression of the gene.

The workhorse of plant genetic engineering is a microbe called *Agrobacterium tumifaciens*. This is a naturally occurring pathogenic bacterium which, when it infects a host plant, releases little parcels of DNA known as *plasmids* that become inserted into the plant's DNA, and are replicated along with it. *Agrobacterium* thereby subverts the plant's biochemical machinery to its own ends. Plasmids can be removed from the bacterial cell and cut open enzymically. DNA sequences responsible for causing disease in the host plant are clipped out, leaving a non-virulent plasmid. Any piece of DNA can be inserted into the plasmid, and the plasmid then returned to the *Agrobacterium* cell. The target plant is then infected with the transformed *Agrobacterium*, which replicates the plasmid and transfers the inserted DNA into the host plant's cell. In practice it is not an intact plant that is infected but a culture of cells or tissues of the target plant. Once a new plant is regenerated from a transformed cell or group of cells, by micropropagation techniques similar to those described in Section 10.2.6, all its cells should possess copies of the introduced DNA.

A particularly exciting prospect made possible by these techniques is the development of virus resistant crops. Many plant viruses consist of a segment of RNA enclosed in a protein coat (in these viruses RNA, rather than DNA, carries the genetic blueprint). It has been shown in the case of the tobacco mosaic virus (TMV), which attacks a range of crop species including tomato and potato as well as tobacco, that the presence of the virus coat protein in a plant cell provides protection against attack by the virus. The segment of RNA in the TMV genome which codes for the coat protein has been identified and sequenced. A corresponding DNA segment has been synthesized in the laboratory and inserted into a plant genome via an *Agrobacterium* plasmid. In the plant this DNA is expressed in the form of TMV coat protein, which confers protection from infection by true TMV.

Transformation technology places no restriction on the source of the introduced gene – it can come from another, unrelated, plant, from an animal or from a microbe. It can even be a synthetic DNA sequence. With the present state of the art, however, there is a restriction on the species to which the gene can be introduced. Only a few species, all of them dicotyledons, have so far been transformed via *Agribacterium* plasmids. They include cotton, tomato, tobacco, oilseed rape and soya bean, together with various vegetable crops and ornamentals. Completely different techniques will probably have to be developed in order to transform monocotyledonous crops such as wheat, maize and rice.

No alternative to *Agrobacterium* as a vector for introducing foreign DNA into plants has been identified which would enable transformation of monocotyledons. However, various vectorless techniques are being tested; some of these offer hope. They include a remarkable 'particle gun' that literally fires DNA-coated particles into intact plant cells.

In many cases it is desirable to insert a gene not into the DNA in the plant cell nucleus but into that of the chloroplasts or mitochondria, as the required protein has to function in one of these organelles. Alternatively, the gene may be inserted into the nuclear DNA but includes the DNA sequence for a 'transit peptide', a protein fragment that assists the main protein in crossing the membranes surrounding the chloroplasts or mitochondria.

Present technology allows only the addition of foreign genes to the genome of a plant, not the replacement of existing genes. This limits what can be achieved through transformation technology. Where the target is to remove some undesirable crop trait such as the presence of a toxic substance in the harvested produce (see Section 12.5), the addition of a gene may not give the desired effect without simultaneous removal of the offending gene.

9.6.3 Regulating the expression of an inserted gene

Success in crop plant transformation will often depend on an ability to regulate the expression of the inserted gene. Just as the natural genes for red colour in the tomato are expressed only in the fruit, and there only when the fruit begins to ripen, it will often be necessary for an introduced gene to be switched on selectively in particular tissues or ogans and at particular stages of development. In the case of genes providing tolerance of environmental stress, pest or disease resistance, or herbicide tolerance, the 'switching on' may need to be specific to the occurrence of the stress, pest, pathogen or herbicide. This is because at other times the expression of the gene may be detrimental.

For this reason, efforts are going on to identify DNA sequences which regulate the expression of particular genes, particularly sequences known as *promoters*, which activate or 'switch on' these genes. It may be possible to engineer a promoter that will respond to a specific chemical inducer by activating a desired gene. For example, in order to activate a gene giving improved drought resistance in a crop, a farmer of the future may spray his crop with a chemical inducer when dry weather is forecast. Similarly, when disease strikes his crop, he may apply an inducer for a natural fungicide gene rather than the fungicide itself.

Summary

1. Most plant-to-plant variation in a species is continuous. The mean value of any parameter such as height or weight in a population can be estimated by sampling, and statistical techniques can be used to measure how precise such an estimate is, as well as the degree of variability that exists in the population. Variability must be measured in field experiments set up to test new varieties, products, machines or techniques, so that variation due to treatment can be distinguished from the background variation due to factors beyond the experimenter's control.

Variation arises from both environmental and genetic causes, but only genetic variation is heritable. Increases in crop yield over the past fifty years are probably due in equal measure to a better environment provided by improved husbandry, and to genetic improvement of the crops themselves by plant breeders. Modern breeding programmes have improved not so much the potential yield of crops grown in ideal conditions, as the realization of that potential through qualities such as reduced lodging or increased disease resistance.

2. Genetic information is present in every living

cell, is transmitted by the gametes from parents to progeny, and is contained within chromosomes which become visible as rope-like bodies in the nucleus during cell division. The information is encoded as a sequence of bases on molecules of DNA. The code is copied on to strands of RNA which migrate out of the nucleus and associate with tiny bodies called ribosomes in the cytoplasm. Here the RNA base sequence dictates the order in which amino acids come together to form proteins. This in turn determines the three-dimensional shape of any protein molecule and gives it any enzyme activity it may possess. A gene is a portion of DNA coding for a single protein.

During normal vegetative cell division each chromosome splits exactly in two in such a way that each daughter cell receives an identical copy of every gene. This process is called mitosis and results in daughter cells having half the DNA complement of the parent cell. The full DNA complement is restored by the synthesis of new DNA molecules which are exact copies of the existing ones. The cell is then ready to divide again.

All cells of a plant are derived by repeated mitosis from a single fertilized egg cell and are therefore genetically identical. They have two full sets of chromosomes, one inherited from the male and one from the female gamete, and are said to be diploid. Their chromosomes can therefore be sorted into homologous pairs. Though the genes on two homologous chromosomes are not necessarily identical they do affect the same characters.

In the formation of gametes a special kind of cell division called meiosis occurs in which members of homologous pairs go into different daughter cells. Each gamete contains only one full set of chromosomes and is said to be haploid. During meiosis the members of different chromosome pairs assort independently and the resulting recombination of genes located on different chromosomes is a major cause of the genetic variability seen in progeny of the same parents. Recombination also occurs between genes on the same chromosome, as a result of crossing over.

3. Genes at the same location on homologous chromosomes may be identical (the homozygous condition) or non-identical though affecting the same character (the heterozygous condition). In the latter case they are alleles of the same gene. New alleles arise by mutation of existing alleles; most mutations are deleterious but a few confer some advantage and are the ultimate origin of all genetic variation. Usually in a heterozygous plant only one allele, the dominant one, is expressed and the other, the recessive allele, is masked. Such a plant is therefore indistinguishable in appearance or behaviour from one which is homozygous for the dominant allele.

Homozygous plants which inbreed by self-pollination are true-breeding, all the progeny being similarly homozygous for the same alleles of every gene. However, self-pollination of a heterozygous plant gives progeny half of which are homozygous and half heterozygous. Thus with repeated self-pollination the proportion of heterozygous individuals rapidly falls and inbreeding crop species such as barley, wheat or peas tend to be highly homozygous. Regular cross-pollination in outbreeding crops such as sugar beet or red clover maintains a high level of heterozygosity in the population. Hybridization of two homozygous inbred lines gives highly heterozygous progeny which, if allowed to self-pollinate, generate enormous genetic variation. This is used by plant breeders as the raw material from which new varieties can be selected combining the best qualities of both parental lines.

Most of the characters in which plant breeders are interested are controlled by polygenes, consisting of several gene pairs. This gives rise to quantitative inheritance characterized by continuous variation in the progeny and makes selection more difficult. Another consequence of polygenes is inbreeding depression, a loss of vigour in a normally outbreeding species which is forced to inbreed, leading to increased homozygosity. The converse of this is hybrid vigour, associated with the high level of heterozygosity resulting from hybridization of two inbred lines. The high yields and great genetic uniformity of such hybrids has led to their development as commercial varieties in a few crops.

4. The domestication and subsequent evolution of crops have been brought about by centuries of intentional and unintentional selection by growers. This process gave rise to land races still containing considerable genetic variation. Modern selection techniques practised by plant breeders aim at the production of varieties which are far more uniform than land races. These varieties have advantages of ease of growing, harvesting and marketing and also predictability of field performance and quality, but are vulnerable to serious damage if infected with a strain of a disease to which they are susceptible.

There are various problems in selection. Most of the important agricultural characteristics of crops such as yield and chemical composition show a high degree of environmental variation. Quantitative inheritance coupled with this means that laborious measurements must be performed on large numbers of plants. Because of genotype-environment interaction it is often necessary to select in a range of environments. When simultaneously selecting for large numbers of characters it is not practicable to reject too many plants on the basis of any one character, thus plants which are far from ideal may have to be selected.

5. Polyploidy, an increase in the number of chromosome sets in plant cells, has given improvement in some crops but not in others. Triploids, pentaploids and to a lesser extent autotetraploids show reduced reproductive fertility but may be useful in crops such as sugar beet which are not grown for their seed. Allotetraploids (e.g. oilseed rape) and allohexaploids (e.g. oats, bread wheat) are no less fertile than diploids and are among our most important seed or grain crops.

6. A series of techniques known collectively as transformation technology or 'genetic engineering' is beginning to have an impact on crop improvement. This technology enables DNA from virtually any source to be inserted into the genome of a plant, animal or microbial species, thereby providing the instructions for making enzymes or other proteins that are totally foreign to that species in nature. Major application for transformation technology are likely to be in the production of crop varieties resistant to a wide range of viral, bacterial and fungal diseases and tolerant of insect pests and herbicides.

10

Vegetative Propagation

Propagation, in its biological sense, means an increase in number, that is multiplication, of individuals of a species, accompanied inevitably by their spread over an area. Most plants achieve this by the production and dispersal of seed, and indeed propagation is one of the main functions of sexual reproduction. But in many species multiplication and spread can be accomplished by purely vegetative organs – roots, stems or occasionally leaves – and we call this *vegetative propagation*.

In its horticultural sense, propagation may mean simply the perpetuation of desirable genotypes of plant, either vegetatively or by seed. Usually, however, the horticulturist using a vegetative method wishes not merely to perpetuate a particularly fragrant rose or delicious strawberry (that is, to ensure its survival) but to multiply it. The method he uses may exploit some natural vegetative propagation system of the species in question, or it may involve artificial techniques such as the taking of cuttings.

10.1 Natural systems of vegetative propagation

10.1.1 Organs of vegetative propagation

Many plants have horizontally growing stems or roots which naturally form daughter plants at some distance from the original parent. Familiar examples include strawberry, the spider plant *Chlorophytum* and creeping buttercup, all of which produce extended stolons, often known as 'runners', from axillary buds near ground level (see Section 6.2.3). The tip of each stolon becomes the upright shoot of a daughter plant and gives rise to a mass of adventitious roots which quickly anchor it and begin to supply water and minerals to the new developing shoot. Initially the daughter is entirely dependent on carbohydrate and other nutrients translocated to it along the stolon from the parent plant, but soon it establishes sufficient photosyn-

thetic area of its own and an extensive enough root system to become quite independent. The stolon then senesces and dies and the link between parent and daughter disappears (Fig. 10.1). Artificial propagation of strawberries utilizes this natural system, the runner being severed and the daughter transplanted once shoot and root systems are well established.

Some woody plants have a readiness to form adventitious roots wherever their stems come in contact with the soil. The blackberry uses this as a natural means of vegetative propagation. Its long stems droop under their own weight and any nodes which come to rest on the soil quickly put down adventitious roots and form daughter shoots from axillary buds. Again the daughter depends on translocation from the parent until it is established but thereafter the connection can be severed with no ill effect.

We have already seen that rhizomes, for example in couch grass (see Fig. 6.19, Section 6.2.3) or coltsfoot can, like stolons, act as organs of extension and hence of vegetative propagation. The apical bud at the tip of the rhizome, or any axillary bud along its length, is capable of growing into an aerial shoot with its own adventitious root system – in effect a complete plant although it may remain for some time connected by the rhizome to other plants. Soil cultivation breaks up the rhizome and stimulates the growth of axillary buds by removing apical dominance. Any small portion of rhizome that carries an axillary bud is therefore capable of giving rise to a new plant. Thus cultivation can inadvertently multiply such weeds.

A similar system of vegetative propagation is possessed by creeping thistle but here the organs of extension are not rhizomes but horizontally growing roots. At any point along them, these roots can give rise to adventitious buds which then grow into aerial shoots. Their capacity for multiplication is even greater than that of rhizomes because the

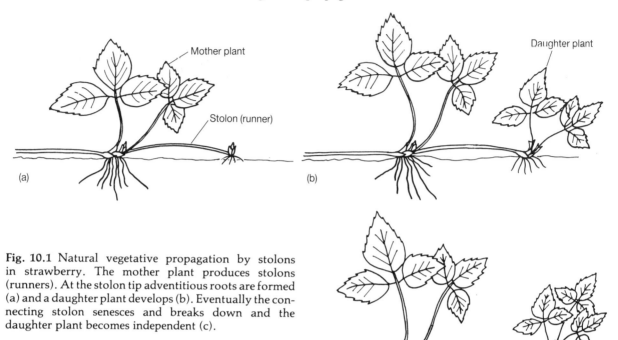

Fig. 10.1 Natural vegetative propagation by stolons in strawberry. The mother plant produces stolons (runners). At the stolon tip adventitious roots are formed (a) and a daughter plant develops (b). Eventually the connecting stolon senesces and breaks down and the daughter plant becomes independent (c).

number of daughter shoots that can arise is not limited by the number of pre-existing apical and axillary buds. Any portion of root can initiate an adventitious bud; it must, however, be large enough to contain sufficient energy reserves to enable the shoot to reach the soil surface. Raspberries and many trees produce adventitious shoots from horizontally growing roots in the same way. These 'root suckers' give rise in nature to thickets and are sometimes used in commercial propagation (Fig. 10.2).

Rhizomes and creeping roots differ from stolons in containing a built-in supply of carbohydrate and other nutrients rather than acting merely as translocation channels between parent and daughter. In most cases they are not only organs of extension and vegetative propagation but also organs of survival over unfavourable periods such as a winter or dry season. As such they usually show dormancy to a greater or lesser degree.

The natural vegetative propagation system of the potato is virtually unique, but is one which has proved highly suited to exploitation as the normal agricultural method of propagating the potato crop. The tuber is, as mentioned earlier (Section 6.2.4), the swollen tip of an underground stolon.

One potato plant may produce many tubers, each capable of growing into a new plant. After a dormant period some of the buds or 'eyes' on the tuber begin to sprout to give rise to aerial shoots. Normally apical dominance ensures that the apical bud develops first and most strongly, but if the apex of the tuber is damaged many of the axillary buds may sprout simultaneously. If a tuber is split up any portion with at least one bud can give rise to a plant. The strength of apical dominance tends to decline during a period of cold storage; tubers sprouted early tend therefore to bear fewer but stronger sprouts than tubers stored longer before sprouting. Each sprout develops its own adventitious root system and eventually becomes an independent plant.

At quite an early stage in the growth of the aerial shoot axillary buds at underground nodes grow into stolons, at the ends of which the new daughter tubers form (Fig. 10.3). The more sprouts that form on the 'seed' tuber, the more aerial shoots will be crowded together, and the more daughter tubers will be borne by these shoots. By appropriate treatment in store the number of sprouts can therefore be varied, and thus the number of daughter tubers per seed tuber. This is especially

Fig. 10.2 Daughter plants arising as 'root suckers' in raspberry; they develop from the adventitious root system of the mother plant. The daughter plants can be severed and transplanted as a means of propagation.

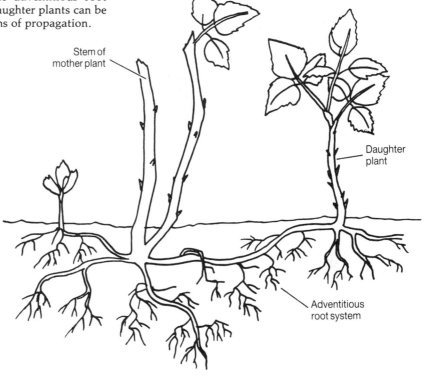

Stem of mother plant

Daughter plant

Adventitious root system

Fig. 10.3 Natural vegetative propagation in potato. Stolons arise from underground nodes below the aerial portions of the shoots and daughter tubers arise at the tips of these stolons.

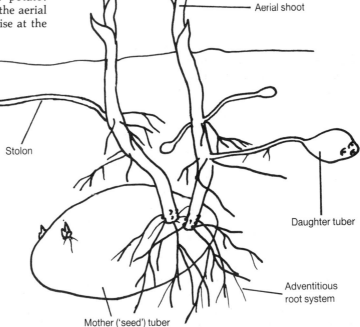

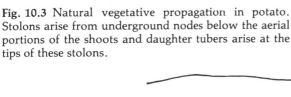

Aerial shoot

Stolon

Daughter tuber

Adventitious root system

Mother ('seed') tuber

important to growers of potato crops intended for use as seed, since they require more and smaller daughter tubers from each tuber planted than do growers producing crops for consumption.

Many equally specialized natural vegetative propagation systems have been described in other species of plant but cannot be fully discussed here. They include the bulbils of wild onion, which are miniature plants formed in place of flowers, the tiny plantlets which form along the margins of the leaf of *Bryophyllum*, and the detachable buds of many aquatic plants which root themselves in the mud at the lake or river bed.

10.1.2 Advantages and disadvantages of vegetative propagation in nature

What advantages does vegetative propagation offer a plant in comparison with reproduction by seed? Vegetative organs are in general much larger than seeds and contain greater reserves of energy. This makes for more rapid, sustained early growth and enables the young plant to become established in the face of intense competition for light, water and minerals from already established vegetation. Thus vegetatively propagating perennials can spread in dense plant communities such as grassland, which cannot easily be invaded by annuals and biennials, whose seedlings cannot tolerate the dense shade and restricted rooting space. Almost all important grassland weeds are perennials and many of them, such as creeping buttercup, stinging nettle and yarrow, spread vigorously by vegetative means.

Another advantage in the wild is that the organ of vegetative propagation will grow only in an environment that is conducive to growth. It will avoid heavily compacted, dry or waterlogged soil. It is therefore 'placed' in a location where the daughter plant has a good chance of establishing itself well – in other words, it is naturally site-selective. By contrast, the dispersal of seeds is a far more random affair.

A third characteristic is that the offspring of vegetative propagation, because they result from purely mitotic cell division, are genetically identical to the parent plant. No genetic recombination takes place. Thus a successful plant, with genetic characteristics well suited to its environment, propagates to produce equally well-adapted offspring, generation after generation.

Against these advantages must be set a number of disadvantages. The same absence of genetic recombination that allows the perpetuation of successful plants leads to reduced variability and therefore reduced adaptability to changing circumstances. So long as the environment remains unchanged, there is no problem, but even a minor change can precipitate wholesale destruction of a plant population. For example, if a plant is susceptible to a particular disease, all its offspring by vegetative propagation will be equally susceptible. This does not matter so long as the organism causing that disease does not occur locally. If, however, the organism is introduced the entire population of plants may be wiped out. There will be no resistant individuals to form the nucleus of a new resistant population. This is why the accidental introduction of a more aggressive strain of the fungal pathogen causing Dutch elm disease has so devastated the English elm trees which until recently were such a prominent feature of the English landscape. This species of elm is almost exclusively vegetatively propagated, spreading slowly by root suckering.

Vegetative propagation in nature can permit only short-range spread. Bracken, for example, spreads by means of its rhizomes at a rate of only about 2 m per year even on the best sites, and on most hill land where it occurs its spread is considerably slower than this. Finally, organs of vegetative propagation are normally produced in much smaller numbers than seeds, hence the capacity for multiplication is in most cases much lower with vegetative propagation than with seed production.

10.2 Vegetative propagation in agricultural and horticultural practice

Crop species such as strawberries, blackberries and potatoes which have well-developed natural systems of vegetative propagation are usually propagated commercially by means of these systems. The advantages that genetic uniformity brings by way of consistency and predictability of field performance and quality of produce have encouraged the development of artificial methods of vegetative propagation for many other species, particularly those which, because of reproductive infertility or seed dormancy problems, are difficult to propagate by seed. Horticulturists have long used such methods to propagate varieties of ornamentals and fruits with rare or desirable combinations of characteristics.

10.2.1 Propagation from underground storage organs

Very often the basis of vegetative propagation is an underground storage organ which, in the case of food crops, is also the harvestable product of the crop. This applies to the potato, the tuber of which evolved partially as a propagative organ, but similar methods are used for such crops as cassava, sweet potato, Jerusalem artichoke and garlic, and for many ornamentals including tulip, narcissus, crocus and gladiolus. In these the storage organs form at the base of the shoot, not at some distance away. In the wild they are merely organs for ensuring survival during an unfavourable period, but in cultivation they are lifted, divided and replanted to multiply the plant.

Cassava and sweet potato have tuberous roots which at the start of a new growing season give rise to adventitious buds, forming the new aerial shoots. Jerusalem artichoke has tuberous rhizomes which sprout like potato tubers from apical or axillary buds. In crocus and gladiolus the base of the vertical stem becomes swollen; this swollen organ, known as a *corm* (Fig. 10.4), is used in propagation. Both corms and the bulbs of tulips, narcissus and garlic produce side-shoots that form small corms or bulbs known as *offsets*, which can be separated from the main shoot and planted individually.

10.2.2 Propagation from basal branch shoots

As was seen earlier in this chapter, many plants bear large numbers of axillary buds at or near ground level. Branch shoots that grow from these buds often develop adventitious roots, making them semi-independent of other shoots from the base of the same plant. This is the case, for example, with the tillers of grasses. Basal shoots with adventitious roots can be readily separated from the main stem. Known variously as slips, suckers or ratoons, these shoots are planted out individually, thus serving as a means of vegetative propagation. This is the principal commercial technique for the propagation of raspberries (Fig. 10.2) and of a number of tropical crops including bananas.

10.2.3 Rooting

In all of the examples we have considered so far, the organ used for propagation either possesses its own root system or produces adventitious roots in the natural course of its development. Other techniques of vegetative propagation rely on the artificially induced formation of adventitious roots on portions of shoot which are not naturally adapted as organs of survival or propagation. This is known simply as *rooting* and its success or failure is

Fig. 10.4 Crocus corm: external appearance (a) and in longitudinal section (b). A new corm is produced above the old one each year.

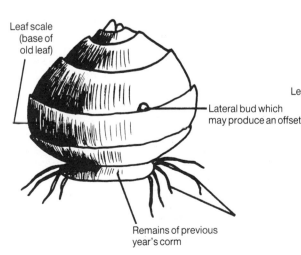

Leaf scale (base of old leaf)

Lateral bud which may produce an offset

Remains of previous year's corm

(a)

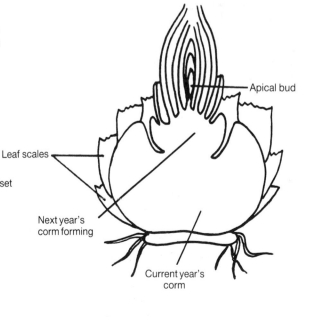

Apical bud

Leaf scales

Next year's corm forming

Current year's corm

(b)

the main factor determining the ease of propagation of a species by these techniques, some of which are illustrated in Fig. 10.5.

Some of these techniques depend on the tendency of stems of certain species to form roots when they are embedded in soil. This happens naturally in blackberries and is used in the artificial propagation of that crop. But branch stems of many other woody species can be bent down and pegged into the ground, where adventitious roots develop. Later the stem is cut between the rooting point and the parent plant, leaving the remainder of the shoot as a new and independent plant. This technique is called *layering*.

It is generally the humidity of the soil that induces rooting. A portion of stem placed in any humid environment will often form adventitious roots. Sometimes wet moss, peat or other medium is placed around the base of a branch shoot and enclosed in polythene while the shoot remains in position, above ground, on the plant. Roots grow out into the moist medium. The shoot may then be cut off the parent plant and transplanted with its new root system into soil. This is the technique of *air layering*. It is similar to ordinary layering in that rooting is induced before the shoot is cut off the parent plant.

These techniques are too demanding of labour to be generally suited to large-scale propagation. More commonly the shoots which are to form new plants are cut off the parent first, and then rooted. These shoots are known as *cuttings*. Cuttings of some species, such as willow, root with remarkable ease when the base is inserted into moist soil. Others are slower to root but the process can be accelerated by dipping the base of the cutting in a solution, paste or powder containing an auxin such as IBA or NAA (see Section 8.7.1) before planting it in the soil. Cuttings of some species, including most conifers, do not root at all and cannot be induced to do so with auxins or other growth regulators.

Fig. 10.5 Some techniques of rooting. (a) Layering, in which the tip of a branch is pegged to the soil, where it develops adventitious roots. (b) Air layering, in which a ring of cuts is made in the bark of a shoot (i), the cut area is wrapped in moss or other water-retaining material and sealed in a polythene wrapping to encourage development of roots (ii), and finally the rooted shoot is cut off the parent plant and is planted in soil (iii). (c) Cuttings, which are removed from the parent plant (i) and placed in sand or light soil in warm, humid conditions (ii) to promote rooting (iii).

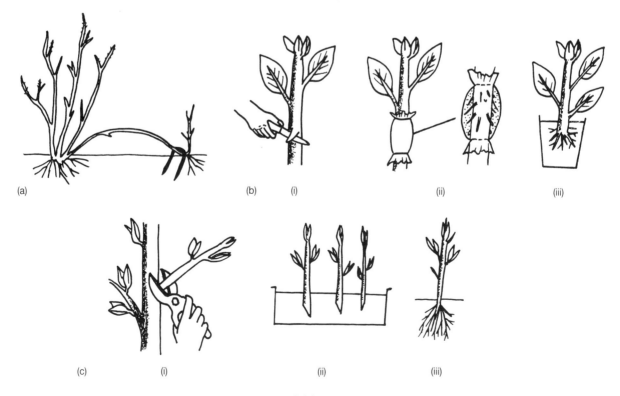

(a) (b) (i) (ii) (iii)

(c) (i) (ii) (iii)

10.2.4 Grafting

Instead of being rooted itself, a shoot cutting may be joined to the base of another plant which already possesses a root system (Fig. 10.6). The joining of parts of plants to promote a physical union between them by growth of tissues across the join is called *grafting* and the join a *graft*. The portions grafted together function as a single plant, but there is no interchange of genetic information between them. Cuttings taken later from either of the components of the grafted plant possess no characteristics of the other component.

The lower portion of a grafted plant, the portion bearing the roots, is known as the *stock* or *rootstock*. Usually it consists of a small piece of stem together with an entire root system but in some cases it may be no more than a small piece of root or no less than a whole plant. The upper portion, which is grafted on to the stock, is called the *scion*. This may be a stem cutting, but very

often it is no more than a bud with a tiny portion of stem tissue attached, in which case the procedure is termed *budding* rather than grafting. Exactly the same principles are, however, involved whether the scion is a stem cutting or merely a bud.

Grafting or budding is a standard procedure in the raising of fruit trees, vines, and many ornamental trees and shrubs, particularly roses. It is performed for a number of reasons. Sometimes it is no more than a means of vegetative propagation, allowing many buds or cuttings from a single desirable plant to be grown on separate stocks. The large root system of the stock enables more rapid growth of the scion than would be possible by growing it on its own as a rooted cutting. It is, however, an operation demanding skilled labour and is more expensive than other conventional propagation techniques.

Grafting also provides a way of combining valuable root and shoot characteristics in a single plant. A particularly attractive rose variety, for example, may have roots which grow poorly or which are highly susceptible to attack by pests and diseases. By grafting or budding it on to a vigorous, disease-resistant stock the grower gets the advantage of the attractive bloom without the disadvantage of poor or unreliable growth. Some specialized uses of grafting are shown in Fig. 10.7.

Probably the most important role of grafting is, however, as a means of controlling the growth and development of the scion. For example, *dwarfing stocks* are commonly used in apples and other tree

Fig. 10.6 Techniques of grafting (a) and budding (b). In grafting, the cut faces of scion and rootstock are placed in close contact (i) and bound tightly together until union occurs (ii). When the binding is removed, a slight swelling may mark the position of the graft (iii). (b) For budding, a T-shaped cut is made in the bark of the rootstock (i). A bud is cut from the scion, always with a piece of cambial tissue at its base (ii). It is then slotted into the cut in the rootstock and bound in position until union occurs (iii).

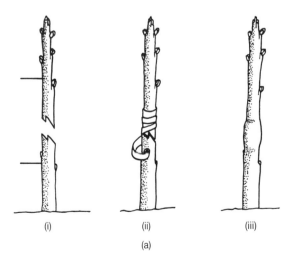

(i)   (ii)   (iii)

(a)

(i)

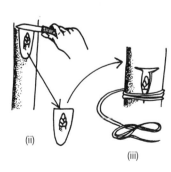

(ii)

(iii)

(b)

Fig. 10.7 Some specialized uses of grafting other than for propagation. (a) Two new rootstocks grafted on to an old scion to preserve it after damage to the original rootstock. (b) Several scions grafted on to an established fruit tree, to enable several different varieties of fruit to be borne on one tree (courtesy of Scottish Agricultural College).

fruits. These retard extension growth in the scion and produce smaller trees. Although the yield of fruit per tree is generally reduced, the yield per hectare may be substantially increased because of the closer spacing of trees that becomes possible when the trees are smaller. Another valuable effect of dwarfing stocks is that they promote the bearing of fruit earlier in the life of the tree. This gives an earlier return on the considerable investment involved in establishment of an orchard. Apple trees on such stocks may come into fruit in as little as two years after grafting, whereas on non-dwarfing stocks it may be up to ten years before any fruit is borne. Delay of returns on investment is a major problem in the financing of all plantation crops.

The dwarfing effect may result from restricted translocation, particularly in the phloem, across the graft. If non-grafted trees have their phloem disrupted by ringing or girdling, they often show dwarfing and early fruiting very similar to that induced by dwarfing stocks.

The graft union is formed by the intermingling of callus tissue produced by both stock and scion in response to wounding – in this case, cutting. As we saw in Section 4.7, wound callus is formed from the vascular cambium, and plants such as monocotyledons which lack cambium cannot therefore be grafted. In budding it is essential that the small portion of stem removed with the bud contains cambium.

The callus tissue is initially undifferentiated but soon cambial cells arise in the callus adjacent to the cambia of stock and scion. These cells then divide to produce phloem towards the outside and xylem towards the inside (Fig. 10.8). The new vascular tissues join those of the stock and the scion, and form the link on which the success of the graft depends. All materials which naturally move in the xylem or phloem, including, it should be noted, hormones and viruses, can cross the graft. To ensure a successful graft the cambia of stock and scion must be placed close together. The graft is bound with tape to ensure that close contact is maintained.

It is not necessary that stock and scion should be genetically identical. Plants do not have antibody defence mechanisms like those of animals, where tissue rejection is a major difficulty in skin grafts and organ transplants except between identical twins. On the other hand, the stock and scion must be genetically similar, either belonging to the same species or to closely related species, otherwise they

Fig. 10.8 A graft union. This consists initially of callus tissue in which new cambial cells form adjacent to the vascular cambium of stock and scion. Later these new cambial cells differentiate to give phloem (P) and xylem (X), which then link the vascular tissues of the scion with those of the stock.

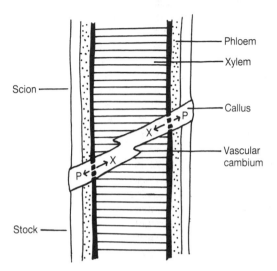

will fail to unite satisfactorily. Roses, for example, are generally budded on to stocks of wild rose species such as *Rosa canina* or *Rosa rugosa*. Similarly, apple varieties are often grafted on to crab apple stocks. Pears, rather exceptionally, are usually grafted on to stocks of quince, which belong not only to a different species but to a different genus from the scion. Plants which cannot form a good graft together because they are too distantly related or for any other reason, are said to show *graft incompatibility*.

In the extreme form of graft incompatibility, the scion dies. But other less extreme forms occur, in which there is weak or abnormal scion growth, excessive growth of callus, or poor mechanical strength of the graft. Graft incompatibility may arise from causes other than genetic dissimilarity, for example the transmission of a virus from a tolerant stock or scion to the other component, which is not tolerant of that particular virus. The dwarfing effect of certain stocks used in fruit tree grafting may result from partial incompatibility.

Sometimes partial incompatibility arising from genetic causes can be overcome by inserting a small portion of stem, with a graft at each end, as a bridge between stock and scion. This bridge, or *inter-stock*, must be fully compatible with both stock and scion even though stock and scion are incom-

patible with one another. The use of interstocks is a common practice with pear-quince grafts.

Other examples are known of more than two different plant varieties being grafted together. Any number of varieties of apple, for instance, may be grafted on to one rootstock to provide a diversity of fruits from a single tree.

Grafts between fully compatible stocks and scions may still fail to 'take' for a number of reasons. We have already seen the importance of close contact between the cambium of the stock and that of the scion has already been noted. Successful grafting also depends on environmental conditions. Callus develops most rapidly at fairly high temperatures, around 25–30°C, and at high relative humidity. High humidity is also desirable to prevent the scion drying out while the graft union forms. Sometimes a wax coat is put around the graft to maintain high humidity, but this may restrict the supply of oxygen which is necessary for respiration and therefore callus growth. A good oxygen supply to the graft is particularly important in the grafting of vines.

10.2.5 Advantages of vegetative propagation in agriculture

Vegetative propagation has certain advantages for the grower, and equally it has certain disadvantages and limitations. Obviously, as traditionally carried out, it cannot be used for annuals or biennials, but only for perennials.

Because the offspring of vegetative propagation are genetically identical to their parents, the quality of tuber, fruit or other harvested part is both predictable and uniform. This is particularly important where the produce must meet certain standards of size, shape, colour or other characteristics before it can be marketed. Most perennial crops, including almost all fruit crops, are outbreeding, so that offspring derived from seed show a wide range of variation from the parent plant and are nearly all inferior to it in quality of produce. Varieties of such crops originate by careful or chance selection from a very large number of plants grown from seed. In this situation, once a suitable offspring has been selected, vegetative propagation is the only way its characteristics can be reliably reproduced through successive generations.

Other advantages of vegetative propagation derive from the fact that the organ used, being large relative to a seed, more quickly generates a sizeable plant. This can shorten the period of growth between planting and harvesting and is a particular advantage in tree crops where the grower may have to wait for years before the first harvest. In some major herbaceous perennial crops like potato, cassava and banana it takes two or more years to produce a full-sized plant from seed, yet they must provide a harvest every year. Only propagation from tubers, roots or other large vegetative organs allows this to be done. Quick establishment also increases the ability of the plant to compete with weeds, especially in the early stages of growth when crops are most vulnerable to weed competition. This is a particularly important advantage in the tropics and is one reason why so many tropical crops are vegetatively propagated.

10.2.6 Rapid multiplication techniques

Very few plants produce more than a small number of vegetative propagules. Thus a major disadvantage of vegetative propagation, until recently, was that it permitted only a slow rate of multiplication. This aspect merits a little more attention, particularly in the light of recent developments, as an increase in the rate of multiplication of a crop is an important agricultural advance.

A given size of plant may be propagated by dividing it into a number of cuttings. The only way to derive more plants from it in a single step is to divide it into smaller pieces. The process can be carried to considerable lengths, for example by cutting each stem into single node sections. But eventually this presents problems. The cutting must develop roots and leaves, and this requires nutrients and energy over a period of time during which water loss is a major hazard. Nutrients, energy and water must be drawn from the cutting's own stores. If the propagule runs out of any of these before its new leaves and roots can provide renewed supplies, it dies. There is thus a conflict between the need to use smaller plant parts to increase the rate of multiplication and the chances of successful establishment.

Some traditional techniques, such as mist propagation, in which cuttings are frequently wetted by a fine spray, in part resolve this conflict by reducing water losses by transpiration until roots develop. The taking of root rather than stem cuttings can also help. Root cuttings of plants such as horseradish and phlox, placed in moist soil, lose little water until the shoot develops, at which time they start to draw water and minerals from the soil.

An instance of a more recent technique which permits rapid multiplication is provided by the narcissus crop. Bulbs are cut up into small portions each containing two thin bulb scales joined at their base by a small piece of stem tissue. Placed in moist peat in a polythene bag these 'twin scales' retain their moisture, draw on their stored reserves of nutrients and in a short time develop a miniature bulb, or *bulbil*, from an axillary meristem between the scales (Fig. 10.9). The bulbil is then planted in soil or compost and eventually develops into a mature plant.

Fig. 10.9 The rapid multiplication of narcissus bulbs. Bulbs are cut into quarters (a) which are then sliced more thinly (b). Each slice consists of sections of leaf bases joined by a piece of stem tissue. The thin slices are then cut into bits (c) each with two leaf scale sections still joined by a piece of stem tissue (d). These 'twin scales' are then placed in a moist vermiculite/peat mixture in polythene bags, after fungicide treatment to prevent them rotting, and left for several weeks until a small bulbil develops between the twin scales (e). The twin scales are then planted out and the bulbils eventually develop into mature plants with full-sized bulbs.

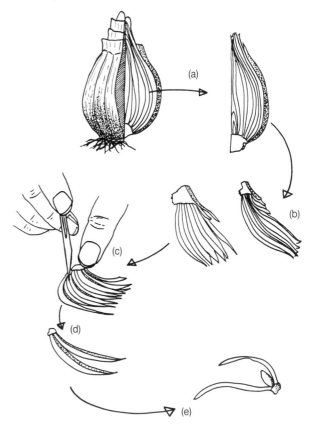

This trend towards rapid multiplication by using ever smaller pieces of tissue can profitably be carried much further. In a series of techniques known as *tissue culture*, a microscopically small piece of the parent plant containing a meristem is taken and placed on an artificial, sterile medium containing minerals, sugar and, where needed, vitamins and hormones, all suspended in a firm clear gel based on agar, a colloidal material extracted from a type of seaweed. The container, holding plant and medium, is sealed, placed under lights and kept at a controlled temperature. Since no moisture escapes, ample organic and inorganic nutrients are supplied in the medium and suitable light and temperature favour photosynthesis, even the smallest propagule need not starve or dehydrate. Precautions have to be taken to prevent moulds and bacteria invading the medium and competing for nutrients or otherwise harming the cultured propagule.

In such techniques, two basic technical problems have been overcome. The limited internal resources of energy, minerals and other nutrients within the propagule are augmented by external supplies of carbohydrate, minerals and vitamins where necessary, in the medium. These can be absorbed by a very small propagule before root development. This limiting factor of nutrition is thus removed, but it does not follow that such minute fragments of tissue can organize themselves into complete plants as they grow, developing the full range of organs, including roots. The advance of technical knowledge, allowing the judicious addition of growth hormones in the correct blend and concentration, enables this to happen, thereby removing the second major limiting factor.

Because of the genetic totipotency of plant cells, the system can be carried to its ultimate length. In theory at least, each cell has the potential to develop into a complete plant and this potential can now be exploited for the reasons explained above. Plant tissues can be broken down into their constituent cells by the use of enzymes that attack the pectin cement between walls of adjacent cells. The individual cells can then be placed in culture media in exactly the same way as whole meristems.

At first a cell in culture divides repeatedly in a disorganized fashion to produce a proliferation of undifferentiated cells called a *callus*, similar to the callus tissue formed by woody plants in response to wounding (Section 4.7). This callus formation can be observed even when meristem culture is employed. Any callus can be repeatedly subdivided to initiate new cultures. Eventually, certain cells in the callus begin to differentiate into phloem and xylem, a more ordered state of affairs develops, and ultimately a small plantlet is produced. In time, this can be planted in the soil.

This entire range of techniques, known as *micropropagation*, can be performed in small laboratories or glasshouses up to the point of transferring rooted plants to the field. Micropropagation allows very rapid multiplication from single plants in a very small area, thereby freeing sizeable areas of land for other uses, while solving the problems of nutrition, moisture loss and tissue organization in small propagules. Furthermore, since it is done indoors in highly controlled conditions it makes propagation more reliable and effectively extends the growing season into cold winters or dry seasons.

A rapid multiplication technique for virus-free potatoes (Fig. 10.10) illustrates some practical considerations as well as the value of micropropagation. From a single sprouted tuber, twenty axillary buds are dissected out and placed in a sterile culture medium for 5–6 weeks. The medium is an agar gel containing sugar as an energy source, salts as a source of essential minerals, and vitamins. Intact plants do not require a supply of vitamins as they make all they need, but tiny portions of tissue detached from a plant cannot make all their own vitamins and must have them supplied in the culture medium. No hormones are added to the medium for potato culture, but they are needed for some species.

Each bud produces a small shoot from which perhaps five buds are once again dissected. This process can be repeated indefinitely, but if done three times it yields in about 15 weeks, from a single initial tuber, about 2500 small plants. These are transferred to pots (Fig. 10.11) and are then transplanted to the field. If each plant produces 12 tubers, at the end of the first year about 30 000 tubers will be harvested. If normal commercial multiplication is then continued for a second and third year, almost five million virus-free tubers will be available for release. To produce this number from a single initial tuber by conventional methods would take at least twice as long.

It is often asserted that, since such techniques are essentially methods of vegetative propagation and only mitotic cell division is involved, the offspring are all inevitably genetically identical and the resulting plants are therefore largely uniform in growth habit and other features. Such is not necessarily the case. In the first place, not all cells of a

Fig. 10.10 A micropropagation technique for potatoes. Twenty axillary buds are cut from the sprouts of one sprouted tuber (a). Each bud is then grown on in a sterile culture medium (b) for 5–6 weeks where it develops into a small rooted plantlet (c). Each shoot is then cut into 5 sections with a bud on each (d) and these are grown on in a similar medium (e). This last stage is repeated three times. Young plantlets are then transferred from the medium to a pot containing compost (f) and then finally planted out in the field.

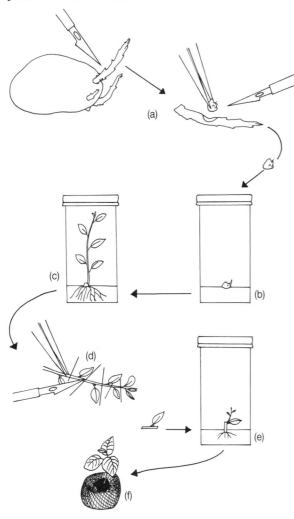

single plant are genetically identical. In some species, for example, many root cells have been observed to be polyploid, and it is very likely that such chromosomal aberrations occur to a greater or lesser extent in cells in other tissues also. Further, it now appears that in the stage of callus formation, which involves the rapid multiplication of cells,

genetic aberrations are not uncommon. It is possible that other factors also influence the situation. Not surprisingly, therefore, it is often observed that genetically atypical plants frequently arise in the micropropagation of plants. These must be eliminated or, in the rare cases where the variant has more desirable features than the norm, selected out.

10.2.7 Plant health problems in vegetative propagation

Another set of problems inherent in vegetative methods of propagation is concerned with the spread and carry-over of plant diseases. Because the offspring of vegetative propagation are genetically identical (once aberrant types have been removed), they are all equally resistant or susceptible to particular diseases and pests. A pathogen able to overcome the defences of one plant can spread through all plants in the crop easily and rapidly. In annually grown, vegetatively propagated crops like potato which at least are harvested and removed from the field once a year, this problem is serious enough, but in perennial plantation crops in which a pathogen can spread continuously over the years that the crop is in the ground, it poses an even greater challenge to plant health.

We must also consider the carrying of infection by vegetative propagules. True seed often carries infection from the parent crop, but this problem is relatively easily controlled by measures for improving the health of the parent crop, by avoiding contaminated seed stocks, or by chemical or other treatments of the seed. Since seed is small in bulk this last method is relatively easy and cheap. Most importantly, for reasons not fully understood, true seed does not usually acquire any virus infections carried by the parent plant. Thus annuals and biennials grown from seed receive a yearly cleansing of viruses.

In contrast vegetative propagules such as bulbs, roots and tubers are relatively bulky and, compared with seeds, they are also succulent. For both these reasons they are much more likely to carry fungal and bacterial infection. As an example, any potato tuber dug out of a field will carry several different pathogenic microbes. Though none may be at a serious level, they are nonetheless carried to any subsequent crop, where they may increase and spread. A further consequence of the succulence of vegetative propagules is that if the material has to

Fig. 10.11 Potato plantlets produced by micropropagation.

be stored between growing seasons, some infections are liable to develop further in store and may even spread from one propagule to others. Where the propagule used is a below-ground organ like a tuber or bulb there is an additional problem of soil contamination. Pathogens and pests dormant or resident in the soil are often carried in soil adhering to organs of vegetative propagation.

These problems, where they have not been controlled, have been sufficient to cause widespread and heavy yield losses in some crops and have even threatened their continued use because of the resulting spread of particular diseases. Past examples include red core in strawberries and wart disease in potatoes.

The most serious threat to vegetatively propagated crops, however, comes from viruses. Since most viruses spread throughout the tissues of their host plants, with the exception of the seed, it follows that cuttings, tubers and other vegetative propagules obtained from any infected plant carry the infection with them. Since viruses normally spread between plants in other ways also, viral infections tend to build up in vegetatively propagated crops. The build-up can be very rapid, and new varieties as well as stocks of older varieties can quickly be rendered commercially useless if steps are not taken to prevent it. Sometimes every single plant of a variety is infected, as was the case with the potato variety King Edward VII, which at one time was 100% infected with paracrinkle virus. In the rhubarb crop until recently many useful varieties were completely infected with several viruses simultaneously.

Control of disease carry-over during vegetative propagation is obtained by a variety of techniques. Before propagating, infected plants may be eliminated by roguing, and heavily infected crops can be avoided altogether. Where infections are very widespread, propagating material may be drawn from a few selected or specially grown crops maintained at high standards of health. If even this is not available, single healthy plants may be selected and propagated by some of the rapid multiplication techniques outlined above.

In situations where every plant carries infection, none of these measures works, and the starting point has to be the infected plant.

Obtaining healthy stocks from infected plants requires different techniques. Where a fungal or bacterial infection is confined to below-ground parts, healthy plants may be obtained by propagating from stem cuttings. This is done, for example, to free potatoes from fungal infections of the tubers such as skinspot, but cannot of course be applied to most viral infections. Fortunately, even with viruses, certain courses remain open. It happens that many viruses are less resistant to high temperatures than their hosts and, in certain cases, if the infected host is grown for several weeks at a much higher temperature than normal, some plants can be freed from infection. Furthermore, meristems are often apparently free of the viruses which infect the rest of the plant and a proportion of plants grown by meristem culture are for this reason free of their parents' infection. These two techniques can be combined. In this way single plants of King Edward VII potato were freed of paracrinkle virus and the entire variety was re-established by multiplying their offspring. The result was a substantial yield increase from this variety. Many other examples could be given. Since vegetatively propagated crops include most fruit crops and a number of crops that are a major part of the staple diet of very large numbers of people, such as bananas, potatoes, cassava and yams, the challenge of producing healthy propagating stock is an important one.

In the case of important vegetatively propagated crops like potato, an entire array of disease controls is frequently combined into a single scheme that ensures the production of healthy, high-quality propagating material. These schemes use the techniques just described, often along with those of micropropagation, to produce initial small stocks of material to meet high health standards. By a combination of measures such as careful site selection to avoid soil-borne infection, and inspection of growing crops and stored produce to eliminate substandard stocks and infected plants, these health standards are then maintained at acceptable levels during the years the stocks are being multiplied up to numbers at which their produce can be used for consumption. Without such schemes, which are usually government-run, and are also used to maintain the varietal purity of stock, many vegetatively propagated crops could not provide the yields required in a developed agriculture.

10.3 The genetics of vegetatively propagated crops

Every cultivated variety of a vegetatively propagated crop such as potato, apple or strawberry is a *clone*; that is to say, every plant of that variety is derived by one or more generations of vegetative propagation from a single plant which itself was the product of sexual reproduction. Thus every plant of one variety is genetically identical with every other plant of the same variety. This leads to great uniformity in a crop – greater even than is possible with inbred lines (Section 9.3.1) or F_1 hybrids (Section 9.3.6) – and is a very useful agricultural feature.

The breeding of new varieties can be accomplished only by sexual reproduction. To breed a new potato variety, for example, it is necessary to collect the true seed (not the so-called seed potatoes which are simply tubers) from the fleshy fruit which follows the flower on some varieties. Some other varieties are never observed to set fruit in normal conditions; these pose problems for breeders wishing to use them as parents for new varieties.

Virtually all vegetatively propagated crops are outbreeding when they do reproduce sexually. If self-pollinated within a single variety, they show inbreeding depression. It is therefore important to the breeder to hybridize at least two distinct varieties to ensure a high degree of heterozygosity in the progeny. The F_1 plants are grown from seed and are subjected to selection over several years; in the case of potatoes this requires repeated cycles of vegetative propagation. Ultimately one or more clones, still genetically F_1 plants, remain as the foundation of new varieties.

Clonal degeneration, a gradual loss of vigour and yield with age of a variety, is a well-known phenomenon in vegetatively propagated crops. Formerly it was thought to be caused by genetic mutations but it is now recognized that the main cause is the spread of virus infection, which, as was seen earlier, tends to build up with each cycle of vegetative propagation.

Mutations do occur in clones, giving rise to aberrant types. In fruit crops *bud sports* are sometimes seen. These are branches or whole shoots which differ from the normal shoots of the variety in the form of leaves, flowers or fruits. Very occasionally, they have certain advantages over normal shoots and can be propagated. Some varieties of seedless grapes, for example, arose as bud sports on seeded varieties.

One of the commonest mutations in potatoes is one that gives rise to 'bolters'. These are plants that have a photoperiodic short-day requirement for tuber initiation. They therefore do not form tubers until very late in the season. The lack of demand for carbohydrates below ground leads to very profuse above-ground growth, often with greatly increased flowering.

Another mutation in potatoes gives 'wildings' with a very bushy growth habit, thin stems and a very large number of small tubers. Tubers from wildings or from bolters, if used in propagation, produce similar aberrant plants. Any potato crop being grown for 'seed', that is the tubers to be planted to produce next season's crop, must therefore be carefully inspected and wildings, bolters and other results of mutation must be rogued out. Aberrant forms also appear in other vegetatively propagated crops such as narcissus, and must be similarly controlled.

Occasionally, mutations have led to new and useful varieties. Examples are the potato varieties Red Craigs Royal and Red King which arose from the older varieties Craigs Royal and King Edward VII by mutations which confer pink rather than white-splashed-pink skin, and the variety Golden Wonder, which is a russet-skinned mutation of the older variety Langworthy.

Summary

1. Many plants have natural systems of vegetative propagation, involving stolons, rhizomes, creeping roots, tubers and other organs. In the wild these systems allow the plants to spread in dense vegetation such as grassland, and provide a less random, though shorter-range, means of dispersal than seed. No genetic recombination takes place in vegetative propagation, therefore successful genotypes can be perpetuated for many generations, but without sexual reproduction the plant population lacks adaptability to changing circumstances.

2. The propagation of crops and ornamentals by vegetative means in agricultural or horticultural practice may utilize natural propagative or regenerative organs, including slips, tubers, bulbs and corms which have roots of their own or produce adventitious roots in the course of their development. Commonly, however, artificial propagation requires techniques such as layering or air-layering to induce rooting, or is accomplished by the taking of cuttings. These are rooted, usually with the aid of auxins, before being transplanted to the field.

Some woody plants are propagated by grafting or budding, in which a cutting or bud is joined to another plant which already has a root system. The cutting or bud is the scion and the rooted part the stock. As well as being a means of propagation, grafting is a means of combining valuable root and shoot characteristics in a single plant, and can produce a dwarfing effect on fruit trees leading to earlier bearing. If stock and scion are not closely related the graft often fails to form successfully.

Vegetative propagation of perennials is often preferred to propagation from seed because of the great uniformity of the offspring it produces and because of the more rapid establishment of plants in the field from large vegetative propagules such as tubers and roots. As traditionally carried out, however, it allows only slow multiplication. By cutting a plant into smaller and smaller pieces the rate of multiplication can be greatly increased, but it becomes necessary to grow the propagules in highly controlled conditions. For tiny portions of tissue or even single cells, artificial culture media containing organic and inorganic nutrients are used. These techniques are known as micropropagation.

A major difficulty in vegetative propagation is the spread and carry-over of diseases, resulting from the succulence of the propagating organ, soil contamination, and the presence of viruses, none of which problems arise on the same scale in propagation from seed. Infections tend to build up with each generation of vegetative propagation, and it is therefore essential to use only healthy, uncontaminated stocks for propagating. If a variety has become completely infected by a virus, heat treatment and meristem culture may produce a few virus-free plants which can then be multiplied by micropropagation techniques to regenerate the variety. Such techniques are often combined with other disease control measures into schemes that ensure a supply of healthy stocks of reliable varietal purity.

3. All plants of one variety of a vegetatively propagated crop are genetically identical. Such a variety is a clone. New varieties can normally be produced only be sexual reproduction. Clonal degeneration, a progressive loss of vigour and yield in a clonal variety, results from virus infection and to a lesser extent from mutations.

Bud sports in fruit trees and aberrant plants such as wildings and bolters in potatoes are also the result of mutations. Occasionally a mutation confers some improved characteristic, such as seedless fruit, and leads to a new and useful variety.

11

Crop Production and Yield

11.1 Meaning of yield

In the preceding chapters of this book the emphasis has been firmly on the individual plant, but here the emphasis has to change. We are now concerned with the *crop*, a whole population of more or less similar plants growing together, interacting with one another as well as with their environment. A crop, as we shall see, is not simply the sum of the individual plants it contains, and this is especially true in any consideration of the *yield* of the crop.

Yield sometimes means the weight of produce harvested from a single plant, for example an apple tree. However, yield defined as quantity of produce harvested per unit of land area is usually a more useful concept and it is in that sense we use it here.

11.1.1 Quality and quantity

In the growing of crops the farmer has two main aims. One is to provide a product of the right *quality*, to use on the farm or to sell on the market. We consider the question of quality of food and fodder crops in Chapter 12. The other principal aim of crop husbandry is to provide a sufficient *quantity* of the product required. What is meant here by 'sufficient' depends on the type of agricultural system practised. In subsistence agriculture the grower has to produce at least enough to feed himself and his family and have a little left over to provide seed to be sown in the following year. In market economies the farmer has to produce a sufficient quantity of saleable commodities to cover the costs of growing his crops and keeping his animals and leave a margin of profit.

The quantity of utilizable or saleable crop produce harvested per unit of land area is the yield of the crop. Note, however, that in the concept of yield it is impossible to separate entirely the ideas of quantity and quality. Consider, for example, a crop of cabbages. Is the yield the weight of all the cabbages produced, including diseased, damaged,

and under- or oversized specimens, or only the weight of those of good enough quality to be sold? Since the profitability of the crop will depend on the weight sold, this second quantity is in this case a more useful measure of yield.

To take a more difficult example, what is the true yield of a crop of grass? The saleable product this time is seldom the grass itself but instead an animal product such as milk, beef or wool obtained by feeding the grass to livestock. If we were simply to measure the total weight of fresh grass produced we would have a measure of yield, but one that has little bearing on animal output because much of what we have weighed is water. A better measure would be the yield of *dry matter*, that is, the weight of all the chemical constituents of the grass (such as cellulose, sugars, protein and minerals) excluding water.

The dry matter yield of grass or any other fodder crop is still not an infallible guide to the likely level of animal output, because the nutritional quality of the dry matter varies greatly between species, with different management treatments and at different times in the growing season, as we shall see in Section 12.1. Ultimately it is the yield of the animal product itself that matters, and this depends almost as much on the quality as on the quantity of grass consumed.

In practice a compromise between quality and quantity usually has to be accepted since higher yields often lead to lower quality. It may be more profitable to produce a relatively small quantity of a high-quality product than a large quantity of an inferior product which is less easily marketable and cannot command such a high price.

11.1.2 Broader definitions of yield

There are other considerations besides direct yield and quality that may influence the farmer's choice of crop and the way he manages it. Many crops give by-products which, though of much lower value than the main product and not usually treated as part of the yield, find a use on or off the farm. For instance, sugar beet tops and cereal straw are fed to animals, and straw is also used as animal bedding.

Crops also have indirect beneficial effects. The unharvested parts of crops, including their root systems, help to maintain the organic matter content of soils which, as we saw in Section 6.1.7, is important for the growth of succeeding crops. An important part of the role of legumes in many rotational cropping systems is the boost they give to succeeding crops such as cereals through the nitrogen fixed in their nodules and later released as nitrate to the soil.

In rotational agriculture it is thus the yield of the rotation as a whole that matters rather than the yield of any one crop in the rotation. As a further illustration of this, certain crops permit interventions by the grower that may be important for the overall management of the rotation. For example, in northern Europe, before chemical methods became available for the control of weeds, pests and diseases it was impossible to grow cereals for more than two or three years in succession on the same land because yields fell off dramatically. By rotating cereals with grass and root crops such as turnips the build-up of weed populations could be prevented and the carry-over of diseases and pests from one cereal crop to the succeeding one could be limited. The grass and root crops gave their own yield but the benefits of growing them were seen in the cereal yields as well. Now that continuous cereal cultivation is widely practised there remains a need for alternative crops which can be inserted into the cereal sequence to allow control of those weeds, pests and diseases that still cannot be controlled in the cereal crops themselves, and perhaps more importantly to help restore organic matter to the soil following years of depletion by cereals. Such crops are known as *break crops*. Ideally they should give an economic yield themselves but this is subsidiary to the yield benefit in the succeeding cereal crops.

In a sense, these by-products and benefits are all yields, since they help increase the overall output of a rotation and, though such benefits may be difficult to quantify, they are important to the farm economy. A related point is that in perennial crops such as grass or soft fruit the major aim is to secure a long productive life, not merely to maximize yield in the first year or two.

11.1.3 Timing and composition of yield

Some crops give their yield more or less continuously. In a banana plantation, for example, the fruit of individual plants is harvested as the plants reach maturity, but the development of plants in the crop is not synchronized and at any one time there are banana plants at all stages of growth. With this type of crop the workload is spread evenly throughout the year and the harvest over a considerable part of it. In temperate agriculture the closest to a continuously yielding crop is grazed pasture, but here the crop is harvestable (by grazing animals) only during the growing season and even within this period there are marked variations in productivity at different times.

Other perennial crops give their yield intermittently at one or more harvests per year. Grass cut repeatedly for conservation as hay or silage comes into this category, as do fruit crops such as apples, blackcurrants, grapes or strawberries. Most agricultural crops, however, are grown as annuals, giving their yield once only at a single final harvest, usually towards the end of the growing season. In these crops yield represents an accumulation of material over the growing season.

The material accumulated as yield in any crop includes organic matter – mainly carbohydrates, proteins and lipids, all derived from the primary products of photosynthesis – together with minerals (including nitrogen, most of which is assimilated into protein) and water. Minerals constitute a small fraction of the weight of a crop, so that even a large variation in mineral content gives rise to an insignificant variation in total yield (though it may have an important bearing on quality).

Water, on the other hand, constitutes a large fraction of the weight of many crop products (for example 80% in potatoes, 88% in swedes, 92% in apples). Relatively small variations in moisture content can therefore produce substantial yield effects. Where the produce is sold fresh, as in the case of vegetables, the increased bulk associated with high moisture content can lead to a higher income, though high moisture content usually means lower quality which may be reflected in a lower price per unit weight. Where the crop is to be consumed by livestock on the farm, water is of little value as a constituent of the yield of the crop. To achieve a given level of animal productivity the farmer has simply to feed proportionately more of a product with a high moisture content. In grain, pulse and oilseed crops, where water constitutes a much smaller proportion of the weight of the produce (commonly 10–15%), variation in moisture content has less effect than in roots, tubers or fruits. However, because these low moisture content products are high-value commodities, merchants are unwilling to pay for even a small increase in weight which is due solely to increased moisture content. The price paid for grains, pulses and oilseeds is therefore usually calculated on the basis of a fixed moisture content. It makes sense therefore to express yield at a fixed moisture content.

For many purposes the yield at zero moisture content – that is, the dry matter yield, the yield of organic matter plus minerals – is the most useful measure. It is especially useful for comparing the yields of different types of crop which may be used for similar purposes, for example wheat and potatoes as staple carbohydrate foods for human consumption, or barley and swedes as stockfeeds.

11.1.4 Harvestable yield and total dry matter production

Since minerals make up such a small fraction of dry matter yield (less than 5% in most crops), dry matter yield can be taken as related to the accumulation of organic matter by photosynthesis over the growing season. The relationship between dry matter yield and photosynthesis is not, however, a simple one, for two main reasons.

First, much of the organic matter produced in photosynthesis is lost again in respiration and photorespiration. There are also losses of dry matter in the death and abscission of redundant leaves, flowers and other organs, and in consumption by other organisms including pathogenic microbes and animals. The dry matter present at harvest time is the accumulated excess product of photosynthesis over these losses.

Second, only part of this accumulated excess is in the harvested organs. A substantial portion is in other organs which may, however, be of value as a by-product or for ploughing back to maintain soil fertility.

Consider as an example a wheat crop in which the total dry weight amounts to 20 tonnes ha^{-1} at harvest. The weight of dry matter in a crop at any given time, expressed on an area basis like this, is the *biomass* of the crop. Though the biomass at harvest is 20 tonnes ha^{-1}, over the course of the growing season the wheat may have photosyn-

thesized twice this amount of organic matter, but respiration, photorespiration, death of plant parts, and consumption by pathogens and predators will have accounted for half the dry matter produced. Of the 20 tonnes ha^{-1} present at harvest time, 6 tonnes ha^{-1} may be in the root system, 6.5 tonnes ha^{-1} in the straw, 1 tonne ha^{-1} in chaff and the remaining 6.5 tonnes ha^{-1} in the grain.

Thus the dry matter yield, in this case 6.5 tonnes ha^{-1}, is a fraction of the biomass at harvest. This fraction, in our present example 6.5/20 or 0.325, may be termed the *utilizable fraction*. The term *harvest index* is sometimes used to mean the same thing but more commonly refers to the ratio of yield to above-ground biomass, that is ignoring roots, the weight of which is very difficult to measure. Thus in this example of a wheat crop, harvest index is 6.5/14 or 0.464. Unfortunately published values of harvest index are often based on fresh rather than dry weights. Whilst harvest index is a useful measure in comparing varieties of one crop such as wheat, or in comparing the performance of one variety under different conditions, utilizable fraction is a more meaningful basis for comparison of different types of crop.

11.2 Plant population and crop yield

Two basic properties of the plant population which constitutes a crop affect the yield of that crop. One is the number of plants per unit area of ground, that is *plant density*, and the other is the pattern of spacing of the plants over the ground. Let us consider first the more important of these two factors, namely plant density, and its effect on yield.

11.2.1 Plant density and the components of yield

It is vital for a grower to achieve the correct plant density, whether by sowing, planting or thinning, to obtain maximum yield. To what extent can yield be increased simply by increasing plant density?

If the crop is very widely spaced, not all the land area is covered by leaves and much of the light available for photosynthesis is wasted. Resources of water and mineral nutrients in the soil are similarly underutilized. In this situation an increase in plant density could be expected to give a considerable improvement in yield. However, as the space between plants decreases, neighbouring plants begin to interfere with one another. Their roots have to forage in the same volume of soil for supplies of water and mineral nutrients that may become depleted, and their leaves may begin to shade one another. This interference between plants is known as *competition* and leads to reduced growth of individual plants. When it takes place between plants of the same species, as in a high-density crop, it is described as *intraspecific competition*.

To understand how intraspecific competition influences yield, consider first a root crop such as sugar beet or carrots, or any other crop, for example cabbage, in which each plant produces a single harvestable organ. Yield in such a crop is the product of two factors: the number of plants per unit area (say a hectare) and the average weight per plant of the harvested part. These are known as the *components of yield* of the crop.

Intraspecific competition results in smaller plants and thus a reduction in the second component of yield, namely the average dry weight per plant of the harvested part. As density increases so competition for available water, minerals and light intensifies and average weight of harvested parts declines, as shown by the example given in Table 11.1.

At relatively low plant densities, light and soil resources are not fully utilized but some intraspecific competition does take place, reducing the

Table 11.1 Decline in average weight of carrots with increasing plant density

	Plants m^{-2}		
	152	494	697
Mean root weight (g)	37	13	8
Yield of roots (kg m^{-2})	5.6	6.4	5.6

From J Salter (1979) *J. Agric. Sci. Camb.* **93**, 431–40.

average size of harvested parts. However, at such densities the effect of increased plant number outweighs the effect of reduced plant size, and total yield increases with increasing density, as shown in Fig. 11.1.

With further increase in plant density, the exploitation of resources reaches a limit at a certain point and any density increase beyond this point is balanced by the resulting decrease in average size

Fig. 11.1 The effect of plant density on the yield of parsnips. Note how total yield of roots continues to increase, even after the number of roots within the marketable size range of 3.8 to 6.4 cm in diameter has started to decline due to a decline in the average size of root harvested. From J K A Bleasdale and R Thompson (1966) *J. Hort. Sci.* **41**, 371–8.

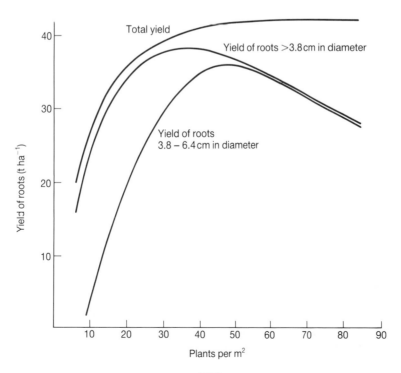

of the organ harvested. Yield, at this point, has reached its maximum.

At even higher densities, a reduction in utilizable fraction is commonly observed. In root crops, for example, intense competition for light induces greater stem production and an increased energy requirement for the production of new leaves as the older ones become shaded by neighbouring plants. There is then less carbohydrate available for accumulation in the organ to be harvested. Consequently there is at these very high plant densities a reduction in yield. This is seen in the example given in Table 11.1.

Such interactions between yield components with increasing plant density are so important that we should look at the principal variations on this theme. Consider a crop in which each plant produces several harvestable structures. In the potato crop we can recognize three yield components: the number of plants per hectare, the average number of tubers per plant and the average dry weight per tuber. Again the first of these components of yield is largely determined by the grower at planting. The second component (tubers per plant) depends on several factors including variety and treatment of the stored 'seed' tubers in the months prior to planting (see Section 11.2.2).

The third and final component of yield, average tuber dry weight, is determined during the period of tuber growth, as carbohydrate, made by photosynthesis in the leaves, is transported to the tubers for storage. Clearly the rate of photosynthesis and the length of time for which it continues are important in determining final tuber size, but so also is the number of tubers among which the carbohydrate has to be shared. In any one plant, the more tubers that have been initiated, the smaller will be the average size of these tubers. In other words, in addition to intraspecific competition between plants, we now also have competition within plants. Effectively the tubers, as sinks, are competing with one another for the limited supply of carbohydrate from the source, the leaves. Increasing the first component of yield, plant density, or the second, say by encouraging the production of more shoots per 'seed' tuber, decreases the third component, tuber size. Again the relationship between yield and density is as illustrated for a root crop in Fig. 11.1.

In crops grown for their seeds, we can recognize a greater number of yield components. In cereals, for example, there are four: the number of plants per unit area (plant density), the number of ears per plant, the number of grains per ear and the average grain weight. Fig. 11.2 shows how, through intraspecific and within-plant competition, these four components interact. The main effect of increasing density is to reduce ear number per plant, though at high densities grain number per ear and average grain weight are also affected. At very high densities, above those shown in Fig. 11.2, total yield of cereals declines, as has been seen in root crops and potatoes.

With most crops, if, instead of harvestable yield, we consider total crop biomass at harvest, we find that this increases with plant density up to a certain point and then stays on a plateau, further density increases causing no reduction. A similar picture is seen with the harvestable yield of certain crops, such as grass, in which virtually the entire vegetative shoot system is harvested. But, as we have seen, where a reproductive structure or storage organ forms the harvested product, a decline in yield is seen at very high densities.

Fig. 11.2 The effect of plant density on the other yield components of barley: number of ears per plant, ●; number of grains per ear, ○; and average grain weight, Δ. From E J M Kirby (1968) *NAAS Quarterly Review* **80**, 139–45.

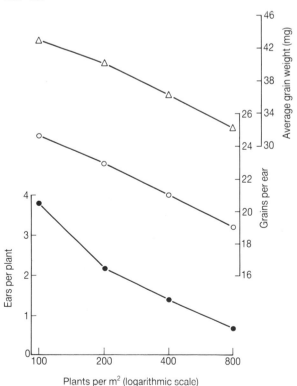

The effect of plant density on yield thus varies considerably between crops, but the crop characteristics which govern the yield-density relationship at low densities are entirely different from those which govern the relationship at high densities. Relatively inflexible crops such as spring oats, which produces few tillers, or maize, which does not tiller, cannot take advantage of unused space around plants in the way that, say, winter wheat can. Their yield therefore increases gradually with increasing density over a wide range of densities. By contrast, flexible crops such as potatoes or winter wheat give maximum yield at relatively low densities, above which yield is scarcely affected until very high densities are reached. Such crops are obviously much more tolerant of wide variations in 'seed' rates than crops whose yield is more dependent on density (using the word 'seed' to cover propagules as diverse as grains and tubers).

A number of characteristics determine whether yield declines gradually or steeply with increasing density at very high densities. As we saw in root crops, the proportion of assimilated dry matter directed towards the harvestable organ often declines at high densities, giving a gradual reduction in yield. In other cases certain catastrophic events, such as lodging in cereals (see Section 4.4) or rapid disease spread, may overtake plants growing in dense stands, causing more severe yield reduction. Wherever vegetables are graded, saleable yield as opposed to total yield declines steeply beyond a certain optimum density, as crowding reduces the average size of harvestable organs. A good example of this was seen in Fig. 11.1.

It should be noted, however, that the tendency of dense stands to undergo self-thinning, whereby the more vigorous seedlings shade out the weaker is so strong that very high seed rates therefore lead to considerable loss of seedlings. The high population densities associated with severe yield loss are therefore seldom encountered in practice.

11.2.2 Control of plant density in crop husbandry

In any crop grown from seed, the amount of seed sown is obviously the main factor affecting plant density. This factor is usually expressed as *seed rate*, the weight (or, in former times, the volume) of seed sown per unit area. Seed rate is not, however, a very accurate measure of the number of seeds sown per unit area, and thus of the potential plant density, because seed of any crop species varies in weight depending on variety and the conditions under which it has been produced.

In winter barley, for example, the thousand-grain weight (see Section 9.1.1) can be anything from 35 to 50 g. In order to give an optimum plant density, around 450 seeds per square metre should be sown; it can be computed that the appropriate seed rate to achieve this is 160 kg ha^{-1} for seed with a thousand-grain weight of 35 g, but as much as 225 kg ha^{-1} for large seed with a thousand-grain weight of 50 g. In some crops large seeds tend to give higher yielding plants but this may not be enough to compensate for reduced plant density if a seed rate appropriate to smaller seeds has been used.

With vegetative propagules, the situation is more complex. Propagules with just one growing point, such as bulbs, behave as seeds, though spacing is usually wider since the propagules are larger. However, many vegetative propagules have several or many buds. The number of buds and the proportion that grow rather than remain dormant varies, and some account of this needs to be taken in the density of planting. The behaviour of potato tubers aptly demonstrates this. Each sprout that develops from a bud on a potato tuber has the potential to become an independent plant, producing its own daughter tubers. But what controls the number of these sprouts?

To answer this, it should be recalled that at the time of harvest of the 'seed' tubers, all buds on these tubers are dormant. Gradually, during storage, this dormancy is lost, first from the apical bud and then progressively from the lateral buds. Thus a tuber which is sprouted after a relatively short storage period will tend to produce only a few sprouts or even a single sprout, whereas after a long storage period many more sprouts per tuber are likely to form. The lower the storage temperature the slower the dormancy is lost, so that, for example, tubers stored at 4°C for 15 weeks will later tend to produce fewer sprouts than tubers stored at 8°C for the same length of time.

Temperature during sprouting is as important as temperature during prior storage. Tubers sprouted at 20°C give rise to a strong apical sprout which rapidly asserts apical dominance over the lateral buds. However, at 12°C apical dominance is not established so strongly and, provided dormancy in the lateral buds has broken, several sprouts develop at once.

Thus by storing 'seed' tubers at 8°C and

sprouting at 12°C the potato grower effectively increases his plant population density even though the number of tubers planted stays the same. Competition among plants derived from the same 'seed' tuber is particularly intense because of their close proximity to one another, and the effect is to reduce average tuber size. This is generally desirable if the crop is intended for 'seed' but not for a ware crop, because tubers used for planting are generally smaller than those preferred for eating.

11.2.3 Pattern of spacing

Obviously, as plant density increases, the average distance between plants, that is the crop spacing, declines, with the various effects on yield components we have illustrated. However, the pattern of distribution of the plants at any one density also has significant effects on yield.

It has been shown experimentally that at a given density, the highest yields of most crops are obtained where the plants are regularly spaced in a square lattice arrangement, as shown in Fig. 11.3. This arrangement evens out competition between plants. It avoids excessive competition at some points and wastage of resources through inadequate exploitation at others. It is an ideal that can be approached in perennial tree and shrub crops, where individual plants can be given attention, but in arable crops practical restraints generally prevent it. Broadcast sowing, that is simple scattering of seed on the ground, either by hand or by machine, results in a considerable degree of randomness in plant spacing. This often leads to appreciable yield loss unless seed rates are

increased to compensate for it. Even crops sown in regular rows by seed drills usually yield less well than those in a square arrangement, because the distance between rows may be too great to prevent wastage of light and soil resources and at the same time strong competition takes place between adjacent plants in the rows. A minimum row width is, however, necessary for access to the crop during the growing season for weed control, harvesting and other operations.

A measure of divergence from the ideal square arrangement is given by the ratio of row width to plant spacing within the row and this is known as the *rectangularity* of the crop. A crop's tolerance of rectangularity depends on the flexibility with which one yield component can increase to compensate for others. Thus sugar beet, in which flexibility is achieved only by increase in root size, shows yield reduction if rectangularity of spacing exceeds about 2:1, whereas a cereal with its ability to tiller may be little affected at rectangularities up to 5:1.

With the advent of selective herbicides, wide rows in crops such as carrots are no longer necessary to allow for mechanical weed control, and such crops are now more commonly grown in beds, where the lower degree of rectangularity has permitted considerable increase in yield.

11.2.4 Crop models and yield components

Modern systems of crop husbandry, particularly cereal husbandry, are based on an understanding of plant populations and of the physiology of the crop at all stages in its growth. The sophisticated grower is already applying to his own crops the latest findings of research into the effects of environmental conditions and management practices

Fig. 11.3 Patterns of spacing in crops: (a) broadcast, (b) beds, (c) rows, (d) square lattice.

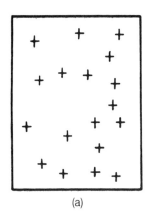
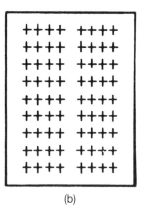
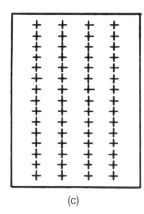
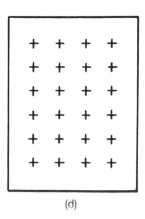

| (a) | (b) | (c) | (d) |

on the various components of yield. When these effects are quantified they can be incorporated into a *crop model* such as that outlined in Fig. 11.4, which enables the grower to predict the likely effect of his management on final yield. At this level the key to successful cereal management is an awareness that the components of yield are determined at different stages in the life of the crop and that environmental or management influences at any one time have their effect on the component of yield being determined at that time. Winter wheat is a good example (Fig. 11.4).

The first component of yield, number of plants per hectare, depends on seed rate (strictly on number rather than weight of seeds sown per hectare), the proportion of seeds which give rise to established seedlings, and the proportion of these which survive to maturity. Most plant mortality occurs at the seedling stage (see Section 5.3), there-

fore number of plants per hectare is determined very early in the life of the wheat crop. Subsequent events have relatively little effect on this yield component.

Seedling establishment can be reduced by low soil temperature, inadequate soil moisture level and poor physical condition of the seed–bed, as discussed in Section 5.3.6. These restraints can be minimized by a suitable choice of soil cultivation system and sowing date and by the use of high-vigour seed. Seedling survival requires careful attention to weed, pest and disease control and the avoidance of waterlogging where possible. Of particular importance to seedling survival in winter wheat is action to minimize winter kill, for example by choosing a hardy variety, keeping seed–bed nitrogen levels fairly low, and using autumn grazing by sheep if necessary to reduce leaf area and hence winter transpiration.

The second component of yield, number of ears per plant, is related to the number of tillers produced, the proportion of tillers which initiate an inflorescence, and the proportion of these which survive to produce a grain-bearing ear. In winter wheat tiller production and inflorescence initiation begin in the autumn and are completed in the early spring. Thus a limit is set to number of ears per plant during the tillering stage, early in the life of the plant.

Tillering is dependent on adequate soil moisture (seldom a problem in late autumn and early spring)

Fig. 11.4 A crop model for winter wheat for eastern England. The final density of ears is the result of a balance between processes that tend to increase ear number, such as emergence and tillering, and processes that tend to decrease ear number, such as death of seedlings and abortion or death of tillers. The ideal numbers of plants, tillers and ears per unit area depend on location. Because of the flexible response of winter wheat to seed rate and growing conditions, the crop can tolerate considerable deviations from the ideal densities of plants and tillers without major yield loss.

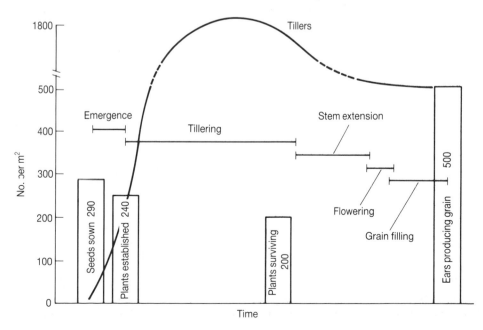

and temperature. This is the stage of wheat growth that is most sensitive to competition from weeds and the benefit of weed control is therefore most often seen in the boosting of ear numbers rather than other components of yield. Some growers apply growth regulators such as chlormequat at around this time to increase the number of ear-bearing tillers per plant.

While the potential number of ears per plant is determined early, the actual number is affected by later events which may cause mortality or failure to develop of flowering tillers. Most tiller mortality occurs during the period shortly after completion of tillering in the spring. Timing of fertilizer application can be very important in maximizing number of ears per plant. Thus an early application of nitrogen fertilizer will increase tiller formation, and a second application after tillering is completed has the effect of improving tiller survival. Control of disease, particularly eyespot, may also be necessary to prevent excessive tiller mortality.

The third yield component, number of grains per ear, depends on the number of florets initiated on the inflorescence, the proportion of florets that develop to carry a flower and the proportion of these which are fertilized and set seed. Fertilization, of course, takes place during the flowering stage but in most cereal crops is close to 100% and has little bearing on variability of yield. Thus the number of grains per ear, like the number of ears per plant, is largely determined not during flowering but much earlier, during and shortly after tillering, when inflorescence development is taking place. Some natural abortion of flowers takes place in winter wheat, but again for the most part this occurs early, before fertilization. Little abortion of fertilized ovules occurs at later stages.

There are considerable genetic differences among winter wheat varieties giving rise to differences in number of grains per ear. Environment and management, however, also play a part in determination of this component of yield. In general those factors which influence number of ears per plant have similar influences on number of grains per ear.

The only yield component likely to be significantly affected by events after tillering is average grain weight. In cereals, very little of the carbohydrate for grain filling comes from reserves, most being produced by photosynthesis taking place after fertilization. This photosynthesis is carried out in the uppermost leaves and in the green parts of the ear itself. Hence these tissues must be maintained in a healthy productive condition if grain filling is not to be restricted and if the yield potential built up in the other yield components is to be realized. In particular, control of foliar diseases, especially in the uppermost leaves, is essential. In a dry season water stress can greatly reduce photosynthesis and thereby limit grain size.

To some extent this final yield component can compensate for low values of the other yield components, but wheat grains, unlike potato tubers or sugar beet roots, cannot grow to indefinite size and their ability to compensate is therefore limited. Thus in wheat and, indeed, in any cereal crop the final yield is strongly influenced by events early in its life. Good management during the seedling and tillering stages of growth is crucial to the development of satisfactory yield.

11.2.5 Plant spacing and size of produce

Both the frequency of spacing, that is plant density, and the pattern of spacing affect size of produce. We have already seen that the average size of harvested organs in root and tuber crops is strongly affected by plant density – the higher the density, the more intense is the competition between plants and the smaller therefore the roots or tubers produced. The same effect is seen on other vegetable crops such as cabbages and onions.

Now market economies require vegetables to be produced within certain size limits that suit the consumer's needs in different markets. Carrots for canning, beetroot and onions for pickling, and potatoes for 'seed' are all cases where the produce must be in smaller grades than when grown for marketing as raw vegetables. Whatever the particular size limits set, all vegetable produce is graded and a proportion will be rejected because it is too large or too small. Vegetables which are too small generally constitute a negligible fraction of the yield and can be dumped without serious financial loss, but oversized produce, even if few in number, may be considerable in weight and may represent significant wastage. If it finds a use at all, such as in the feeding of livestock, its value to the grower is still very low.

It is therefore important that as large a proportion of the produce as possible is within the required size range. This demands not only control of the overall plant density but also uniform spacing between plants and uniform growth rate of plants. Any gaps or plants showing retarded growth will allow neighbouring plants to grow larger and thereby increase the proportion of harvested vegetables that will exceed the size limit. To minimize gapping it is essential to use high quality, reliable seed and as far as possible to protect the crop, particularly in the seedling stage, from stresses that cause mortality and checks to growth.

Formerly, seed rates were aimed at producing seedling densities much higher than the optimum required. Seedlings were then thinned or singled to the correct spacing by hand or machine. The high seed rate gave a safety margin that could absorb any losses of plants that took place up to thinning. However, thinning is an expensive procedure and is seldom accurate enough to give the precise control of plant spacing that is necessary.

Now seed drills are used that plant seeds at carefully pre-set spacing, and if this *precision seeding* is used with vigorous, reliable seed the problems are much reduced. However, the remaining population of young plants has little margin of safety if appreciable mortality occurs, and soil conditions may not be uniform enough to allow even germination and growth. Sometimes seedlings emerge in two distinct flushes, the first shortly after sowing from seed that lands in a moist situation, and the second during a later spell of favourable weather from seed that found a less ideal site. Thus in a field of cauliflower, to take one instance, two or more plant populations of different ages may be discerned by the experienced eye. Raising seedlings in a seed–bed where there is better control of conditions during germination, and transplanting them later to the field, can reduce this variability, but as explained in Section 6.1.4 checks to growth persist due to disturbance of root systems during transplanting. Hence a further advance is to rear seedlings in preformed blocks of soil or other growing medium and plant these out at precise spacings.

11.3   Determination of yield

We have seen that dry matter yield can be expressed as the utilizable fraction of biomass at harvest. Let us leave utilizable fraction aside for the moment and consider the determination of biomass at harvest. This depends on the length of the growing season and the rate of dry matter production at different times during the growing season. Again we can return to consider the length of the growing season later and concentrate at present on the rate of dry matter production. This, as we have

seen, is the rate of accumulation of organic matter by photosynthesis after accounting for losses due to respiration, mortality and other causes.

The rate of photosynthesis depends on the intensity, quality and daily duration of light falling on the crop, the proportion of this light which is intercepted by the leaves and the efficiency with which the intercepted light energy can be converted to organic matter. Of key importance is the interception of light, which depends on the *leaf area* of the crop.

11.3.1 Leaf area

The ratio of the total area of the leaves of a crop to the ground area occupied by the crop is known as the *leaf area index*, or *LAI*. Thus if 1 m^2 of a wheat field carries wheat leaves totalling 3 m^2 in area, the LAI of that crop is 3.

Early in the life of an annual crop, LAI is very small. Interception of sunlight by the crop is poor, most of the light falling on bare soil. As the season progresses, LAI increases, light interception improves and the rate of dry matter production goes up.

However, as new leaves continue to be produced at the top of the crop canopy they begin to shade the older ones lower down. These older leaves, receiving less light, photosynthesize less but continue to respire as before. With further increase in LAI the shading of the lower leaves becomes so great that photosynthesis no longer exceeds respiration in these leaves, which then cease to be net producers of dry matter. When this stage is reached, the lower leaves usually begin to senesce and die, so that they no longer form part of the productive leaf area of the crop. The continued production of new leaves at the top of the canopy balances the loss of old leaves at the bottom and LAI levels off (Fig. 11.5).

In most crops LAI levels off at a value which gives about 95% interception of light and this value is known as the *optimum LAI* for the crop. In most dicotyledonous crops, the leaves of which tend to be broad and flat and held more or less horizontally, the optimum LAI is around 3. Cereals and grasses have narrower leaves which are held more vertically so that light penetrates further into the canopy. They tend to have higher values of optimum LAI, commonly around 4 to 5.

Optimum LAI for most crops is greater when the angle of the sun is higher, as light from overhead penetrates further into a canopy than light from

Fig. 11.5 Development of leaf area index in a swede crop. After about twelve weeks from sowing, the production of new leaf area at the top of the canopy is balanced by the loss of leaf area through senescence at the bottom. After eighteen weeks, new leaf area production is not enough to balance loss by senescence and leaf area index declines.

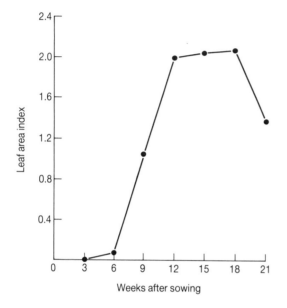

an oblique angle. Similarly, high-intensity light penetrates to lower levels in a canopy than does low-intensity light. Thus optimum LAI is greater in tropical than in temperate latitudes, and in temperate latitudes it is greater in midsummer than earlier or later in the season.

Though LAI is usually taken as a measure of the area available to intercept light for photosynthesis, it should be remembered that structures other than leaf blades contribute to the photosynthetic area. For example, in maize the stem and tassel (the male inflorescence at the top of the plant) each contribute about 9% of the total photosynthetic area of the crop but are generally ignored in measurements of LAI. In temperate cereals the ear is usually very important in photosynthesizing carbohydrate for grain filling. In barley, for instance, anything from 24 to 84% of the carbohydrate in the grains has been manufactured in the ear. Indeed by the grain-filling stage LAI is rapidly declining (Fig. 11.5) through leaf senescence and it might therefore be thought that LAI is largely irrelevant to grain yield. However, the size of the ear and hence its ability to photosynthesize depends on earlier photosynthesis by the leaves.

For this reason there is in cereals a close correlation between yield and the maximum LAI reached.

In crops such as sugar beet, which maintain LAI at around optimum levels for longer periods of time than cereals (Fig. 11.6), yield is related not simply to LAI at any one time but to the product of LAI and the time for which it is maintained. Some crop scientists have related yield to a quantity represented by the area under the curve for LAI against time, as shown in Fig. 11.6 (in mathematical terms, the integral of LAI from crop emergence to harvest), and this quantity is known as *leaf area duration*, or *LAD*.

One approach to the improvement of crop yield is to increase LAI or LAD. This can be achieved in various ways. Early sowing permits the establishment of a high LAI by midsummer, when days are long and light intensities are high. This is a major reason why autumn-sown cereals outyield spring-sown cereals. Increased sowing density leads to the optimum LAI being reached earlier in the season, as does an increased use of nitrogen fertilizer, which encourages vigorous foliar growth. Varieties within crops may differ in optimum LAI; those whose optimum LAI is greater tend to be higher-yielding, probably because the incident light is spread over a larger leaf area. For example, varieties of perennial ryegrass with erect leaves have a greater optimum LAI than varieties with a more prostrate habit of growth, and are in general higher-yielding. The use of such varieties is thus another way of improving productivity.

Action can also be taken to minimize those influences that reduce LAI. Foliar diseases, insect and other animal pests and competition from weeds all reduce LAI and their control can therefore result in improved yield.

11.3.2 Net assimilation rate

LAI is thus a most important factor, determining the rate of dry matter production per unit ground area in a crop at any time. This rate is often called *crop growth rate (CGR)*, and is the product of LAI and the rate of dry matter production per unit leaf area, or *net assimilation rate (NAR)*. Crop scientists investigating genetic, environmental or husbandry factors affecting yields often use techniques of *growth analysis*, one of the most useful of which can be expressed by the simple formula CGR = LAI × NAR.

Just as an increase in LAI can give improved yields so, theoretically, can an increase in NAR.

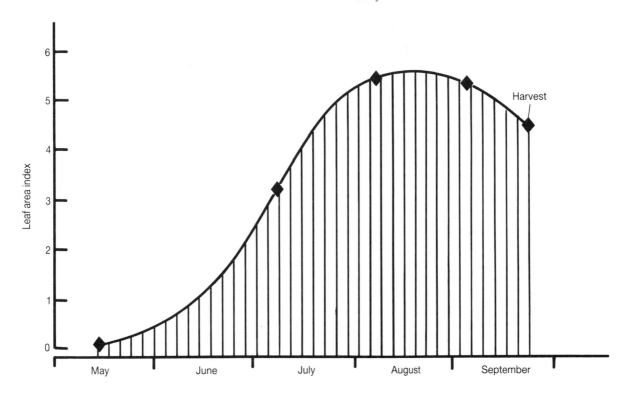

Fig. 11.6 Development of leaf area index in sugar beet. The shaded area under the curve is a measure of leaf area duration (LAD). From R K Scott *et al.* (1973) *J. Agric. Sci. Camb.* **81**, 339–47.

This might be achieved by increasing the amount of light incident on the crop, for example by artificial lighting in glasshouses in winter or by the elimination of shading. Tall-growing weeds in a low-growing crop can cast considerable shade, thereby reducing NAR. Water stress reduces NAR because stomata remain closed to conserve moisture and the gas exchange needed for photosynthesis is severely restricted. Thus in dry situations NAR can be improved by irrigation.

An alternative approach is to improve the efficiency with which the incident light is converted to organic matter. Plant breeders have put considerable research effort into trying to improve NAR in crops, primarily by selection of plants with high NAR. This has not, however, proved easy, since there is relatively little genetic variation in NAR within crop species. Surprisingly, it has been found that many high-yielding crop varieties actually have lower NAR than their wild relatives.

11.3.3 Length of growing season

A high CGR, then, whether resulting from high LAI or high NAR or a combination of both, tends to lead to high yield, but only if it is sustained over a long period. Thus the length of the growing season is another important factor affecting yield.

The growing season is often limited by unfavourable weather conditions. In cool temperate climates, the start of the season can be delayed by low soil temperatures in spring and the end is often curtailed by low air temperatures in autumn. Frost-sensitive species cannot be sown until the danger of spring frosts is over, and they have to be harvested before the first frosts of autumn. In climates with a pronounced dry season the growing season begins with the first rains and ends when the dwindling moisture reserves in the soil can no longer support growth.

In many cases, however, the growing season is limited by the crop itself. Many crops come to maturity well before the end of the available season, for example early potatoes, winter barley and winter oilseed rape. The early harvest offers the grower several advantages, such as good weather conditions at harvest and a spreading of the harvest workload over a longer period, and the produce can often be marketed at a higher unit price. The remaining portion of the growing season is not wasted if it is used for the establishment of another winter crop or the production of a quick-growing catch crop such as stubble turnips.

Biennial root crops such as carrots, sugar beet and swedes tend to bolt if sown too early. In this case it is the start of the growing season that is delayed by an attribute of the crop.

The growing season can be lengthened, and hence yield improved, by various means. The use of varieties which are hardier, less sensitive to frost or less susceptible to bolting is one type of improvement that has been made in many crops. Where the growing season is limited by water availability, irrigation will extend it. Some crops are started off during the late winter under protection of glass and transplanted to the field in spring. Finally, the use of later-maturing varieties will generally give higher yields but, for reasons already mentioned, it may be preferable to accept a rather lower yield earlier.

11.3.4 Utilizable fraction

The three factors so far discussed, namely LAI, NAR and length of growing season relate not strictly to yield but to the biomass of the crop at harvest time. For yield to be high it is also important to have a high utilizable fraction or harvest index (Section 11.1.4).

One of the great successes of plant breeding, especially in cereals, has been the steady improvement in harvest index. Indeed, if we compare a modern wheat variety with one released fifty years ago there is likely to be little difference between them in total above-ground biomass at harvest. But the modern variety will have a greater proportion of that biomass in the grain and less in the straw and chaff.

Management of the crop can also influence the size of utilizable fraction. An excessively high sowing density, for example, causes intense intra-specific competition leading to increased stem growth. This lowers the utilizable fraction both in cereals and in root crops. In a crop such as flax, however, where long straight stems are desired for their long fibres, high sowing density is advantageous to yield.

Very high levels of nitrogen fertilizer application encourage vigorous stem and leaf growth. Where the crop is grown for above-ground vegetative parts, as in the case of grass or kale, heavy nitrogen applications will boost both biomass and utilizable fraction. With cereals, root, tuber and fruit crops, however, it is important not to apply too much nitrogen as the effect is to reduce the utilizable fraction, though total biomass may be increased.

The amount of utilizable fraction can, within limits, be manipulated by the use of certain growth regulators. For example, chlormequat, ethephon and mepiquat chloride are used to shorten cereal straw. The main justification for this is to improve standing ability, especially where fairly high levels of nitrogen are being applied (see Section 4.3.2). Even in situations where no lodging occurs, and improved standing ability therefore has no direct benefit, the same growth regulators often give increased yield through an improvement in harvest index.

11.4 Effects of weeds on crop yield

We saw in Section 11.2.1 that individual plants in a crop compete more intensely with one another at high sowing densities, and that this intraspecific competition leads to a lower yield per plant. However, the greater number of plants per hectare will compensate partly or totally for this and the result is that yield per hectare is in many crops little affected by sowing density over a wide range of densities.

But what if we have more than one species of plant competing with one another? This is known as *interspecific competition*, and it occurs in two types of situation. The first is where only one of the competing species provides the economic yield and the others are weeds. The second, which we shall return to later, is where two or more crop species are grown together, each contributing usefully to the economic yield.

Weeds cause losses to farmers in various ways, for example by contamination of produce with their seeds or other parts and by interfering with harvesting operations, but the principal way in which they cause loss is by competing with crops for light, water and mineral nutrients, thereby reducing yield. Where a weed is tall in relation to the crop, it shades the crop and the resulting competition for light is at least partially responsible for any yield loss. Less tall weeds, though their leaves are shaded by the crop, may still compete strongly below ground, removing from the soil water and mineral nutrients which would otherwise be available to the crop. Probably in no crop-weed situation is competition purely for one resource such as light or nitrogen; it is more likely that several resources are being competed for simultaneously. The situation is further complicated by the fact that almost always there are several weed species competing with the crop and with each other.

11.4.1 Factors affecting yield loss due to weed competition

It should not be taken for granted that if a crop contains weeds its yield will suffer as a result. Since it costs money to control weeds, whether by mechanical means or chemically by the use of herbicides, it is worth doing only if the benefits of weed removal are greater than the costs. What, then, determines whether or not weeds are reducing yield and if so by how much?

Crop species. Some crops are more susceptible than others to yield damage by weed competition. In general cereals suffer less than wide-spaced row crops. In Great Britain, for example, a typical weed infestation, if uncontrolled, will reduce wheat or barley yield by 10–15% but sugar beet or swede yield by 50–75%. This is because the cereal crop itself is highly competitive and suppresses weed growth, whereas the large open spaces that persist for a long period in a row crop permit luxuriant weed growth which in turn can smother the crop.

Crop variety. In some crops the choice of variety may affect the level of yield loss caused by weeds. Where tall-growing weeds are a problem, a tall variety, for example of peas or kale, will be less affected by competition for light. In the early stages of grass establishment prostrate, densely tillering grass varieties give faster ground cover than erect varieties and hence restrict weed growth.

Crop density. At high sowing densities the crop covers the ground more quickly and again gives less opportunity for weed growth. The better a crop can suppress weeds, the less it suffers itself.

Weed species. In any given crop some weed species are more damaging to yield than others, but the relative effects of different species are not the same in all crops. In general, the more similar a weed is in growth habit to the crop in which it is growing the more damaging it tends to be. The combination of a closely related crop and weed often shows particularly intense competition. Thus weeds of the same family as the crop, such as wild oats in cereals, fat hen in sugar beet, and charlock in oilseed rape, are all highly detrimental to yield. They are also more difficult than other weeds to

control with selective herbicides, because of the difficulty of finding a herbicide that will kill the weed but cause no injury to the closely related crop.

Weed density. Not unexpectedly, the degree of yield loss increases with increasing weed density. The relationship, however, between weed density and crop yield is usually not linear. For example, in barley various experiments have indicated that yield is inversely proportional to the square root of the density of wild oats (Fig. 11.7). This comes about because as weed density increases, weed plants increasingly compete with one another as well as with the crop. Individual weeds are smaller when growing at high density, thus the competitive influence on crop yield of 10 wild oats m^{-2} is not ten times as great as that of 1 wild oat m^{-2}. Probably in most crop-weed situations the degree of crop yield loss is directly related to the biomass of weeds present.

Relative time of emergence of crop and weed. If weeds emerge before or along with the crop, they are much more damaging to yield than if the crop gets a head start. In cool temperate areas, root crops such as swedes and carrots are sown in late spring, a time which coincides with the peak of weed seedling emergence, but spring cereals are sown earlier and are already established before the weeds start to emerge. This is another reason why root crops are more prone to yield loss from weed competition than cereals.

Fig. 11.7 The relationship between yield of barley and density of wild oats, based on results of several field experiments (courtesy of E B Scragg, North of Scotland College of Agriculture).

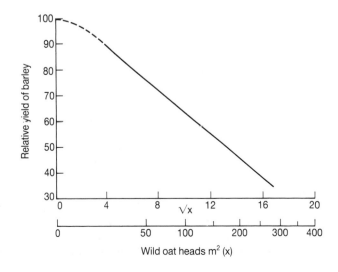

Environment and husbandry. The crop and its associated weeds are not identically adapted to the environment in which they grow. They will differ in optimum temperature or soil pH for growth; their rate of response to fertilizer, drainage or irrigation will also differ. The nearer that conditions approach to the ideal for a species, the more competitive that species will be. Thus corn spurrey, a weed encouraged by low soil pH, is more damaging as well as more abundant on acid than on neutral soils. Crops in general respond better to nitrogen fertilizers than do weeds, therefore weeds often do less damage where heavy applications of nitrogen are made. Exceptions to this include fat hen in sugar beet and docks in grassland, both of which become more competitive on nitrogen-rich soils. Chickweed can grow at temperatures below the minimum for grass growth, thus it can be especially damaging in autumn-sown grass, growing during brief spells of favourable weather throughout the winter; by spring it may have smothered the grass.

11.4.2 Time of weed-crop competition

Experiments in a wide variety of crops have shown that yield-damaging weed competition often occurs only for a limited period quite early in the life of the crop. In one set of experiments, if onions were kept weed-free for just 2 weeks, from 5 until 7 weeks after crop emergence, the yield was as high as if the crop was kept weed-free for the whole season. Weeds emerging with the crop thus do not compete before the 5th week, while weeds that emerge after the 7th week never compete sufficiently with the crop to cause yield loss. The minimum period during which the crop must be kept free of weeds to prevent yield loss is known as the *critical period* for weed competition.

In some crop situations the critical period is so short as to be effectively a point in time. In beetroot, for example, experiments have shown that a single, well-timed weed removal, around 4 weeks after crop emergence, is sufficient to prevent yield loss. If the weeds are removed earlier than this time, subsequently emerging weeds are able to compete later in the season. If, on the other hand, weed removal is delayed until after this time, those weeds that emerged along with the crop will already have competed sufficiently to cause yield loss at harvest time.

A third situation has been observed in spring-sown broad beans, where again a single removal of weeds was enough to prevent yield loss, but in this case the timing of the operation was less critical. Weed removal once only, any time from 1½ to 4 weeks after crop emergence, was found to give as high a yield of beans as keeping the crop weed-free all season.

In some crops the time of yield-damaging weed competition depends on the particular growing season. In a wet year when weed growth is more prolific, root crops not only suffer greater yield loss if weeds are not removed, but also show less flexibility in the time at which weeds must be removed to prevent yield loss. Sugar beet in a wet season has been shown to have a longer critical period for weed competition than in a dry season. In swedes a single removal of weeds has been found to be sufficient to prevent yield loss, but in a wet year the timing of this operation was highly critical at about 6 weeks after sowing, whereas in a dry year weeding once only at any time from 4 to 7 weeks after sowing was adequate (Fig. 11.8).

Such findings are of great relevance to the practice of weed control. Mechanical weed control, for example by hoeing or soil cultivation between the crop rows, should be carried out at the time which will allow a full yield to develop or, if there is a critical period for weed competition, at least twice, at the beginning and end of the period. Selective herbicides which are applied to the growing crop, that is, *post-emergence herbicides*, should similarly be applied at the correct stage to minimize weed competition.

However, the timing of application of a post-emergence herbicide is often determined not by considerations of weed competition but by the susceptibility of the crop itself to damage by the herbicide, which tends to diminish as the crop gets older. Alternatively it may be determined by the susceptibility of the weeds, which also declines with advancing growth. In recent years emphasis has been placed on the development of herbicides which will safely control wild oats in wheat or barley at advanced stages of growth, but the need is really to control wild oats early, before any yield-damaging competition starts.

Many herbicides are applied to the soil before sowing the seed or immediately after sowing but before emergence. These *pre-emergence herbicides* act by killing weed seedlings as they germinate or during their early growth. The activity of pre-emergence herbicides persists for some time after they have been applied. The duration of this persistence must be long enough to prevent weed

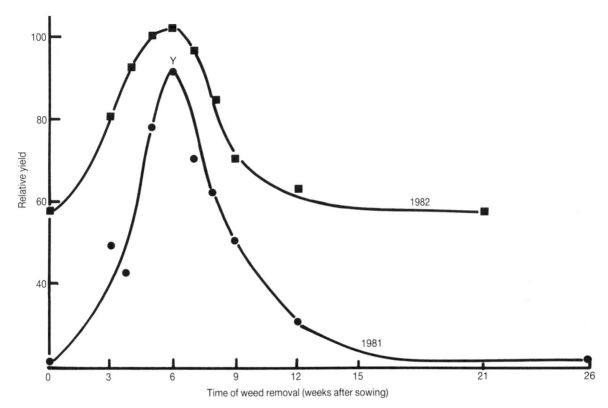

Fig. 11.8 Effect of time of weed removal on yield of swedes (relative to yield in the total absence of weed competition) in a wet year (1981) and a dry year (1982). Significance of point Y explained in the text. From J C Forbes (1985) *Ann. Appl. Biol.* **106**, 505–11.

emergence before the point marked Y on Fig. 11.8.

There are other justifications for weed control besides preventing yield loss, and timing of herbicide application need not always be related to the time of weed-crop competition. For example, a late application of a post-emergence herbicide is often made to sugar beet long after the end of the critical period, in order to ensure a weed-free crop to facilitate harvesting. The herbicide glyphosate is applied to ripening cereal crops for control of couch grass and other weeds, not only to speed up harvesting but to reduce the burden of weeds in the succeeding crop.

11.4.3 Allelopathic and parasitic weeds

Competition is by far the most important way in which weeds cause yield loss in crops. Certain weeds, however, may reduce yield through other effects.

We saw in Section 6.1.5 that some plants produce exudates which retard or suppress the growth of other species growing in close proximity. This is known as *allelopathy*. Pot experiments have shown that a number of weed species produce such exudates and that certain crops grown on soil in which these exudates are added show reduced growth. In the field, however, it is virtually impossible to separate the effects of competition and allelopathy, and we do not know if allelopathy plays any significant part in the reduction of crop yield caused by weeds.

A few weed species, including dodder and broomrape, are directly parasitic on crop plants. Dodder, for example, has no chlorophyll and cannot photosynthesize its own organic matter. Its slender stem twines around its host plant, producing from each node specialized adventitious roots which grow into the tissues of the host, drawing out the products of the host's photosynthesis to meet its own requirements. Clearly all growth of the parasitic weed must be at the expense of growth of the host crop, and yield is inevitably reduced.

11.5 Yield of mixed crops

Farmers commonly wish to grow two or more crops on the same land in the same year in order to make better use of the limited land area available to them. This can be achieved in various ways.

11.5.1 Multiple cropping

One approach is to take two or more crops sequentially within one year from the same area, sowing one immediately after, or in some cases even before, the previous one is harvested. This is normal practice in tropical or subtropical climates with no pronounced dry season or where irrigation can be practised during the dry season. For example, in the Canary Islands it is possible to grow three crops of potatoes in one year.

In temperate climates with a cold season this approach is more difficult, but even in Great Britain it is possible to grow a main crop which is harvested in mid to late summer, followed by a secondary, or *catch crop* to utilize the remainder of the growing season. The further north one goes in Britain the shorter the growing season becomes and the less time is available for catch cropping. Any catch crop must be fast-growing; to save time it is often direct-drilled into the stubble of the preceding

cereal crop without ploughing. Most catch crops are grazed by animals in late autumn or early winter. Some commonly used are Westerwolds ryegrass, a very fast-growing annual form of Italian ryegrass, and various brassicas such as forage rape and stubble turnips.

An alternative approach to multiple cropping is to grow two or more crops simultaneously on the same land. The crops may be quite different from each other and harvested separately. Commonly annual and tree crops are grown together (Fig. 11.9). In the tropics, coconut plantations are sometimes underplanted with pineapples or cocoa, whilst in southern Europe a common sight is a field of lucerne (alfalfa) or barley with regularly spaced olive or almond trees, thus again yielding two quite separate products.

A recent study in North Dakota showed that a mixture of wheat and linseed could give higher economic return than separate wheat and linseed crops grown on the same total land area. Practical difficulties of growing such a mixture have been eased by the advent of herbicides which control

Fig. 11.9 Multiple cropping involving an annual crop (sugar beet) grown under a tree crop (poplar for timber) in Belgium.

weeds selectively in both crops, and by harvesting equipment which can separate the two kinds of seeds.

Various mixtures of cereals and legumes, for example oats or barley with vetch or peas, are sometimes grown as a mixed forage crop, the legume enriching the protein content of the feed and the cereal supplying the bulk of the carbohydrate and supporting the weak-stemmed legume in the field. Other advantages of mixed cropping include the spreading of risks from pest and disease attack and, especially in tropical agriculture, the reduction of soil erosion.

However, two crops cannot occupy the same ground at the same time without competing for light, water and mineral nutrients. The yield of each crop will almost certainly be lower than if it were grown on its own. One exception to this is tea, which in some areas yields better in the shade of trees than in the open where it tends to suffer from overheating of the leaves. In general, the practice of mixed cropping is worth while only if the combined yield of the two crops is greater than the yield of either grown alone.

In glasshouses various forms of multiple cropping are practised. Tomatoes or chrysanthemums are sometimes interplanted with lettuce, which grows fast and is removed for selling before the main crop has produced a full leaf cover. This technique effectively increases leaf area index in the early part of the season and is similar in effect to catch cropping.

In the humid tropics, the various traditional systems of shifting agriculture, in which clearings made in the forest or bush are cropped for a few years before being abandoned, almost always employ mixed cropping, usually with quite a wide range of species. Where year-round rainfall permits, series of overlapping crops can maintain almost continuous complete ground cover. The resulting increases in yield over those achievable by growing the same crops individually can be considerable. For example, two hectares of a mixture of maize and beans can yield 35% more than one hectare of maize and one hectare of beans. The yield advantage of similarly mixing sorghum and beans is as much as 55%. In 'hand tool' farming such increases in productivity per hectare are important because the alternative, which is the cultivation of a larger area, is usually not possible with the limited labour available. Another benefit is the reduction of risk. If one crop fails totally, through attack by insects, disease or other causes,

the other crop or crops in the mixture will still provide a yield from each cultivated hectare.

In northern Europe, a cereal (usually spring barley or oats) is often used as a cover crop for grass sown around the same time or slightly later. This practice of *undersowing* grass allows earlier utilization by animals than if it is sown after the cereal has been harvested, and ensures greater grassland productivity in the year after sowing. To understand the benefits of undersowing cereal crops with grass, it is necessary to look at production over two years. If grass were sown in spring, without a cereal cover crop, it would produce in the year of sowing one silage or hay cut followed by regrowth (the aftermath) for grazing. With a cover crop, however, the most that could be expected would be light grazing in the autumn. By the following year the grass in either case would be well established, but the grass sown without a cover crop would almost certainly perform better than the undersown grass as it would be better tillered and have a better developed root system. However, the yield of undersown grass in its second year is considerably greater than that of grass sown alone in its first year.

As far as the cereal is concerned, what is often not realized is that it also suffers by competition from the undersown grass. This competition may result in a loss of grain yield of up to 0.5 tonne ha^{-1}. However, it may be worth sacrificing some grain yield in order to obtain a higher grass yield the following year. Some agronomists argue that undersowing is a bad practice because the grower makes neither a good job of the cereal nor a good job of the grass. Among the disadvantages of undersowing is a great restriction on the choice of herbicides that can safely be used for weed control in the cereal crop.

11.5.2 Mixtures of similar crops

Since almost all grasslands, most forests and a large proportion of tropical crops consist of mixtures of species, it can be seen that mixed cropping is the rule rather than the exception. In temperate agriculture, however, most arable crops are grown on their own, all plants in the crop normally being of the same variety. Such a crop is known as a *monoculture*, though this term is also sometimes used to mean the growing of the same crop year after year. From time to time interest is shown in growing crops of different species, but similar growth habit and end-product, together in a mixture. An example is barley and oats grown for stockfeed. Mixtures of varieties of one species can be thought of in the same way as mixtures of different species.

It is sometimes claimed that barley-oat mixtures show greater stability of yield than either cereal on its own, because if a particular growing season is less favourable to one component species of the mixture the other will compensate. It is certainly true that attacks by disease are less devastating in mixtures than in monocultures.

But how do mixtures compare in yield with monocultures? To answer this, we must first recall that in any crop the individual plants compete strongly with one another. The question is whether the competitive effect of species B on species A is greater or less than or equal to that of species A on itself, and likewise whether the competitive effect of species A on species B is greater or less than or equal to that of species B on itself. Various possibilities are illustrated in Fig. 11.10.

In example 1 the response of either A or B to the presence of neighbouring plants is unaffected by whether the neighbours are A or B. In this case interspecific competition is no different from

Fig. 11.10 Yields of monocultures of species A and B (solid lines) and of mixtures of A and B in varying proportions (broken lines). Only in example 4 does the yield of a mixture give a higher yield than a monoculture of A, because in this example both species suffer less from interspecific than from intraspecific competition.

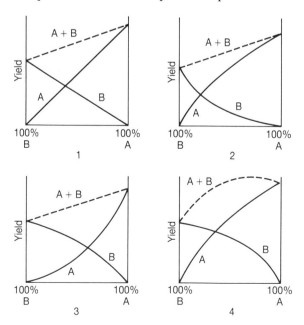

intraspecific competition, and A and B are said to compete *equivalently* with one another.

In examples 2 and 3, A and B do not compete equivalently. One species is affected more, and the other less, by interspecific than by intraspecific competition. The increased yield per plant of one component of the mixture is entirely at the expense of the other component, so that the total yield of the mixture is, as in example 1, intermediate between the yields of monocultures of A and B at the same total plant density.

In example 4, both species are less affected by interspecific than by intraspecific competition. In this situation the mixture is higher-yielding than either monoculture. This does not mean that somehow the presence of one species stimulates the growth of the other. It simply means that the mixture is exploiting the environment better than monocultures. It may be, for example, that one species is deeper-rooting than the other, so that the components of the mixture are utilizing substantially different reserves of water and mineral nutrients. They are thus in less direct competition than are plants of the same species.

In mixtures of cereal varieties competition has usually been found to be equivalent and the yield of the mixture to be intermediate between the yields of monocultures of the varieties in question. Where two cereal species such as barley and oats or wheat and rye are mixed the situation is as in example 2 or 3 of Fig. 11.10, but which species of the two is more competitive depends on soil or weather conditions. For example, in a barley-oat mixture the oats are more competitive and tend to replace the barley on soils of pH lower than 6, whereas on soils of pH higher than 6 the barley generally has the upper hand. Similarly, in a wheat-rye mixture wheat is the more competitive of the two components in a hot, dry season, but rye is more competitive in a cool, moist season. In no situation, however, is the yield of the wheat-rye mixture greater than that of the higher-yielding monoculture.

One type of crop that is more commonly grown as a mixture than as a monoculture is grass. Here it is often observed that the yield of the mixture is greater than that of a monoculture of any of the component species (cf. example 4 in Fig. 11.10). Grassland is, in most cases, an extremely complex community of many species and varieties and its yield is further complicated by the fact that it is harvested by animals, who are selective grazers. For simplicity, therefore, let us take an initially half-and-half mixture of two grass species, cocksfoot and timothy, and measure the yield of digestible dry matter from this mixture and from cocksfoot and timothy monocultures (Fig. 11.11).

In a dry situation, the deeper-rooted cocksfoot performs better than timothy. In the mixture, cocksfoot contributes considerably more than half the total digestible dry matter yield, though initially the two species were present in equal amounts. The yield of the mixture is intermediate between the yields of the monocultures, but is closer to that of cocksfoot than to that of timothy. In a wet situation, the relative yields of cocksfoot and timothy are reversed and timothy contributes more than half the yield of the mixture. In a situation which is neither excessively dry nor excessively wet the yield of the mixture will probably be greater than that of either monoculture.

Fig. 11.11 Yields of monocultures and mixtures of two pasture grasses, cocksfoot and timothy. In dry situations the contribution of cocksfoot is greater than that of timothy, whilst in wet conditions timothy contributes more than cocksfoot.

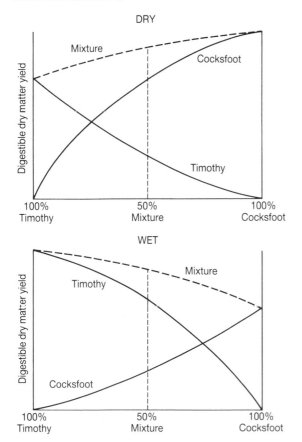

How does this come about? Partly it is because of the difference in rooting depth which gives cocksfoot access to reserves of water and nutrients not tapped by timothy, while allowing timothy to exploit the upper layers of the soil more fully. It is also partly due to differences in the seasonal pattern of growth of the two species. Cocksfoot is earlier to commence growth in spring than timothy, thus when sown in a mixture the early growth of cocksfoot is little affected by competition from timothy. But cocksfoot is also earlier to flower, a process which greatly lowers its digestibility (see Section 12.1.4), and it is digestible rather than total dry matter yield that is important in a grass crop. While the digestible dry matter yield of cocksfoot is falling in early summer, that of timothy is still high. Thus the cocksfoot-timothy mixture effectively uses a greater part of the growing season to produce digestible dry matter than either species does on its own.

This example also provides a good illustration of the benefit of mixtures in providing stability of yield. While a monoculture of cocksfoot or timothy will show marked yield variation between wet and dry years, the mixture can be expected to vary much less in yield.

11.6   Impact of stress on yield

In agricultural practice the full yield potential of a crop on any site in any season is probably never realized. Periods of unsuitable weather may cause drought, waterlogging or frost damage. Insect pests or parasitic fungi may attack the crop and reduce its yield.

These are examples of what may be called the negative influences on crop yield. Where any such influences are injurious to the health of a crop, we say that the crop is subject to *stress*, using the word in its widest meaning. Some biologists are not entirely happy with this use of the term 'stress', but it is widely used for lack of a better one. To make matters clearer, we use the term here to denote a negative stimulus in the plant's environment, but not the plant's response to it. (In Chapter 4 we use the terms stress and strain with the more restricted meaning derived from civil engineering, covering only mechanical stress and the plant's response to it, respectively.) Much of the effort that goes into the management of a crop is aimed at reducing stress and its effects, and we should therefore spend some time considering the impact of stress on yield, as it can be very substantial.

Broadly, stresses can be classified into those caused by living, or *biotic*, agents, and those caused by non-living, or *abiotic*, agents. Abiotic factors include adverse weather or soil conditions, and toxic chemicals such as may occur in atmospheric pollution. Biotic factors include large animals, especially grazing animals such as cattle and sheep, and the numerous types of smaller invertebrate animals that feed on plants, including insects, mites and nematodes. Weeds are also biotic factors, but, as we saw in Section 11.4, their influence on yield is primarily through competition for resources rather than direct stress on crop plants. Also within the category of biotic factors are the various microbes that invade plants. Chief among these are the plant-pathogenic fungi, but there are also many species of bacteria which are important in this area. Viruses, although often not thought of as living organisms, are normally included as biotic factors, and have a major influence on the yield of many crops. One source of stress of increasing significance arises from man's attempts to control weeds, pests and diseases in his crops, though it is not clear whether we should class this as biotic or abiotic. An example is herbicide damage. Here the grower, in removing one stress, that arising from weed competition, may be unwittingly inflicting another.

The conditions required for a crop to produce maximum yield, together with some examples of negative influences, are laid out diagrammatically in Fig. 11.12. These conditions seldom even approach attainment in practice, despite the best efforts of the most skilled growers to provide them.

Fig. 11.12 Conditions required for a crop to give maximum yield (inner circle) and examples of stress factors which can lower yield (outer circle). A stress factor may lower yield either through its own direct impact or through its influence on another stress factor. For example, frost reduces yield through its direct impact but also assists entry by infectious microbes. Excessive rainfall has little direct effect but causes damage by inducing waterlogged soil conditions. Particular atmospheric and soil factors influence the spread of pests and diseases.

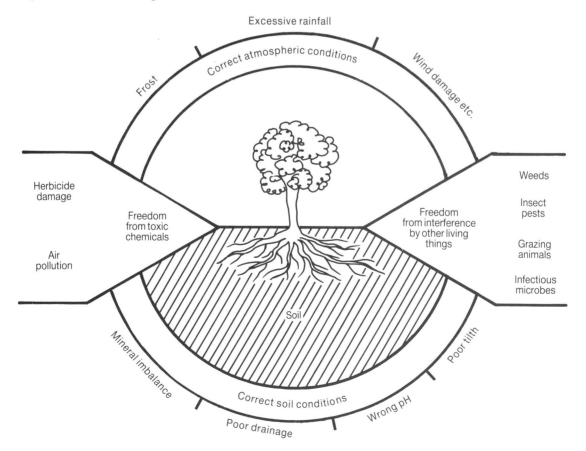

Weather conditions in most climates do not remain ideal for more than very short periods, and the same is generally true of soil conditions. Every plant in even the best managed crop can be shown, by careful examination, to carry a range of insect and other pests, and to be affected, at least to some degree, by several plant pathogens. The 'healthy' plant, that is a plant entirely free from such negative influences, may exist as a concept, or occasionally in the laboratory, but not in practical agriculture, which has to be satisfied with much lower standards. In practice, health is a relative rather than an absolute term, and the combined impact of negative factors, both biotic and abiotic, on the yield of crops remains considerable.

11.6.1 Types of damage caused by stress factors

It is not possible to produce a neat classification of the types of damage to plants caused by stress. There are many kinds of effects and many overlaps between them. Nonetheless, it is helpful to note some of the main types.

The most obvious form of damage is where tissue is removed or destroyed, simple examples being the activities of grazing animals or leaf-eating insects such as caterpillars. Many pathogenic microbes cause destruction of tissue by rotting it. Examples include the fungus *Phytophthora infestans* which rots the foliage and tubers of potatoes to cause the disease late blight, and bacteria of the genus *Erwinia* that cause soft rots of many vegetables and fruits, in which the tissues of the invaded organs collapse into a liquid purée. Frost and hail damage are instances of comparable tissue destruction by abiotic factors.

Certain insects such as aphids and leafhoppers extract juices from the plant by inserting their long, hollow, needle-like mouthparts through the cuticle and epidermis into the phloem, and allowing the sap to flow through them into their digestive tract. In other words they remove material from tissues without removing or destroying the tissues.

Other organisms act by releasing toxins into the translocation system of their host plant, the toxins damaging those parts they reach. They may thus cause tissue destruction at some distance from the site of attack. Toxins also play a role in the effects of certain abiotic factors – for example, the damage caused by waterlogging is partially caused by toxins produced under anaerobic conditions in the soil entering the plant.

Whereas many parasitic organisms act like *Phytophthora infestans*, simply destroying plant tissue as they attack it, others are more subtle in their impact, avoiding tissue destruction but causing various alterations in the physiology of their hosts. Some viruses, for instance, cause little obvious damage to host tissues but the host plant is nonetheless stunted as a result of physiological and metabolic changes that follow infection. Similarly, some of the most highly evolved fungal pathogens, such as the powdery mildew fungi that infect many crops, do not cause extensive tissue damage. The host tissue remains alive even when heavily infected. However, soon after infection has become established, it is observed that, among other effects, the products of phytosynthesis are induced to flow more towards the site of infection, where they are utilized by the fungus, rather than towards those parts of the plant where they are needed. The fungus thus creates a sink which competes with grains, fruits or tubers for carbohydrates and other materials, thereby reducing yield.

Another type of effect induced by some stress factors is alteration in the growth pattern of the plant. Viruses, for example, frequently cause leaf distortion as well as stunting of the whole plant. In some virus infections quite profound changes occur, such as the partial transformation of petals into leaves. Growth distortion is also a characteristic effect of the damage caused by some herbicides if wrongly applied, and even by some air pollutants. Some insects, bacteria and fungi cause the development of outgrowths known as *galls* on susceptible plants. Galls are usually produced by the over-enlargement and rapid, repeated division of cells to form a mass of poorly organized tissue, as occurs in the roots of brassica crops invaded by the fungus *Plasmodiophora brassicae*, giving the disease known as club-root (Fig. 11.13).

However, by far the most common effect of stress is simply to reduce growth rate and thereby, in most cases, yield. Examples of stresses which have this effect are low temperature and shortage of soil moisture, nitrogen and other minerals, notably phosphorus.

Throughout a growing season, many stress factors come into play at different times to limit growth. The yield achieved reflects the cumulative impact of these factors.

Plants are able to adapt or acclimatize to certain types of stress in such a fashion that damage is reduced. This is particularly evident with drought or cold. If maturing plants of many species are

Fig. 11.13 Club-root disease of a brassica plant. Invasion by the causative fungus causes development of root galls which greatly restrict the ability of the root system to absorb water and mineral nutrients, leading to severe yield loss.

subjected to drought, individuals raised under conditions of relative water scarcity tend to be less affected than those raised in a regime of plentiful water supply. This may be partly because when water is scarce root systems grow deeper to explore greater volumes of soil and partly because of changes such as greater deposition of cuticular wax to reduce transpiration.

In cool temperate areas, plants with poor cold resistance are often raised under glass until the arrival of warmer weather allows them to be planted outside. Generally they are not transferred directly from glasshouse to field, but via an intermediate stage in which they are grown for a week or two at lower temperatures than in the glasshouse, though still with some protection. This process, known as *hardening off*, often takes place

in cold-frames and allows the plants to develop greater resistance to cooler growing conditions.

In general, the more sudden and unpredictable the onset of stress, the less likely it is that the plant will already be acclimatized or that it can rapidly acclimatize in response to it. Even so, it is possible that we may have underestimated the ability of plants to adapt to stress. There is some evidence, for example, that certain trees can pre-adapt, and increase their resistance, to insect attack by sensing emissions of organic substances induced in nearby trees that have already been attacked.

11.6.2 Timing of stress and the crop's response

Stress can be particularly damaging when it coincides with critical stages or processes in the crop's development. As we saw in Section 5.3, the seedling stage is a particularly vulnerable one to all kinds of stresses, but there are also later stages of development where stress can have severe effects. For example, in an experiment in peas, one day of soil waterlogging just prior to flowering reduced yield by one-third.

Indeed in most crop species the period from just before flowering through to and including seed set is a vulnerable one to many types of stress. To some extent, this is due to the comparative delicacy of flowering parts, such as petals, ovules, anthers and developing pollen. Frost, for example, can damage flowers at temperatures at which leaves are untouched. The early stages of organ formation are also vulnerable to stress. Even short periods of stress during initiation or early development of any organ are likely to have highly damaging and long-lasting effects on that organ.

From the above, it will be apparent that the impact of a stress depends at least as much on the stage of development reached by the plant as on the nature and severity of the stress. Some simple experiments in which watering was withdrawn from equal numbers of barley plants in pots, at different stages in their growth, demonstrated this well (Table 11.2). The impact of the imposed drought depended largely on what the plant was doing at the time. Drought at any stage of growth reduced yield to some extent, though stress applied at the earliest stages had the least effect. This was to be expected, since young plants have the opportunity to compensate for reduction in some components of yield such as number of ears per plant by increases in others such as average grain weight at a later stage. Predictably, straw height was most

Table 11.2 Response of barley in pots to withdrawal of watering at different stages of its growth

Timing of drought period	Total grain weight per pot(g)	No. of ears per pot	No. of grains per ear	Average weight per grain (mg)	Straw height (cm)
Well watered throughout	14.4	19	16.8	45	94
Drought at seedling stage	12.1	17	14.2	50	79
Drought during stem extension	12.2	27	10.1	45	46
Drought just prior to and just after head emergence	6.3	28	6.8	33	64
Drought during seed filling and ripening	6.7	18	14.0	27	94

From Agricultural Development and Advisory Service, *Report of Field Experiments, East Midland Region, England, 1969.*

reduced when drought occurred during the period of stem extension. In contrast, drought during and just after ear emergence, when much of the stem extension was completed, but when flowering parts were still in an early stage of development, affected straw height much less, but gave a much more severe depression in yield. Examination showed this was due to abortion of grains at the tips and bases of the ears, and this is apparent from the column in Table 11.2 showing the number of grains per ear. The latest period of drought applied during seed filling and ripening caused an equally severe yield reduction, but for different reasons. The grains were established and could not be aborted, hence the number of grains per head remained high, but the grains simply did not fill, as is apparent from the column showing the weight per grain. Naturally, no effect on straw height occurred, since that was already fully developed.

Some of the more subtle implications of the relationship between timing of a stress and its effect on yield are demonstrated by studies in Britain of the effect of infection by *Erysiphe graminis*, which causes powdery mildew, on the yield of winter barley (Fig. 11.14). If the infection develops in the autumn when the barley plants are young, the main effect is to stunt the growth of the roots. This results in more plants being killed by winter conditions as they become more vulnerable, particularly to frost, in their weakened condition. A reduced number of ears reach maturity, partly because of the increased winter kill and partly because fewer tillers form on those plants that survive. As a result yield may be substantially lowered.

Late attacks of the disease cannot affect earlier stages of crop development such as tillering, but

Fig. 11.14 The fungus *Erysiphe graminis* growing on a leaf of barley, causing the diease powdery mildew. Long chains of spores can be seen forming (photograph courtesy of S J Wale, Scottish Agricultural College).

reduce the supply of carbohydrate to the developing grains. This is particularly serious when infection spreads to those organs that are primarily responsible for grain filling, which in barley are the uppermost two or three leaves and the awns (the long bristles on the ear). Thus yield in barley is affected both by early and by late disease attack, though for quite different reasons.

However, the response of crops to stress is often such as to reduce or even nullify the effect on yield. Plants commonly produce organs in excess of their needs, giving them a margin of safety that can be called upon if necessary, a form of insurance if it helps to regard it in that way. Cereal plants, for example, always have a large number of potential tillers in the form of axillary meristems. Some of these develop but many are never activated. Of those that develop as vegetative tillers many subsequently die without bearing ears. If early stress reduces the number of vegetative tillers produced, a higher porportion of these survive to produce grain, and there may be no observable effect on yield.

If the number of ear-bearing tillers per plant is reduced by stress at a slightly later stage, the cereal plant still has flexibility of response. It may produce more grains per ear, or larger grains, or both. The increased number of grains per ear is achieved not by more flowers being initiated but by fewer aborting, since these too are produced greatly in excess of need. This is also seen in many non-cereal crops. Cotton, for example, forms more flower-buds than can ever produce cotton bolls. Where stress reduces the number formed, fewer abort or remain inactive, and again there may be no loss in yield. In general, crops with an indeterminate flowering habit, such as peas, are even more flexible in their ability to respond and compensate in this manner than crops with a determinate habit, such as cereals.

As an example of compensatory response to stress at a later stage in development, many fruit trees such as apples set more fruit than they can ever ripen, shedding the excess shortly after setting. In this situation stress may cause obvious damage such as reduced fruit set without any yield loss resulting. Such examples of 'slack' in the plant's organization should not be seen as inefficiency. This slack has evolved to give the plant margins of safety to cope with numerous varying stresses throughout its life.

Another situation where the effect of stress on yield is minimized is where the plant simply has time in hand in which to repair the damage done. For example, barley sown in the autumn may enter the winter carrying considerable leaf infection by various fungal pathogens, but control of these diseases in the autumn often produces no measurable increase in yield the following year. The explanation seems to be that the plants need only grow sufficiently in the autumn to establish themselves before the winter, and provided the disease stress does not persist, as we saw in the case of powdery mildew, any loss of foliage due to fungal attack can be readily restored in the spring. Time to accomplish this is not a limiting factor.

The plant's ability to recycle minerals and other nutrients from redundant or damaged tissues, which was discussed in previous chapters, also gives some flexibility to its response, though the nature of the damage caused by stress in some cases, such as loss of tissue to grazing or insect attack, precludes such recycling. Where damage is done to organs important for supplying products of photosynthesis to the harvestable parts, for example the uppermost leaves of cereals which make major contributions to the swelling of the grain, there is usually enough flexibility in the plant's organization to allow other organs to take over their function at least in part, and to reduce the potential yield loss.

It is noticeable that crops often sustain considerable damage to both photosynthetic and root tissues without measurable yield loss. Part of the explanation lies in the fact that these tissues are relatively transient and expendable. Roots, for example, are constantly subject to attrition by numerous insects, nematodes, fungi, bacteria and other organisms in the soil. The loss of roots by this means over a growing season is very considerable. Plants therefore produce these organs on a renewable basis .

Inevitably, the capacity for flexible response, by whatever mechanism, is progressively lost as the plant ages, and more and more yield components such as number of ears per plant and number of grains per ear become established and thus fixed. Not only is the number of flexible factors reduced, but the margins of safety on the remaining ones are probably smaller.

11.6.3 Interactions between stresses

It has been pointed out that under normal growing conditions plants are frequently subjected to more than one stress at any one time. It follows that interactions between stresses and the crop's response to multiple stress are significant matters influencing yield.

At the simplest level, there may be no interaction between stresses. Their overall impact on the crop is simply additive. The combination of frost and hail occurring within a short space of time is an example of two additive stresses.

More subtly, one stress factor may act synergistically with another; that is to say, their total impact is greater than the arithmetic sum of their individual impacts if they occurred separately. Air pollutants often occur in mixtures, and these may interact synergistically. A clear instance of this is the interaction of sulphur dioxide and nitrogen dioxide, which damage plants when presented as mixtures at concentrations which, separately, would not induce symptoms.

Very commonly, two forms of stress are linked in their occurrence. For example, one stress may, by its impact outside the plant, induce another. Partial drought not only causes water stress in crops but also, as was noted in Section 3.4.2, greatly reduces the plant's ability to take up minerals, including those derived from applied fertilizers. Thus nutrient stress is added to water stress and yield may be reduced by both. Furthermore, drought tends to occur during long periods of continuous sunshine. These are not only the hottest periods, but are also the periods when leaves are least subject to the cooling effects of transpiration, because of the shortage of water. Thus drought stress, especially in the tropics and subtropics, tends to induce additional stress due to overheating of leaves.

Another type of interaction is where one stress increases the susceptibility of a crop to another stress. Any factor which inflicts mechanical wounds on a plant, such as hail damage or rough handling of fleshy storage organs at harvest, creates openings in the plant's defences through which parasitic microbes or insects may enter. Drought makes many trees more susceptible to invasion by fungi and insects. Overheating of tree trunks by direct sunshine inhibits the production of substances in the tree that defend it against pathogenic fungi. Many other examples of cause and effect relationships between stresses exist.

Whatever the mechanism of interaction, it appears to be general practical experience that crops can often tolerate damage from one stress without appreciable yield loss, but not by two or more coming together. For example, crops affected by frost or by herbicide damage often recover and show no yield loss, but not if affected by both at the same time. It may be that one can be accommodated within the safety margin of the crop but that two take it across the threshold of damage that cannot be outgrown.

11.6.4 Levels of yield loss due to stress

The impact of stress on yield thus depends on three things. It depends firstly on the nature of the stress – to be more precise the type of damage inflicted: its extent, its duration and its timing. Secondly, it depends on the reaction of individual plants and of the crop as a whole, and thirdly on how the stress interacts with other stress factors.

The effect of stress on the overall performance of the crop can range from slight stunting to death, either of the organs attacked or of whole plants, but it is difficult to relate the damage caused to the ultimate impact on yield, so complex are the processes involved and the reactions of the plant. Few biotic or abiotic stresses have only one effect, as we saw in the case of drought stress. Not only yield, but also quality is often affected. Table 11.3 illustrates the impact on yield and malting quality of barley due to one disease, leaf blotch, revealed when the disease was controlled by spraying the crop with a protective fungicide.

Sometimes the relationship between the stress and its impact on yield is obvious. These tend to be situations where yield loss is total or nearly so. This chiefly occurs when what is affected is some critical part of the plant, or some critical process or phase in its development. Decay by fungi or bacteria of harvestable organs, such as the tubers of potatoes or yams or the fruits of apples or strawberries, often leads to total yield loss because it directly affects the harvested product. Even slight blemishes to such parts can have major impacts on saleable yield when, for example, they reduce the eye appeal of fruit or flowers to potential buyers. Frost, when it kills the flowers of crops such as coffee, is also in the category of stress factors causing catastrophic losses. Total yield loss also sometimes results from stem or root rots which, once the causal fungi or bacteria have rotted through the entire diameter of the stem or killed the root system, inevitably cause the death of all parts above the site of infection. The fungus *Sclerotinia sclerotiorum* causes this kind of damage to the stems of a wide range of crop species. A number of wood-decaying fungi and bacteria similarly

Table 11.3 Effects of control by the fungicide thiophanate methyl of leaf blotch of barley (caused by *Rhynchosporium secalis*), on the quality and quantity of yield. Spraying at other growth stages influenced straw length, average grain size, and the nitrogen content of the grain. Nitrogen content is critical for the grains' usefulness for malting. Germinative energy is a test used by maltsters to test speed of germination.

	Total infection at treatment (% of green leaf area)	Reduction in infection (%)	Yield of grain (t ha^{-1}) at 15% moisture content	Increase in yield (%)	Thousand-grain weight (g)	Increase in thousand-grain weight (%)	Germinative energy (%)	Nitrogen content (% of dry matter)
Control (no fungicide treatment)	180.3	—	3.97	—	33.0	—	84	1.79
Sprayed with fungicide in the early stages of grain ripening	122.9	32	5.07	28	35.9	8.8	94	1.76

From *Research Investigations and Field Trials*, North of Scotland College of Agriculture, 1979–80.

destroy the stems of fruit and timber trees.

When, on the other hand, a stress causes no measurable yield loss, it may have simply been so slight in its effect that no permanent damage was done to the plant. More often, some of the various recovery or compensation mechanisms we have outlined account for the lack of yield loss.

Most commonly, it is not a case of all or nothing, but a gradation of impact on yield that is correlated with the degree of damage inflicted. To understand this, it should be recognized that a crop, as a population of plants, behaves differently from individual plants in its response to stress. A crop draws on the resilience of individual plants within it to increase their growth to take advantage of reduced competition from injured neighbours. As a result, the recuperative power of a crop is substantially greater than that of individual plants.

For example, in experiments on sugar beet, removal of up to 10% of the plants at the seedling stage caused virtually no loss of yield (Fig. 11.15). Even removal of half the young plants reduced yield only by about 10%. The effect is exactly like that of reducing sowing density, in that the individual roots grew larger to compensate wholly or partially for the smaller number of plants. The sugar beet root, with its many concentric vascular cambia acting as areas of potential growth, is flexible enough to respond to reduced competition in this way (see Section 6.1.9). In contrast, as we saw in Section 11.2.1, the potato crop has two levels of flexibility – in number of tubers per plant and in tuber size. Plant removal experiments in

Fig. 11.15 The effect of plant removal (open circles) and defoliation (solid circles) at the four-leaf stage on final root yield of sugar beet. From F G W Jones *et al.* (1955) *Ann. Appl. Biol.* 43, 63.

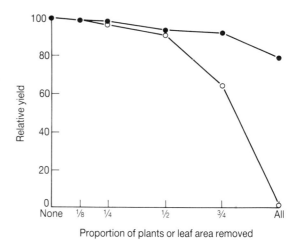

potatoes suggest that compensation in this crop is primarily by increasing tuber number rather than tuber size.

Returning now to our sugar beet example in Fig. 11.15, if instead of removing whole plants, half the foliage of seedling beets was removed leaving the plant density unchanged, yield was again reduced by about 10%. This time the average size of harvested beets was reduced but their numbers were not. The response of sugar beet to infection by the virus that causes the condition known as sugar beet yellows was in marked contrast. If only 12% of plants became infected early in the season, significant yield loss resulted. Why should disease reduce yield more severely than partial defoliation or removal of whole plants?

The answer is that artificial removal or defoliation of plants at one point early in the season still allows the surviving plants or plant parts to respond thereafter in a 'healthy' manner, and recoup losses. In practice these treatments correspond to single event stresses such as frost or hail damage, or loss of seedlings through insect attack. Many crops probably have substantial safety margins built into their seed rates to allow for such stresses, especially in developing countries where the seed is often of poor quality and may not be treated with protective chemicals.

However, once a plant is infected with the sugar beet yellows virus, the disease continues to affect the host's performance throughout its life-span. The stress is sustained, not transient. In addition, the pathogen spreads to more plants during the season. Perhaps the most significant point, however, is that healthy plants around those with the disease do not grow much more to compensate for the yield loss of their neighbours, which are only partially stunted. The diseased plants each grow to occupy an area almost equal to that of a healthy plant, but they utilize it less efficiently. In short, the effect of diseased plants on the yield of the crop is rather like that of weeds; they compete with the healthy plants but give little useful return themselves.

This association between apparently quite low levels of stress on a crop and substantial impacts on yield is typical of sustained stresses. In particular, it is typical of diseases such as sugar beet yellows in which an altered physiology of the host plant rather than crude tissue destruction is the primary effect. Other examples of such diseases include powdery mildew of barley. In experiments in Great Britain it was found that if this disease affects only 4% of leaf area at the completion of ear emergence, it causes 5% yield loss, and if 10% of leaf area is affected there is 16% yield loss. In potato crops, potato virus X usually causes symptoms barely noticeable to the untrained eye, yet, before it was controlled, it reduced yields of potatoes in many areas by around 13%. Thus how spectacular the damage is to a crop may be little guide to its impact on yield.

It should also be noted that damage to plants is often not evenly distributed throughout a crop.

Pest and disease attacks, and problems such as poor drainage, are often locally concentrated. Within such intensely damaged patches compensatory growth by neighbouring plants is not possible except around the margins of the damaged areas. The effect of the non-random distribution of damaged plants on yield loss is shown in Fig. 11.16.

With single event stresses, as shown by artificial removal or defoliation, the greater the proportion of the growing season that remains for recovery, the less damage is done to final yield. On the other hand, with sustained stresses like virus diseases or prolonged drought, the plant's performance is permanently impaired, and the earlier the onset of the stress the greater is the proportion both of the growing season that is inefficiently utilized and of the potential yield that is lost.

Late blight of potatoes demonstrates these points in a slightly different way. The causal organism, *Phytophthora infestans*, simply destroys its host's foliar tissue as it invades it, but unlike the powdery mildew fungus in barley, it does not alter its host's physiology in any profound way. Defoliation experiments have shown that a substantial part of the leaf area of the potato crop must be destroyed

Fig. 11.16 The effect of systematic, random or localized plant losses on yield of sugar beet. From F G W Jones *et al.* (1955) *Ann. Appl. Biol.* **43**, 63.

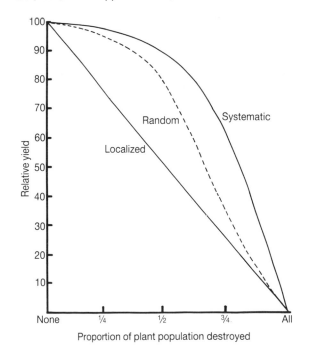

before yield is greatly affected, and thus the destruction of foliage by late blight has less effect on yield than might be expected. Only when about 75% of the foliage is destroyed does tuber growth fully stop, but if the infection continues to spread because of favourable weather conditions this level of damage is ultimately reached and dry matter accumulation in the tubers ceases. Thus although no basic alteration in the host's physiology is involved, the stress is sustained and the earlier in the season that foliar damage reaches the critical level, the greater again is the yield loss.

Perennial crops such as fruit and timber trees and herbage grass, while following the general trends discussed above, bring new variations to the theme. In the first place, yield effects may be delayed in time. For example, if stress reduces the number of fruit buds laid down in one year, the effect on yield will not be evident until the following year. Again, differences in flexibility of reponse are evident. If deciduous trees are defoliated by insects, they can replace the leaves in that season, but with evergreen conifers such as spruces and firs this reaction is delayed until the following year. The advantages of the immediate response of the deciduous species for rapid recovery are obvious.

Second, perennials start the growing season with substantial reserves of energy and minerals, which permit greater resilience in their response. Evergreen conifers survive defoliation by insects in the spring before new foliage has fully emerged, but they may die if this happens after full leaf emergence when, presumably, reserves have been depleted. The significance of stored reserves applies also to herbaceous perennials. Part of the skill of managing grassland is in cutting and grazing in such a way as to leave adequate carbohydrate reserves in the plants for rapid regrowth.

Third, the potential growth of a woody perennial in any one season is, to a great extent, dictated by the size of the 'mainframe' (trunk, branches and twigs) on which the growth takes place. It follows that any check to growth of a woody perennial, by retarding the development of the mainframe, affects potential yield for several subsequent seasons. In other words, because growth is cumulative, stress effects in woody perennials tend to be cumulative in their impact on yield. In timber crops the effect is not so much to reduce final yield as to delay maturity and thereby set back the date of cutting, with equally important economic consequences.

Fourth and last, where perennials are grown in mixed stands of several species, as in grass swards or mixed woodland, stresses which act preferentially against one or more constituents of the mix can have rather more subtle effects on yield. By weakening the competitive ability of affected species, or by simply eliminating susceptible individuals, the gradual demise of these species is accelerated or their occurrence reduced, usually under competitive pressure from the less affected species. In this manner, over a period of years, the species composition of a grassland sward tends to drift, often to a less agriculturally desirable mix. Over a longer period, similar deleterious changes can be induced by stress in mixed forests.

11.6.5 Effect of stress on yield stability

Putting even rough figures on the total losses of crop yield due to stress is difficult. The substantial gap that exists between maximum potential yield and average yields in practice is only partly due to the impacts of negative factors. Estimates of annual losses from pests and diseases of 10–20% for crops in temperate areas, and 20–30% in tropical areas, must be treated with the caution that all such estimates deserve, but the figures are probably of the right order.

Despite the very great impact of stress on quantitative yield, it is important not to underestimate its significance for other aspects of yield. We have mentioned certain effects of stress on quality of saleable produce and also the phenomenon of delayed yield in perennial crops, but another aspect, the impact on stability of yield, should not be ignored.

Maize in the USA brings out this point nicely. In the eastern part of its range, it receives considerable annual rainfall, and yields are high. As its range is followed west, it receives decreasing amounts of rain, and the yield of unirrigated crops declines. In addition, annual yields become increasingly unstable, rising and falling in response to variation in rainfall from year to year. Irrigation in these areas, by reducing drought stress, both increases and stabilizes maize yields.

The impact of stress on the stability of yield is often underestimated in its wider importance. In market economies the effect of even slight overproduction of a commodity which has an inflexible demand, and may also be highly perishable, is to cause a fall in price out of all proportion to the degree of overproduction. Similarly, slight under-production can boost prices disproportionately, perhaps beyond what many potential consumers can afford. The effect is most serious in the case of staple items of the diet that are prone to these erratic variations. Good examples are potatoes and yams.

In subsistence farming, year-to-year variations are often even greater than in market economies, due to erratic climatic variations in many of the areas where it is practised and to the lower level of technology in use. These circumstances can produce periods of feast and famine. The latter have disastrous effects on the survival and stability of the communities involved, and indeed on the political stability of entire nations. Thus, in subsistence agriculture, stabilizing yield is often a more important goal than simply boosting yield. Land races of crops that have evolved in these systems of cultivation are usually more tolerant of stress than the modern varieties grown in the developed world, and are better at producing at least some yield in the face of considerable adversity. Even in highly developed economies with intensive systems of agriculture, market instability due to variations in yield is still a central problem. Not surprisingly, therefore, much of the effort of crop management goes into alleviating the problems of stress rather than just raising yield.

11.6.6 Manipulation of stress to increase yield

In some circumstances, stress can have beneficial effects on crop yield. It is never likely to increase total biomass production per hectare but it often has the effect of enhancing the development of one part of a plant at the expense of others. Stress of various kinds may be deliberately applied in a controlled way in order to produce this type of effect.

The most common application of this kind of technique is where the balance of vegetative and reproductive growth is affected. These two facets of growth tend to be in competition with one another and, where the product harvested is a seed or fruit, too much vegetative growth may therefore depress the yield. In lucerne (alfalfa), for instance, the highest yields of forage are obtained where plenty of water is available, but if the crop is being grown for seed, seed yield may be greater where lesser amounts of water are available.

This phenomenon is even more clearly seen in perennial crops, where again excessive vegetative growth tends to take place at the expense of flower production. The first flush of vegetative shoots in certain varieties of raspberry, growing each year from the base of the plant, is often killed with a non-translocated herbicide to prevent them from competing with the fruit for nutrients. A later, less vigorous flush of vegetative shoots is then allowed to grow to provide the following year's fruiting canes.

Maturing orchards of apples and pears are undersown with grass to restrict vegetative growth of the trees and thereby encourage fruiting. In other situations individual trees or shrubs may have their roots pruned to restrict vegetative shoot growth. Certain ornamentals, such as some species of geranium, produce adequate numbers of flowers only if given a low level of mineral nutrition.

Stress in the form of grazing is used in grassland to encourage grass tillering. This probably works by allowing more light to reach the base of the plant, which in turn induces more dormant buds to develop as tillers. This creates a dense sward more resistant to the ingress of weeds and with a longer productive life.

11.7 Pre-harvest losses of realized yield

Even after yield has been produced in the growing crop, losses continue up to and beyond harvest. Post-harvest losses are dealt with in the next section; here the prime causes of loss of realized yield before and at harvest are briefly considered.

Pre-harvest losses of realized yield occur from the standing crop due to various causes, especially pests, diseases and weather. Cereals can lose sizeable amounts of grain by the wind shaking it out of their ripe heads on to the ground. The act of harvesting itself is of very variable efficiency, and a considerable amount of the crop may simply be left in the field. For example, up to 3 tonnes ha^{-1} of sugar beet are commonly left unharvested as roots are missed by the harvesting machine in one way or another. An average 2-3% of the yield of cereals cut by combine harvester is lost through inefficiencies in the threshing processes of the machine. With mechanically harvested leguminous crops, harvesting losses are usually very high, sometimes reaching 30-40% in peas.

In extreme cases, severe weather conditions at harvesting time may totally prevent harvest of a crop, which is then left to rot in the field. The risk of such losses is greatest where the crop produce is highly perishable, as in the case of many fruits and vegetables.

However, the most common losses caused by

harvesting operations are those which do not become evident until later. As will be seen in the following section on post-harvest losses, crop damage through inappropriate or careless harvesting technique can greatly accelerate deterioration of crop products in store. Likewise, since the storability of many crops is highly dependent on their physiological condition at the time of harvest, harvesting at an inappropriate time, especially at too early a stage of maturity, can also result in serious post-harvest losses.

11.8 Post-harvest yield losses and their control

Yield losses continue after harvest. Indeed some of the worst losses occur at this stage, often due to misunderstanding or ignorance on the part of handlers of products. Many crops are only half way through their life-span at harvest. This section therefore considers the crop from harvest onwards. It includes storage, preparation for marketing, transport and sale. Preparation for marketing can entail various processes such as grading by size, removal of diseased or otherwise undesirable material from the bulk, and packaging. Fig. 11.17 shows a flow chart of these processes. Other procedures in storage and marketing such as processing and preservation are considered in Section 12.4. Since the impacts of post-harvest

handling on yield are often inseparable from effects on the quality of the produce, these are considered here also, though the subject of quality is covered in much greater detail in Chapter 12.

The grower should understand that his role is important for all post-harvest processes. He has, for the most part, charge of harvesting and often also of storage. The function of storage is vital, especially with staple items of diet. Crop production is highly seasonal, outside certain restricted

Fig. 11.17 Flow chart of post-harvest processes for agricultural products. 1. Marketing without storage, as occurs commonly in peasant economies (note that there may be long period of domestic storage in the consumer's home). 2. Marketing without storage in the case of highly perishable products such as green salads (note the very short period of domestic storage). 3. Marketing direct from farm to caterer leading to simultaneous sale to, and consumption by the final consumer. This is an increasingly important route. 4. Sale to retailer, who sells to consumer. 5. Sale to merchant, who sells to retailer. Some products may pass through a chain of merchants before being retailed, especially if they are exported or marketed at a distance. The storage life of the product (—)lasts from harvest until the product is moved off the farm after first sale. Its shelf life (----)lasts from the end of storage until sale to caterer or final consumer, and domestic storage (~) is the period thereafter during which the product is held pending use by the caterer or final consumer.

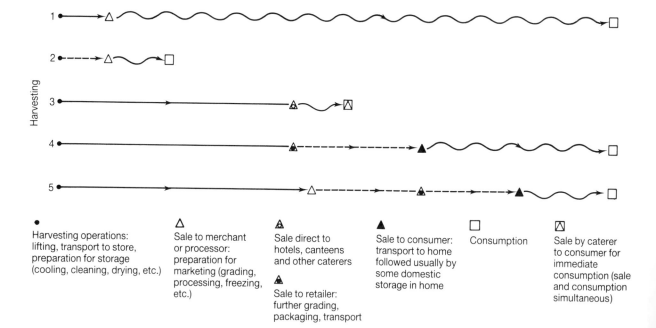

•	△	⚠	▲	□	◩
Harvesting operations: lifting, transport to store, preparation for storage (cooling, cleaning, drying, etc.)	Sale to merchant or processor: preparation for marketing (grading, processing, freezing, etc.)	Sale direct to hotels, canteens and other caterers	Sale to consumer: transport to home followed usually by some domestic storage in home	Consumption	Sale by caterer to consumer for immediate consumption (sale and consumption simultaneous)
		⚠ Sale to retailer: further grading, packaging, transport			

tropical areas, but human appetite is not. Effective storage evens out the inconsistencies between supply and demand and avoids a feast-and-famine cycle. Moreover, although in any developed market economy the grower is seldom involved in all the production processes from sowing to consumption, his handling of the earlier stages profoundly affects the quality and storability of the product at later ones. The husbandry of the crop in the field, at harvest and in store decides whether it is fit to be marketed, although the impact of faulty husbandry at these earlier stages may not become apparent until marketing or processing is well advanced. For example, internal blemishes and off-flavours or other defects in quality initiated by bad handling months earlier may not be detected or even expressed until the products are cooked or eaten. With the trend to long-distance marketing and increased processing of foods, both of which demand tighter quality control, the significance of effects of husbandry in the growing crop on storability and marketability is further increased.

Although much effort is expended on growing a crop, at least as much again and more is often expended on it before it is consumed. At each stage of post-harvest handling, with successive inputs of time, energy and materials, the value of the crop increases, but so also does the scope for losses or even disaster through mismanagement of the product. Examples of post-harvest wastage in common crops are given in Table 11.4, and these give some idea of the quantity of yield losses often experienced. In underdeveloped agriculture these losses may be much greater, often of the order of 40%. Losses in quality, such as the development of blemishes or reductions in vitamin content, are less spectacular, but may be equally important.

Throughout storage and marketing, in striving to reduce these losses, it is the botanical characteristics of the plant part in use that largely decide how the product is best handled. Table 11.5 shows the array of products that must be considered and their general groupings. Even within one group there may be large differences in storability among individual products, for example between top fruit such as apples and pears, and soft fruit such as strawberries and raspberries. There are, of course, individual plant products that do not conveniently fit into any of these categories, such as rhubarb. Seed storage is a special case and has been dealt with in Section 5.3.5.

Despite the very wide range of types of plant product, there are underlying principles that guide the handling of them all. After harvest, every product ages and tends ultimately to decline in quality to a condition in which it is no longer marketable. Skilful handling maximizes the length of time for which it can be retained in a condition acceptable for consumption. This maximum time available for storage and marketing is called the *post-harvest life* of the product, and is usually divided into the *storage life*, denoting the period of time that the product can be maintained in an acceptable condition in store, and the *shelf life*, denoting the period of time that it can be maintained in an acceptable condition during

Table 11.4 Post-harvest losses in a range of fruits and vegetables at retail and consumer level, in New York (USA) marketing area, as measured by various researchers

Commodity	Cause of loss (%)			
	Mechanical damage	Parasitic disease	Non-parasitic disorder	Total loss (%)
Apples	1.1	0.2	0.4	1.7
Cucumbers	1.2	3.3	3.4	7.9
Grapes	4.2	0.4	0.9	5.5
Lettuce	5.8	2.7	3.2	11.7
Oranges	0.8	3.1	0.3	4.2
Peaches	6.4	6.2	—	12.6
Peppers	2.2	4.0	4.4	10.6
Potatoes	2.5	1.4	1.0	4.9
Strawberries	7.7	15.2	—	22.9
Sweet potatoes	1.7	9.2	4.2	15.1

From J M Harvey (1978) Reduction of losses in fresh marketing fruits. *Ann. Rev. Phytopath.* **16**, 321–41.

Table 11.5 The main types of harvested plant product, with examples, post-harvest life, and main causes of limits to storage life

General group	Subgroups	Examples	Approximate post-harvest life*	Common causes of deterioration limiting storage life
Dormant vegetative storage organs	Roots	Parsnips	4–5 months	Loss of moisture, developing fibrousness, sprouting
	Tubers	Potatoes	6–9 months	Loss of moisture, sprouting
	Bulbs	Onions	10 months	Sprouting, root growth
	Swollen buds	Brussels sprouts	3–5 weeks	Loss of moisture, leaf senescence
Fruit	'Soft' fruit	Raspberries	3–5 days	Fungal decay, overripening
	Top fruit	Apple	3–5 months	Overripening
	Tropical and subtropical	Watermelon	2 weeks	Overripening, storage rots
Growing shoots	Salads	Lettuce	2–3 weeks	Loss of moisture, leaf senescence
	Herbs	Mint	1 week	Loss of moisture, leaf senescence
	Cut flowers	Roses	7–9 days	Overdevelopment of flower
	Leafy vegetables	Celery	2–4 weeks	Loss of moisture, leaf senescence
Seeds	Cereals	Wheat	} years	Very gradual loss of nutritional value over years
	Pulses	Peas and beans		
	Nuts	Cashew		
Roughages	Hays	Grass or herb-rich mixtures with grass	} years	Very gradual loss of nutritional value over years
	Silages			

*Allows for maintained quality during period of subsequent marketing.

marketing. These last two terms are often used loosely and interchangeably. Also, a product does not necessarily have both phases. For example, highly perishable commodities such as salads go straight from harvest to marketing and have only a shelf life.

11.8.1 Causes of post-harvest deterioration

For several reasons, post-harvest deterioration is inevitable. First, a harvested organ is cut off from its source of the products of photosynthesis. Yet, unless it is killed and preserved by drying or other means, it continues to respire, using up its limited energy store. Second, it is also cut off from its water supply in the soil, but continues to lose water by transpiration or other routes. Third, a harvested organ still tends to continue its normal processes of development, which may render it unsuitable for storage or consumption. Fourth, harvested produce remains susceptible to the attacks of pests and pathogens, and indeed, as will be seen, storage increases its vulnerability to some of these.

Despite provision of the most favourable storage and marketing conditions, one or more of the

trends outlined above ultimately cause deterioration. The aim of good husbandry during storage and marketing is simply to reduce the speed with which these trends proceed.

Three major factors govern the speed of post-harvest deterioration. These are the type of plant material being handled, the condition of the product on entry into storage or marketing, and the environmental conditions it experiences during these processes. To be easily understood, these factors and their interactions need to be elucidated in a little more detail.

11.8.2 Effects of type of plant material on post-harvest deterioration

Compare two general groups of product for durability. Products consisting of young, actively growing tissues such as growing shoots, as in salad cress, or young flower buds, as in cut flowers or broccoli, or developing seeds, as in peas and beans harvested before maturity, have a short post-harvest life – perhaps only of days – unless they are specially treated. In contrast, storage organs which are dormant or quiescent when harvested, such as tubers, roots, bulbs or mature seeds, have a long post-harvest life, often of months, or, in the case of many seeds, years. Many fruits also have a fairly long post-harvest life if picked unripe and kept in storage conditions that retard ripening. These differences between products in post-harvest life are principally due to differences in their rates of water loss, their general levels of metabolic activity and their susceptibility to infection. Even differences among varieties of one crop species in features such as moisture content and disease susceptibility can significantly influence post-harvest life.

It can be appreciated that products in the first group, particularly those bearing leaves, will tend to lose moisture more rapidly through their greater surface area per unit volume than more compact organs such as storage roots or tubers. In addition, storage organs often have suberized skins that effectively reduce water loss. The suberized periderm of a tuber is more waterproof than the cuticle of green tissue, but even cuticles differ in this respect, depending on their thickness and structure. Some products, such as citrus fruits, are sometimes coated with wax during marketing, partly to decrease water loss. The rate of water loss and its control are considered more fully in Section 11.8.4.

Young actively growing shoots also have much higher metabolic rates than dormant storage organs. The initial respiratory rate of the product going into store or marketing is a very good indicator of its potential post-harvest life. At 15°C, for instance, broccoli may have fifteen times the respiratory rate of potato tubers, and an appropriately shorter shelf life. This is partly because more rapid respiration uses up energy reserves more quickly, but also because the higher metabolic rate is associated with more rapid growth and development which continues after harvest. With growing shoots such as cut flowers or salad greens, this is an immediate problem. Cut flowers may blossom and start to senesce, salads may wilt, broccoli may start to flower, before these products even reach the shops.

This is in marked contrast to dormant or quiescent storage organs such as roots or tubers, which may cease growth for months. These are often mistakenly assumed by growers to be inert or unchanging after harvest, but this is a fundamental error. Growth may have halted, but metabolic activity, development, including the initiation of leaves, flowers and roots, and physiological aging continue as the organ prepares for the next burst of seasonal growth that it would produce under

natural conditions. Further, these organs are highly responsive to the environmental conditions in which they are maintained post-harvest, and temperature in particular influences the pattern and rate of development. This, too, is considered in more detail in Section 11.8.4.

Finally, leafy shoots and flowers are more readily attacked and decayed by pathogens after harvest than storage organs such as sweet potato tubers or onion bulbs, which have more effective defences against invasion by fungi and bacteria. Storage organs do, however, suffer attacks by pathogenic microbes. In general, the higher the moisture content of an organ and the greater its concentration of readily utilizable carbohydrates, the more vulnerable it is to attack.

11.8.3 Effects of condition of produce on post-harvest deterioration

The second major factor influencing post-harvest life, the condition of produce entering storage or marketing, is critical. This is particularly true regarding the presence of pests and diseases in the produce, and the level of mechanical damage inflicted on it. Many post-harvest diseases are simply the further development of infections that were already established in the field. Foliar fungal infections, such as downy mildews of green salad crops like lettuce or leaf spots on cut flowers, are good examples of this, and normal environmental conditions in storage or marketing can accelerate their development. In addition, pathogens may be introduced as contaminants on the crop, for example in soil adhering to tuber or root crops, or as spores on the foliage of leafy crops. These may later infect the stored material. Quiescent infections, already established before harvest, may also be activated by storage and handling conditions. Thus the health of material entering storage or marketing is important, and attention must be given to the husbandry of the crop prior to harvest to reduce levels of infection and contamination. Additional measures may also be taken, such as the cleaning of stores before use, and the application to produce of fungicides and bactericides as dips, sprays or fumigations during or shortly after harvest.

It is inevitable that some degree of physical damage is inflicted on a crop by harvesting and handling it. This gives rise to another important aspect of condition in plant material after harvest. In addition to the cut ends of stems, the numerous wounds inflicted by machinery during harvesting, grading, transport and packaging are an important problem. Even apparently tough tissues such as roots and tubers are more easily damaged than is commonly realized, and are very susceptible to wounding by mechanical handling.

Such wounds have several effects. They greatly increase the rate of water loss by the crop. The bruised and damaged areas and their surrounding tissues develop a raised respiratory rate. Moreover, many fungal and bacterial pathogens gain entry by wounds. Good examples of such *wound pathogens* are the fungal parasites *Phoma exigua*, causing the rot known as gangrene on potatoes, and certain *Penicillium* species which cause moulding of citrus fruits. Most mature plant organs have lost the ability to heal wounds, but certain roots and tubers including potatoes, sweet potatoes and yams retain this ability. Hence these products are often 'cured' by being subjected to periods of 10–14 days in store in which high humidity and raised temperatures are maintained, to prevent such wounds gaping through drying out, and to encourage them to heal quickly by the rapid formation of a cork barrier, thereby excluding wound pathogens. Elementary precautions such as padding sharp or abrasive surfaces on machinery where these might damage plant products, and designing machinery so that the product does not fall over sizeable drops or on to hard surfaces during processing, also reduce wounding.

Now the manner in which pre-harvest treatments affect wounding is enlightening. With roots and tubers, the degree of maturity of the crop at harvest, often strongly influenced by sowing date and other cultural measures, has a significant effect. Mature roots, tubers and bulbs develop toughened outer skins which are less easily damaged. Going even further back in the history of the crop, the selection of site has a bearing on the matter. On stony soils, for instance, the stones, if agitated with roots and tubers during lifting, inflict considerable mechanical damage. Another important influence is the level of fertilizer application. High levels of nitrogen application, in particular, increase the susceptibility of crops to mechanical damage by increasing their moisture content and thus making them softer.

In fact, the mineral nutrition of the crop affects post-harvest life in other ways, too. High levels of nitrogen fertilizer application increase the susceptibility of the foliage of leafy vegetables to field infections that may then develop further during storage and marketing. Many crops in store develop disorders that are related to calcium deficiency. These disorders can be prevented or reduced by the application of calcium salts during growth, or even directly to the product after harvest. Examples include bitter pit in apples, end spot in avocado and soft nose in mango. Lack of other minerals such as potassium and boron can also give rise to specific post-harvest disorders.

From all this, including the need to manage crops to reduce infection and contamination by pathogens and infestation by pests in plant produce, what clearly emerges is how strongly pre-harvest husbandry influences the durability of the crop during storage and marketing and its suitability for its target markets.

11.8.4 Effects of environmental conditions during storage and marketing on post-harvest deterioration

This leaves the third major factor influencing post-harvest deterioration to be considered, namely the influence of environmental conditions during storage and marketing. Two factors in the environment exert a dominant influence on the fate of the crop post-harvest. These are the temperature of the product and the composition of the air trapped in the stored bulk or packaged product.

The influence of these factors must be seen in the context of certain problems created by the very act of bulking materials for storage or marketing. While a crop is growing in the field, its respiratory products of carbon dioxide, water and heat are freely dissipated into the surroundings. Once it is bulked this cannot easily happen, and water vapour, carbon dioxide and heat, plus certain volatile plant products, tend to accumulate within the bulk. Simultaneously, oxygen tends to be depleted through crop respiration. There is thus an inherent tendency in bulked plant material for relative humidity, temperature and carbon dioxide concentration to rise, oxygen to be depleted and, in extreme conditions, for semi-anaerobic, excessively warm conditions to develop.

A second result of bulking crop products is that it makes them particularly prone to epidemics of various infections and infestations. Spread of pests and diseases between closely packed hosts is facilitated, especially if the bulked organs touch one another as in the case of roots or bulbs in bulk stores. Further, if environmental conditions in the bulked crop are not monitored and controlled and correct ventilation practised, the raised levels of temperature and humidity and semi-anaerobic conditions which develop are very conducive to the growth and multiplication of many pests and pathogens.

Temperature strongly influences all the processes that lead to deterioration. The tendency of harvested produce to continue its growth and development has already been mentioned. Plainly, raised temperatures will quickly encourage green growing shoots such as those of broccoli to sprout, or raise their respiratory rate undesirably. But the sensitivity of dormant or quiescent storage organs to environmental influences is often underestimated. Such organs retain the ability to respond, often surprisingly rapidly, to conditions around them. With potato tubers, as described in Section 11.2.2, raising the temperature will accelerate first the loss of dormancy and then the loss of apical dominance, and hence alter the sprouting pattern of the tuber, giving, if it is to be used for planting, more plants per tuber in the field. It is the condition of the material, rather than how long it has been stored, that matters – its physiological age in this case, rather than its chronological age.

Obviously, as temperature is lowered from, say, 30°C to 0°C, the rates of respiration and other metabolic processes within the product decline, but not uniformly over the whole temperature range. The rates of these processes fall more and more sharply as temperature declines. Thus a fall of as much as 10°, from 30°C to 20°C, may have comparatively little effect on metabolic rates and therefore on the post-harvest life of the crop, but a decline of even 1° from, say, 5°C to 4°C has a very marked effect, as Fig. 11.18 makes clear. Consequently, many crops such as onions, root crops, leafy vegetables and soft fruits like raspberries are stored at temperatures approaching 0°C and rapidly deteriorate if held at higher temperatures. They cannot be stored below 0°C because this would result in freezing injury.

Exclusions to the general practice of refrigerating perishable products are crops of tropical and sub-tropical origin, including sweet potatoes, tomatoes, aubergines and peppers, which must be stored above 5°C to avoid chilling injury. It is a common practice to extend the period over which fruit may be marketed by picking it unripe and slowing its further development by cooling. In non-climacteric fruit (see Section 7.5.3), ripening is simply slowed down by cooling, but in climacteric fruit it is stopped entirely, and the fruit is brought to higher

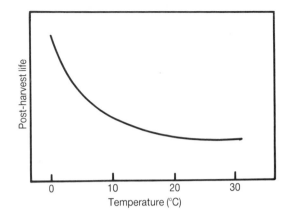

Fig. 11.18 Effect of temperature on post-harvest life of stored moist products. From R H H Wells *et al.* (1981) *Post Harvest*. Granada.

temperatures to ripen when it is desired to market it.

The influence of temperature on other physiological processes is also important. An array of commercially significant features known to be influenced by storage temperature of potato tubers is shown in Fig. 11.19 and demonstrates how critical this factor is even in a quiescent organ.

Many storage organs store their energy principally as starch, but some sugar remains in the cells, strongly influencing flavour. The ratio of sugar to starch is not fixed. It is a dynamic equilibrium which changes with temperature, and below a critical temperature that depends on the species, the equilibrium shifts in favour of sugar accumulation. For potatoes, the critical temperature below which there is a net conversion of starch to sugar is 10°C, but for sweet potatoes it is 51°C. Sugar accumulation may or may not be desirable. In sweetcorn and peas a high sugar content gives good flavour and these crops are harvested immature, when sugar content is still relatively high. But sweet flavour is not desired in many starchy vegetables. In potatoes, not only is it undesirable, but tubers with high sugar content have a poor texture when boiled, and over-brown when fried due to caramelization of the sugars.

Not surprisingly, considerable effort is put into bringing products to a suitable temperature for storage and marketing and then maintaining it. A central problem is that newly harvested material nearly always comes in from the field much warmer than it can be successfully stored. With products having a high intrinsic metabolic rate,

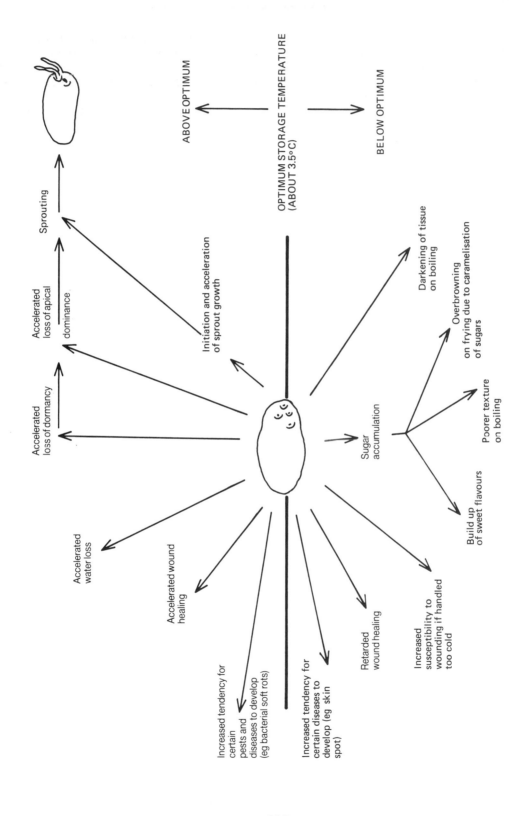

Fig. 11.19 Some effects of storage temperature on the keeping quality of potato tubers.

Sprouting

Accelerated loss of apical dominance.

Accelerated loss of dormancy

Initiation and acceleration of sprout growth

ABOVE OPTIMUM

OPTIMUM STORAGE TEMPERATURE (ABOUT 3.5°C)

BELOW OPTIMUM

Accelerated water loss

Accelerated wound healing

Increased tendency for certain pests and diseases to develop (eg bacterial soft rots)

Increased tendency for certain diseases to develop (eg skin spot)

Retarded wound healing

Increased susceptibility to wounding if handled too cold

Sugar accumulation

Build up of sweet flavours

Poorer texture on boiling

Overbrowning on frying due to caramelisation of sugars

Darkening of tissue on boiling

and thus a shorter post-harvest life, like soft fruit and leafy vegetables, removal of this *field heat* must be done quickly. Various techniques of cooling prior to storage and marketing, such as forced ventilation by air passed over ice, or standing the produce in a cooled room, are employed. Products with an intrinsically low metabolic rate and long post-harvest life can be cooled with less haste, and this is usually done in the store itself by judicious ventilation with cool air. In either case, it is extremely important that the temperature of the product be monitored and controlled throughout storage and marketing.

Turning to the effects of the gaseous composition of the air surrounding the produce, it is water vapour that is of most frequent concern. The most direct and obvious effect of relative humidity is on the rate of water loss from the product. Mention has already been made of how such factors as surface area to volume ratio, types of surface coatings on the product, and mechanical damage all influence water loss, but relative humidity plays a crucial role. Since most products are sold by weight, evaporation of water causes direct loss through loss of marketable yield, but the effects on quality are often more important: wilting, or simply loss of crispness, may develop in salads; cut flowers may also wilt. Products differ in the water loss they can sustain before they become unacceptable. Only 3% water loss, resulting from a very few hours in the wrong conditions, may bring matters to this point in lettuce, but at the other end of the scale, up to 10% water loss may be acceptable in onions.

For most products, water loss can be greatly reduced by maintaining relative humidity above about 97%. Unfortunately above 95% relative humidity many pathogenic bacteria and fungi are highly active, and there is therefore a conflict of demands. In most cases this can be resolved by storing at low temperature as well as high relative humidity. This not only reduces the activity of pathogenic microbes, but also makes it easier to maintain the relative humidity at the desired high level. This is because cool air has a much lower capacity to hold water vapour than warm air. Thus fully saturated air (that is, air at 100% relative humidity) contains a much lower absolute concentration of water vapour at 5°C than at 15°C. A result of this, as explained in Section 2.4.2, is that air at a given relative humidity has much lower capacity to absorb moisture from a stored product at 5°C than at 15°C.

Weekly rates of water loss of 0.5–1% of the weight of stored products are common. Cooling and maintaining high relative humidity can reduce these losses by perhaps half, and lengthen storage life. In one experiment carrots after six months in store at 0.5–0.7°C, ventilated by air humidified and cooled by passing it over ice, maintained acceptable quality for a further ten months, while those in conventional storage conditions at 2–3°C showed greater water loss as well as obvious root and shoot growth.

In practice, the relative humidities that are employed in storage and marketing are a compromise between the conflicting demands of decreasing water loss, the activity of pathogens, the temperature sensitivity of the product and other factors. For fruit, a relative humidity of around 90% is commonly used, but for leafy vegetables and some root vegetables like carrots, which are prone to rapid water loss, 98–100% relative humidity is necessary. In other groups of products, the compromise relative humidity can be very different. Onions and cucumbers, for example, are stored at 65–70% relative humidity, to prevent excessive decay by pathogens and, in the case of onions, to discourage root growth.

The accumulation in store of volatile compounds given off by the product itself is another aspect of storage atmosphere that can create problems. Ethylene, a plant hormone (see Section 8.4.4), is the most prominent of these compounds. Many fruits, as they develop and mature, become more sensitive to ethylene and respond to traces of it by ripening faster. Even fumes from internal combustion engines, when allowed to leak into a fruit store, can accelerate ripening because of their ethylene content. Where concentrations of ethylene from stored fruits build up in store, ripening is accelerated and storage life thus shortened. Accumulation of ethylene and other volatile compounds can also cause storage disorders.

Accumulation of carbon dioxide and depletion of oxygen do not necessarily have deleterious effects. It has long been known that both increasing the carbon dioxide and lowering the oxygen concentration of the atmosphere around plant tissues independently slow down respiration and other metabolic processes. This has usefully been applied in prolonging the storage life of products where this is normally limited by physiological changes within the products, such as loss of chlorophyll during senescence, as in spinach, or toughening in vegetables such as asparagus. To achieve the desired effect, modification of the composition of the storage atmosphere has to be substantial, for example a hundredfold increase in carbon dioxide concentration to around 3%, or reduction in oxygen concentration from 20% to around 3–4%. In such *controlled atmosphere storage*, these two methods are often used together, combined with low temperature. Below about 2% oxygen concentration, however, anaerobic conditions begin to develop in parts of the plant tissues, leading to disorders. In addition, the plant's ability to defend itself against pathogens is much reduced due to inhibition of its biochemical defences, and storage rots may spread rapidly.

Less sophisticated but nonetheless effective methods are more commonly used to control post-harvest environmental conditions. Storage buildings are insulated to protect products from frost and excessive heat. The grower intervenes through ventilation to prevent the build-up of heat, carbon dioxide and moisture and, where necessary, volatile compounds generated by the stored product. Even for storage of dormant organs with low metabolic rates such as tubers and bulbs, it is important to encourage the free movement of air around them.

11.8.5 Effects of packaging on post-harvest deterioration

Packaging for marketing not only increases ease of handling and enhances the presentation of the product to the potential buyer. It also protects the product from dirt and mechanical damage during transport and handling. Packaging that completely encloses the product, such as cardboard boxes or coverings of clear plastic film, also protects it from temporary exposure to suboptimal temperatures and lowers the rate of water loss by maintaining a higher relative humidity around the product. Figure 11.20 illustrates a variety of methods of packaging for marketing.

Packaging must be skilfully designed. For example, plastic film wrappings reduce water loss and thus delay deterioration, yet if excess mois-

Fig 11.20 A range of products packaged for marketing, showing (a) peaches in cardboard cartons wrapped in plastic film and packed in wooden boxes, (b) carrots washed and put in string net bags, (c) lettuces individually wrapped in polythene bags and packed in cardboard cartons, and (d) spring onions tied in bunches and packed in cartons.

ture accumulates within the package, or oxygen becomes excessively depleted due to respiration of the product, rotting and decay are accelerated. To avoid these disadvantages, the correct type and thickness of film must be used to allow sufficient diffusion of gases through it, and often ventilation holes are provided to assist in this process. Where containers are stacked, they must be arranged in such a way that ventilation penetrates to those at the centre of the bulk.

11.9   Efficiency of crop production

The term 'crop production' is commonly extended to include the marketing as well as the growing of crops. Here its meaning is confined to the biological process of accumulation of harvestable dry matter by plants in the field. Yield is the end-product of the dry matter production process, which requires such inputs as energy, water, nitrogen, minerals and propagating material. As important as absolute yield is yield relative to the quantity of these inputs, that is, the *efficiency* of crop production.

The productivity of a crop can be judged with regard to any input. For example, with increasing input of nitrogen fertilizer the yield response per unit of input declines, as shown in Fig. 11.21. The levelling off of productivity reflects a declining efficiency in the utilization of nitrogen by the crop.

Fig. 11.21 Response of a highly productive tropical grassland to nitrogen fertilizer. Note that at low nitrogen rates there is a strong yield response but that at higher rates little further increase in yield is obtained. From G W Cooke (1982) *Fertilizing for Maximum Yield*. Macmillan, New York, NY.

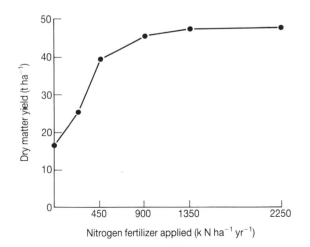

All inputs are subject to the law of diminishing returns in this manner. The high costs of nitrogen fertilizers in recent years have brought the efficiency of utilization of nitrogen more to the fore.

In multiplying stocks of propagating material from very small numbers, for example, of cuttings taken from a single virus-free plant, the speed of multiplication of these stocks is of paramount interest. Thus efficiency of production in this case may be considered as the number of propagules (for example tubers, bulbs or seeds) harvested per propagule planted or sown.

In semi-arid areas where the provision of water is a limiting factor to productivity, yield per unit of water supplied may be the most important measure of agricultural efficiency.

For most purposes, however, the most appropriate input against which to measure efficiency of crop production is energy, because most of the inputs and outputs can be expressed in units of energy.

The yield of a crop can be thought of as a quantity of energy, fixed as organic matter in the process of photosynthesis. This energy can be released again if the organic matter is oxidized back to carbon dioxide and water, as happens when it is consumed as food or burned as fuel. On average the oxidation of 1 kg of dry plant material releases about 18.5 MJ (megajoules) of energy in the form of heat. This indicates that in the production of 1 kg of dry matter by the plant 18.5 MJ of solar energy has been fixed.

Now the energy we get out of any system, whether it be a motor car, a power station or an agricultural crop, can be compared with the energy that has to be put in. No system can yield more energy than it takes in. Indeed the conversion of one form of energy to another (petrol to motion, coal to electricity, sunlight to organic matter) is always accompanied by irretrievable energy losses. Thus the energy obtained from the system is only a fraction of that put in. This fraction is a measure of the *energetic efficiency* of the system.

The energetic efficiency of plant agriculture is thus the ratio of the energy obtained as yield to the energy supplied to the crop in order to produce that yield. In this case there are two distinct energy inputs: the radiation energy of sunlight falling on the crop, and the energy supplied in various forms to the crop by the farmer. There is, for example, the energy expended by the workforce on the crop, the energy of fuels (mainly diesel fuel) required to drive the machinery and perhaps dry the crop, and the very considerable energy which has been used to make the fertilizer and other chemicals applied to the crop.

11.9.1 Photosynthetic efficiency of crop production

Let us consider first of all the energetic efficiency of crop production in relation to the solar radiation energy input. *Photosynthetic efficiency* in a crop can be defined as the fraction of incoming radiation which is fixed as crop biomass.

What is the nature of this incoming radiation? The sun emits radiation over a broad spectrum of wavelengths from ultraviolet (short wavelength) through visible light to infrared (long wavelength). Virtually all the energy is emitted in the band of wavelengths between 0.3 and 2.0 μm. Visible light is radiation between about 0.4 and 0.78 μm.

Gases in the atmosphere selectively absorb certain wavelengths, so that the spectrum of radiation falling on plants is modified from that emitted by the sun. The main absorbers are carbon dioxide and water vapour, filtering out long-wave radiation in the infrared, and, most importantly, ozone, filtering out most of the dangerous ultraviolet as well as some of the infrared.

Most of what remains is in the visible wavelengths, so that solar radiation reaching the earth's surface can, without serious inaccuracy, be referred to as 'light'. Not all the wavelengths of sunlight are, however, equally absorbed by chlorophyll and its accessory pigments, Blue light, of wavelengths around 0.45 μm, and red light, around 0.65 μm, are more efficiently absorbed than green light, around 0.55 μm (Fig. 11.22). A greater proportion of green than of red or blue light is reflected or transmitted by plant leaves, which is why they appear green.

About 45% of the radiation energy received at the surface of the earth from the sun is of wavelengths that can be used in photosynthesis. This 45% is known as the *photosynthetically available radiation*, or *PAR*.

The amount of PAR per unit area received by a piece of land in the course of a year will clearly be an important factor determining the agricultural productivity of that land. It depends very much on latitude, for the farther from the equator the land is, the more obliquely the sun's radiation strikes the surface. A given amount of radiant energy is thus spread over a greater area of land and the incoming energy per square metre is, for this reason, less at

Fig. 11.22 Absorption of light of different wavelengths by two of the principal pigments of the chlorophyll system, chlorophylls *a* (broken line) and *b* (solid line).

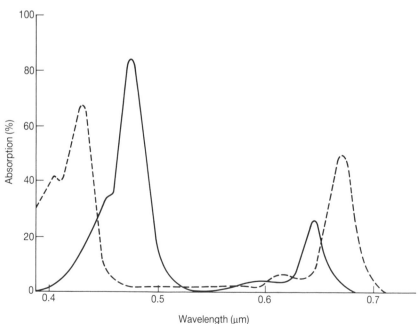

60°N than at 30°N (Fig. 11.23). In addition, radiation arriving obliquely has to pass through a greater depth of atmosphere and a greater proportion of it is therefore absorbed before it reaches the ground.

PAR input depends not only on latitude but on local climate. Cloud reflects back to space much more radiation than clear sky, thus an area with a moist climate characterized by frequent cloud cover has a lower input of PAR than one at the same latitude with a drier, sunnier climate.

Differences in cloud cover can in fact outweigh the effect of latitude. The relatively cloud-free, semi-arid belts to the north and south of the equator receive greater light energy input than areas closer to the equator with heavy rainfall and associated cloud cover. Typical annual PAR inputs for semi-arid regions of the tropics are in the region of 3500 MJ m^{-2}, compared with around 3000 MJ m^{-2} in equatorial regions. These semi-arid belts, because of their high light energy input, have the highest potential crop productivity in the world, but yield is severely limited by lack of water. It is in areas like these that irrigation can therefore bring the greatest rewards.

Further from the equator, latitude and local climate can have major effects on light energy input

Fig. 11.23 The effects of latitude on the intensity of solar radiation at the earth's surface. At higher latitudes, the earth's surface is at a more oblique angle to incoming radiation and a given amount of radiation is spread over a wider area (a_2) than nearer the equator (a_1). Further, the radiation pathway through the atmosphere at higher latitudes is longer (d_2) than it is nearer the equator (d_1), resulting in more solar radiation being absorbed by the atmosphere and thus less reaching the ground.

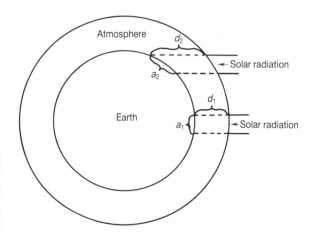

even over relatively short distances. In Great Britain, for example, annual PAR input ranges from about 1750 MJ m^{-2} in the south to 1250 MJ m^{-2} in the extreme north. If we exclude from the calculation PAR arriving during the winter period when soil temperature is below 6°C (an arbitrary but commonly accepted limit to crop growth), the PAR input during the growing season is about 1650 MJ m^{-2} in southern England and only 1050 MJ m^{-2} in Shetland. A mean value for the principal agricultural regions of Great Britain would be around 1500 MJ m^{-2}.

How much of this is actually fixed by a crop? The highest recorded values of above-ground dry matter production in Britain are around 20 tonnes ha^{-1}, for intensively managed grassland. Below-ground production might bring the total to around 30 tonnes ha^{-1}, or 3 kg m^{-2}. If each kg represents the fixation of 18.5 MJ of solar energy, a total of $3 \times 18.5 = 55.5$ MJ m^{-2} will have been fixed in the course of the growing season. This represents 55.5/1500 or about 3.7% of the PAR received.

Few crops, however, achieve an annual rate of dry matter production approaching 30 tonnes ha^{-1}. Furthermore, as we have shown only a proportion, and in many crops it is a small proportion, of the accumulated biomass is harvested as yield. Thus the efficiency of PAR energy conversion to yield in agricultural crops is generally much lower than 3.7%. For example, the UK national average dry matter yield of maincrop potatoes is less than 6 tonnes ha^{-1}, representing an efficiency of PAR conversion of only 0.7%. The national average wheat yield, again expressed as dry matter, is under 4 tonnes ha^{-1}, representing an efficiency of less than 0.5%. Even the all-time British cereal yield record was produced with an efficiency of PAR conversion of less than 1.5%.

What are the prospects for improvement in these apparently very low efficiencies? It is known that under ideal conditions when leaf area index is at its optimum value and photosynthesis is not being limited by water deficit or mineral deficiency, a crop can, over short periods of time, achieve an efficiency of conversion of up to 10% of PAR. Efficiencies of this order have been recorded for such disparate crops as pearl millet in the humid tropics of northern Australia, maize in the eastern USA and sugar beet in Great Britain.

Studies of the photochemistry of energy fixation in plants suggest a theoretical maximum photosynthetic efficiency of around 20% after allowing for losses in respiration and photorespiration. Even under ideal conditions, however, about 10% of PAR falling on a crop is reflected from the leaf surface, 5% is transmitted to the soil and about 10% is absorbed by non-photosynthetic structures such as cell walls. Thus, at most, 75% of the PAR received by a crop is available for absorption by the chloroplasts and the theoretical maximum efficiency of fixation of incident PAR is therefore 20% of 75%, or 15%.

No crop, however well managed, could be expected to sustain a level of PAR conversion efficiency of even 5% for prolonged periods, far less for a whole season. Various crop scientists have suggested that the maximum sustainable photosynthetic efficiency by arable crops in a cool temperate climate is of the order of 3%. This is in fact achieved by coniferous forest plantations in Great Britain and it is not unreasonable to suggest that crops such as potatoes, sugar beet and cereals could match this. A theoretical dry matter yield equivalent to 3% of average PAR input in Great Britain would be around 24 tonnes ha^{-1}. Calculated values such as this, applying to a geographical area but not to any particular crop, are sometimes taken as estimates of *potential crop production*. Very substantial further improvements in harvest index and the rate of establishment of leaf area will be necessary before such yields can even be approached.

11.9.2 Cultural energy efficiency of crop production

Unfortunately, high photosynthetic efficiency is possible only where limiting factors to crop growth other than light are overcome. This usually requires the use of large quantities of fertilizer, especially nitrogen, and the consumption of considerable amounts of fuel in soil cultivation and crop spraying. In some cases irrigation may also be necessary. All this represents a large input of *cultural energy* to the crop.

Until the twentieth century, virtually all the cultural energy input to agriculture was in the form of human hand labour, work by horses, oxen or other animals, and human and animal manure. All of this energy was ultimately supplied by food. It was essential that the crops grown at the expense of this energy yielded more food than was consumed in their production.

With increasing mechanization, the replacement of the horse by the tractor and the greatly increased availability of inorganic fertilizers, agriculture in the developed world has come to depend less and less on human labour and draught animals and more and more on the energy of fossil fuels, especially oil. The inorganic nitrogen fertilizers that now support so much of the world's crop production are, in energy terms, very expensive to produce. But until the energy crises of the 1970s and 1980s fossil fuels and products such as fertilizer made using the energy from such fuels were cheap and abundant, so that efficiency in the use of cultural energy was not an important consideration in developed agriculture.

Once again, however, energy from fossil fuels is an expensive resource and there is need for systems of crop husbandry which can maintain high levels of productivity but with reduced dependence on high cultural energy inputs. Direct drilling in place of ploughing, frequent small fertilizer dressings to minimize wastage and the use of chemical desiccants to reduce drying costs are examples of techniques that can help achieve this.

At present, cereals such as barley and wheat in Great Britain yield about 2.5 times as much energy in grain as they consume in the form of cultural energy in their production. Sugar beet is a little more efficient, producing about 3.5 units of energy for every unit consumed, while potatoes are less efficient, returning only about 1.5 units of energy for every unit consumed.

In British cereals, about 50% of the cultural energy input is in the form of fertilizer and 15% is in the fuel used directly in the growing of the crop. A further 20% on average is used in drying the grain. The manufacture and maintenance of machinery acounts for about 10% but transport costs are low – about 3% of energy input. Potatoes and sugar beet do not incur drying costs but transport costs per unit of dry matter production are considerably greater because of the lower dry matter content of the produce. Some specialist crops such as soft fruit incur high energy costs in the form of chemicals for pest, disease and weed control and growth regulation.

In spite of the high energy cost of producing the food we eat, it is worth noting that this is only about 20% of the energy cost of getting the food to the dinner table. The remaining 80% is incurred in processing, packaging, transport to and from market and preparing the food for the table.

Summary

1. Yield is the quantity of utilizable or saleable crop produce harvested per unit land area. Optimizing yield is a major aim of crop production, but equally important is ensuring that the produce is of the required quality. Other aims are to maintain soil fertility, to make the most of by-products, to secure a long productive life in the case of perennial crops and to improve the yields of other crops in the rotation.

In most crops, the yield represents an accumulation of organic matter, minerals (including nitrogen) and water in the course of a growing season. Because the moisture content of crop produce is highly variable, yields are best expressed in terms of dry matter, of which organic matter generally accounts for more than 95%. Dry matter yield is related to biomass at harvest, which is the total dry weight of all parts of the crop and represents the accumulated excess of the products of photosynthesis over losses in respiration, photorespiration, abscission, consumption by pests and mortality throughout the growing season. Utilizable fraction is the ratio of yield to biomass at harvest; harvest index usually means the ratio of yield to total above-ground weight.

2. Crops such as sugar beet or cabbage in which each plant produces one harvestable organ have two yield components: the number of plants per hectare (plant density) and the average dry weight per plant of the harvested part. Increasing plant density leads to more and more intense competition between plants, resulting in a reduced dry weight per plant. Thus over a wide range of densities yield is scarcely affected.

Potatoes typify a crop that produces more than one harvestable organ per plant. In potatoes there are three yield components: plant density, number of tubers per plant and average dry weight per tuber. Tuber number per plant depends on the number of tubers initiated, which in turn depends on variety, storage conditions and sprouting conditions of the parent tubers. Potatoes can compensate for reduced plant density or reduced tuber number by increased tuber size.

Yield in cereals is made up of four components: plant density, ear number per plant, grain number per ear and average grain weight. As in other crops plant density is determined by seed rate and seedling mortality. Ear number depends on number of tillers produced, the proportion of tillers which initiate an inflorescence and the proportion of these which survive to bear grain. This yield component is determined during the tillering period, quite early in the life of the crop. Grain number depends on the number of florets initiated on the inflorescence and the proportion of these which develop flowers. This is also determined during and shortly after tillering, as the later process of fertilization and seed set in cereals is seldom much below 100%. Only grain weight is likely to be significantly affected by events after tillering.

The spacing of plants in the field can be used to regulate the size of produce of vegetable crops. For uniformity of size, even spacing, vigorous and reliable seed and protection of young plants from disease and other stresses are all important.

3. The amount of photosynthesis that can take place in a crop to build up its biomass depends largely on leaf area index (LAI), the ratio of total leaf area to the ground area covered. LAI is very small early in the growing season but rapidly builds up to an optimum value when lower leaves are shaded to the point where they cease to be net contributors of carbohydrate to the plant. Thereafter the loss of lower leaves by senescence and abscission is balanced by the production of new leaves at the top of the canopy and LAI levels off. Optimum LAI is around 3 for most temperate broadleaved crops and 4–5 for cereals and grasses. In cereals, where synchronous leaf senescence sets in shortly after maximum LAI is reached, yield is closely correlated with maximum LAI, but in crops such as sugar beet, which maintain optimum LAI for longer periods, yield depends not only on maximum LAI but on the time for which this is maintained.

The second major factor affecting yield is net assimilation rate (NAR), the rate of dry matter production per unit leaf area. NAR is relatively little affected by variety or husbandry and in practice is a less important determinant of yield than LAI.

Length of growing season, determined by climatic factors and by the crop's response to temperature, daylength and moisture, affects yield by its effect on the duration of photosynthesis. It may be extended by transplanting from glasshouses, by irrigation or by the use of varieties which are more tolerant of adverse conditions early in the growing season or are later maturing.

Much of the improvement in crop yields has come from the introduction of varieties with a better utilizable fraction or harvest index.

4. Weeds reduce yield by competing with crop

plants for light, water and nutrients. The degree of yield loss is dependent on crop species, variety and density, weed species and density, whether weeds emerge before, along with or after the crop, and environmental and husbandry factors which affect the relative competitive ability of crop and weed. Yield-damaging weed competition often occurs for a relatively short period of time quite early in the growing season. Thus pre-emergence herbicides need not have very long persistence in the soil and there is frequently no yield benefit to be had from mid-season or later application of post-emergence herbicides. Some weeds also reduce crop yield by parasitism or allelopathy.

5. Competition also takes place among the constituents of mixed crops. In many cases, however, the yield of the mixture is greater than that of monocultures of the individual species that comprise it. In cereals undersown with grass both crops suffer yield reduction through competition between then, but the advantage of earlier establishment of the grass makes the practice worth while. Mixtures of cereal species or varieties seldom show a yield advantage over monocultures, but a mixed grassland sward generally outyields a grass monoculture because different grass species show different seasonal patterns of growth and utilize soil resources in a complementary fashion.

6. In practice yield is almost always affected by stress factors, which damage crops in various ways: tissue removal or destruction, extraction of sap, formation of toxins, physiological changes, growth distortion, or reduced growth rate. Stress factors may be biotic (pests and disease) or abiotic (for example mineral imbalance, frost, over-heating, drought or waterlogging). Their effect on yield is not necessarily related to the severity of the damage, but instead depends on whether the stress is short-lived or sustained and on the stage of crop growth when the damage occurs.

In general, short-lived stresses are most damaging to yield when they occur late in the season, when the crop has little time for recovery, or when they coincide with a critical or vulnerable stage in the development of the crop, such as flowering. Sustained stresses tend to be most serious when they occur early, before the crop has had a chance to accumulate harvestable dry matter. Most crops, however, have substantial built-in safety margins to allow them to respond flexibly to stress and thereby minimize yield loss.

In the course of a season, a crop has to cope with many different stresses, some of which interact in simple or complex ways. The control of stress factors is important not only to improve average yields but to give greater stability of yield from year to year and in particular to avoid occasional catastrophic yield losses.

Certain types of stress may be deliberately applied in a controlled way in order to enhance the growth of one part of a plant at the expense of others, and thus increase harvestable yield.

7. Even when part or all of the potential yield of a crop has been realized, substantial yield losses may continue. Weather and other factors can cause losses in the standing crop, and inefficiencies in harvesting operations account for further losses.

8. The grower's role in ensuring that products are fit for storage, marketing and processing is vital. The crop increases in value as it goes through these stages but also becomes more vulnerable to mismanagement and consequent loss. Post-harvest deterioration is inevitable because the harvested plant parts eventually utilize their limited stores of energy and water, are subject to attack by pests and pathogens, and tend to continue their growth and development. The speed with which these trends proceed is dependent on the type of plant material, its condition on entering storage or marketing, and the environmental conditions during storage and marketing. Post-harvest deterioration can be greatly delayed by good husbandry throughout the growth of the crop in the field, including measures to reduce its infection and contamination by pests and pathogens, by careful handling during harvesting, storage and marketing, to reduce mechanical damage, by pre-storage cooling, and by control of relative humidity, temperature and ventilation in store.

9. Efficiency of dry matter production by crops can be measured with respect to any input but is most commonly expressed as energetic efficiency. The main energy inputs are solar radiation energy, about half of which is available for photosynthetic activity, and cultural energy. Energetic efficiencies of up to 3% fixation of photosynthetically available radiation (PAR) can be sustained by high-yielding crops over a growing season, but average yields represent efficiencies of conversion of less than 1%.

The main forms of cultural energy required in crop production are nitrogen fertilizer, which is, in terms of energy, expensive to produce, and the fuel used in machinery. Major arable crops in Great Britain yield 1.5–3.5 times the energy consumed in their production.

12

Plants as Food

The main concern of agriculture is the production of food for mankind, through the cultivation of crops and the husbandry of animals. Agriculture is not the only source of food; a significant contribution to our nutrition is made by commercial fisheries, together with a small contribution from the hunting and gathering of wild animals and plants.

Not all agricultural products are edible or are intended to be eaten by man or his domestic animals; note, for example, the commercial importance of timber, vegetable fibres, tobacco and rubber. Furthermore, not all edible plant products are actually used as food; apart from seeds and tubers saved for sowing or planting the next season's crops, many potential foodstuffs are processed for industrial products such as lubricants, starches and alcohol.

The aim of this chapter is to consider the *quality*, in a broad sense, of plants and plant products as foods for man and his livestock. In considering each group of nutrients the major factors that influence the food value of crops for those nutrients are described, and this variability is also considered more generally in terms of the effects of growing conditions (Section 12.2). Some effects of post-harvest handling on the quality of crops have already been discussed in Section 11.8.

Like plants, animals need a continuous supply of water, energy and the raw materials necessary for body construction and maintenance. Unlike plants, however, animals are not capable of photosynthesis, so that energy and raw materials can be obtained only by the consumption of food. Water is obtained by drinking and to a lesser extent by the consumption of food.

12.1 Plants in human and animal nutrition

What are the essential raw materials that animals, including humans, get from food? It will be remembered that a plant's requirements are very basic – carbon dioxide, water and minerals, including nitrogen mainly in the form of nitrate ions. Animals, however, have no enzyme systems capable of extracting carbon from carbon dioxide or nitrogen from nitrates. They must get their carbon and nitrogen already built into organic molecules such as carbohydrates, lipids and proteins. In most cases they break these down into relatively simple molecules – sugars, fatty acids and amino acids respectively – before reassembling them into their own body construction materials.

Certain organic substances which are required in small amounts for the functioning of some enzymes in both animals and plants, cannot be synthesized in animal cells from simpler molecules, although plants build them from scratch. These are the *vitamins*, another set of essential ingredients in the animal's diet (except, as will be seen, where microbes living symbiotically in the animal's gut can manufacture them on the animal's behalf).

Finally, animals require mineral elements for reasons similar to those of the plant and for building bones and teeth. Most of their mineral supply is in their food, although some portion of their requirement may be contained in their drinking water.

12.1.1 Water in plant foods

Water is every bit as essential to animal life as it is to plant life. The human body, for example, is approximately 62% water, and the adult human must consume about 1 litre of water per day to replace losses in urine, faeces and perspiration (more than this in warm conditions or with heavy exertion). A lactating cow requires about 30 litres per day depending on milk yield.

Many plant foods have a very high moisture content and indeed most fruits and vegetables have a higher moisture content than milk. The human digestive system could not, however, cope with the bulk of food that would be necessary to supply the daily water requirement; humans have to drink as well as eat to maintain a correct water balance. Cattle or sheep fed on fodder roots such as swedes or mangels may derive enough water from their food not to require drinking water, but most feeds are not rich enough in water to enable animals to do without a supply of additional drinking water. 'Concentrate' feeds such as grains and pulses and oilseed cakes are particularly low in water. Table 12.1 shows the water content of a range of common foods and animal feeds.

Water content and dry matter content are opposite sides of the same coin. In calculating animal feed rations it is dry matter content that is important, water being regarded simply as a diluent. The higher the dry matter content, the cheaper and easier a feed is to transport and store. The dry matter intake of cattle fed *ad lib* (allowed to eat as much as they want) on grass silage is often reduced if the silage has a low dry matter content, but this does not appear to be a direct consequence of the greater intake of water. There is no conclu-

Table 12.1 Moisture content of various foods and animal feeds of plant origin

	Typical moisture content (%)
Foods for human consumption	
wheat flour	13.0
oatmeal	8.9
Brussels sprouts	88.1
lentils	12.2
peas (picked green)	78.5
potatoes	75.4
sweet potatoes	70.0
tomatoes	93.4
Concentrate feeds	
barley grain	11.4
soya beans	10.0
maize grain	13.5
rapeseed meal	6.4
Roughages	
timothy hay	12.3
lucerne hay	10.0
oat straw	9.9
ryegrass (fresh)	75.7
grass silage	59.2
Fodder roots	
fodder beet	87.0
turnips	90.7

sive evidence of a similar depression of appetite with low dry matter roots or fresh green forage.

12.1.2 Energy in plant foods

Animals require energy for metabolism, temperature regulation, growth, reproduction and lactation. Dietary energy standards for farm livestock are usually calculated by adding together the energy required just to keep the animal at its present body weight, that is, for *body maintenance*, and the energy required for *production* at the desired level (the amount of additional energy needed for pregnancy, lactation, egg production, growing or fattening).

Dietary energy requirements are normally expressed in megajoules (MJ). A joule is equivalent to about 0.24 calories, the original metric units of heat energy. The 'calorie' of dieticians and weight-watchers is really a kilocalorie; 1 megajoule is equivalent to 240 kilocalories.

If a sample of food is burned in a calorimeter to carbon dioxide, water vapour and ash, the heat produced is a measure of the *gross energy* content of the food. Most plant materials have a gross energy content in the region of 18.5 MJ kg^{-1} dry matter, but materials rich in oil or other lipids (e.g. oilseeds) have higher gross energy contents because of the greater amount of energy locked up in lipids than in the same weight of carbohydrate or protein.

Routine chemical analysis of animal feeds (Fig. 12.1) partitions the dry matter of a sample into five broad categories of material, known as the *proximate constituents*. The first of these, *crude protein*, is a rough estimate of protein content derived by measuring the total nitrogen content of a sample. The average protein contains 16% by weight of nitrogen; the measured nitrogen content of the sample is therefore multiplied by $100/16 = 6.25$ to give crude protein. By no means all the nitrogen in a feed is in the form of protein, but the true protein content is very difficult to determine and for most purposes crude protein is an adequate approximation. Of the other proximate constituents, *ether extract* consists mainly of oils, fats, waxes and other lipids; *crude fibre* is predominantly cellulose and other cell wall materials, including lignin; *nitrogen-free extract* contains a wide range of materials, the most important of which are the non-structural carbohydrates such as starch and sucrose; and *ash* consists of the oxides of the mineral elements (excluding nitrogen) contained in the feed.

Fig. 12.1 Proximate analysis of an animal feed. As explained in the text, crude protein is a measure of all nitrogen-containing substances, not only true protein. Ether extract contains lipids such as oils and waxes. Crude fibre consists mainly of cellulose and other cell wall materials. Ash contains minerals. Subtraction of these four constituents from the total dry weight of the sample gives nitrogen-free extract, which consists largely of non-structural carbohydrates such as starch and sugars.

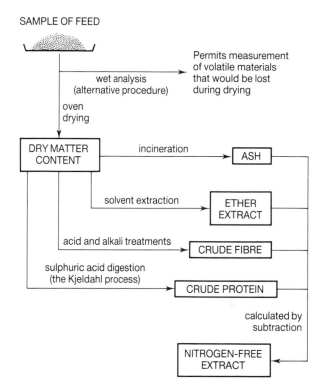

The four organic fractions, but not the ash, contribute to the gross energy content of the feed. Proximate analysis and calorimetric determination of gross energy content of a wide range of plant materials have shown that a close estimate of gross energy content can be derived from the formula:

$$GE = \frac{2.3CP + 4.1EE + 1.9CF + 1.8NFE}{100}$$

where gross energy (GE) is expressed in MJ kg^{-1} dry matter and crude protein (CP), ether extract (EE), crude fibre (CF) and nitrogen-free extract (NFE) are expressed in g kg^{-1} dry matter. Table 12.2 gives typical proximate analyses of some common feeds and their calculated gross energy contents.

Table 12.2 Proximate analysis of some common animal feeds

| | Dry matter (DM) (g kg^{-1}FW) | Analysis of dry matter (g kg^{-1}DM) | | | | | Gross energy (MJ kg^{-1}DM) |
		CP	EE	CF	NFE	ash	
Concentrate feeds							
barley grain	886	135	24	84	720	37	18.6
oat grain	910	133	53	132	647	35	19.4
maize grain	865	105	46	21	814	14	19.4
soya beans	900	421	200	56	272	51	23.8
sugar beet pulp	910	100	7	209	644	40	18.2
Roughages							
lucerne hay	900	184	22	298	402	94	18.0
timothy hay	877	87	26	332	493	62	18.2
grass silage	243	132	32	348	417	71	18.5
barley straw	860	37	16	488	392	66	17.8
ryegrass (cut, fresh)	243	165	41	218	420	156	17.2
ryegrass pasture (close-grazed)	200	265	55	130	445	105	18.8
Fodder roots							
fodder beet	130	123	8	69	685	115	16.8
turnips	93	140	22	118	624	97	17.6
potatoes	246	89	4	20	850	37	17.9

FW   fresh weight
CP   crude protein
EE   ether extract (oils, fats and other lipids)
CF   crude fibre
NFE   nitrogen-free extract (mainly non-structural carbohydrates)

12.1.3 Digestibility and metabolizable energy content of plant foods

Not all the organic matter or the energy it contains can be digested by an animal. The material which is not digested is excreted as faeces. A comparison of the proximate analysis of a feed with that of the resulting faeces reveals what fraction of each of the four organic constituents has been digested. This fraction is the *digestibility coefficient* for that constituent. Thus if 10 kg of crude protein in a feed is eaten by an animal and only 2 kg are retrieved in the faeces, the crude protein digestibility coefficient for that feed and that animal is 0.8.

Foods in which a high proportion of the energy is contained in the crude fibre component are known as *roughages*. They include grazed and conserved grass, kale and other green forage brassicas, green vegetables and straw. Roughages are not a suitable energy source on which to base a pig or human diet but are ideal for ruminants.

A measure of ruminant digestibility of a feed which is particularly useful in describing roughages is the *D-value*. This is the total digestible organic matter expressed as a percentage of the dry matter. With the onset of maturity in grasses, for example, the lignin content of the shoot increases. This lowers the digestibility not only of the crude fibre but of other components as well, because the contents of lignified cells may not be fully exposed to the digestive enzymes. The result is a progressive fall in D-value as the herbage matures (Fig. 12.2).

Grass for silage is cut earlier than hay and therefore produces a more digestible feed.

Subtracting the gross energy content of the faeces from that of the food gives a measure of the *digestible energy* content of the food. This unfortunately cannot be determined simply from the D-value because the energy content of digestible ether extract (DEE) is about two and a quarter times that of digestible crude protein (DCP), crude fibre (DCF) or nitrogen-free extract (DNFE).

Dietary energy standards for farm animals in the United States are based on the *total digestible nutrients* (TDN) system whereby the TDN content of a feed is DCP + 2.25DEE + DCF + DNFE. In Britain the *metabolizable energy* system is used.

Metabolizable energy (ME) differs from digestible energy in taking account of energy losses from the animal in urine and, in the case of ruminants, methane gas which is produced copiously during microbial fermentation in the rumen. Just as gross energy content of a feed can be estimated from proximate analysis, ME content can be estimated from the digestible proximate constituents by a similar formula. For ruminants the formula is:

$$ME = \frac{1.5DCP + 3.4DEE + 1.3DCF + 1.6DNFE}{100}$$

where ME is expressed in MJ kg^{-1} dry matter and DCP, DEE, DCF and DNFE are expressed in g kg^{-1} dry matter. ME contents of some feeds are given in Table 12.3.

Fig. 12.2 Changes in D-value (% digestible organic matter in total dry matter) of perennial ryegrass (variety S24) with maturity.

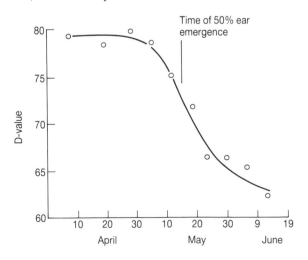

Table 12.3 Metabolizable energy content of some animal feeds

	ME (MJ kg^{-1}DM)
Pasture grass	
very leafy	10.8
leafy	10.6
early flowering stage	9.7
late flowering stage	8.7
Marrowstem kale	11.0
Sugar beet	
roots	13.7
leaves	9.9
Turnips	
roots	11.2
leaves	9.2
Oat grain	11.5
Wheat grain	14.0
Maize grain	14.2
Peas	13.4

Since the human digestive system can derive no metabolizable energy from cellulose, roughages are utilized less efficiently by humans than by ruminants. On the other hand, foods low in fibre such as potatoes and wheat grain are more efficiently metabolized by humans than by ruminants because of the much lower energy wastage in the form of methane.

An adult human male of average height and weight pursuing a reasonably active life requires something in the region of 12 MJ of metabolizable energy in his diet each day. Normally this would be satisfied by a combination of plant and animal foods, but a purely vegetable diet, judiciously chosen, could supply the necessary energy. For instance, the following each provide 1 MJ: 60 g cane sugar; 55 g wheat flour; 125 g bread; 250 g potatoes; 40 g peanuts; 25 g olive oil. Many plant foods, however, are very low in metabolizable energy, and for this reason are popular with slimmers. To get 1 MJ you would have to consume about 2 kg of apples, 1.5 kg of turnip or 10 kg of rhubarb.

12.1.4 Factors affecting energy content of plant products

Clearly the higher the dry matter content of plant material (that is, the lower the water content) the greater is its gross energy content, expressed on a fresh weight basis. Of more interest, however, is the energy content expressed on a dry matter basis. We have seen that plant products, except for those rich in oil, have a broadly similar gross energy content of around 18.5 MJ kg^{-1} dry matter. There are, however, big differences between products in metabolizable energy content, resulting mainly from differences in digestibility.

Some important factors which affect the digestibility and therefore the ME content of ruminant feeds are listed below.

Part of plant. In general, seeds (even those without a high lipid content) are richer in ME than roots or tubers, which in turn are richer than leaves. Stems tend to be low in both digestibility and ME content, mainly because of the relatively high proportion of lignin in the crude fibre component. Table 12.3 compares the ME contents of a range of seeds, roots, leaves and stems. Where both stem and leaf are eaten, as in kale, forage rape and grass, digestibility and consequently ME content decrease with increasing proportion of stem to leaf.

Species of plant. As Table 12.3 also shows, there are differences between plant species in the ME content of any one part of the plant. Oat grain has, for example, a lower ME content for ruminants than wheat grain. Here we are not strictly comparing like with like, for the wheat grain is a simple fruit with only relatively thin 'bran' (the pericarp and seed coat) containing much indigestible fibre, while the oat grain has a fairly massive fibrous 'husk' (the lemma and palea) surrounding the fruit. Grass species differ greatly in the digestibility of their dry matter, as a result of differing proportions of stem to leaf at any given time as well as differences in lignin and silica content both of stems and of leaves.

Variety of plant. Varietal differences in digestibility and ME content are probably of little significance in seeds and roots, but can be quite marked in herbage species. In particular, tetraploid ryegrass and red clover varieties tend to be of higher digestibility at a given stage of growth than their diploid relatives. There are also considerable differences in digestibility of straw between barley varieties, with D-values ranging from 35 to 50.

Stage of growth. This is the most important single factor affecting the digestibility of grassland herbage. Spring grass, consisting almost entirely of leaf, typically has a D-value of 70–75 and an ME content of 11–12 MJ kg^{-1} dry matter. Digestibility falls as the flowering stem begins to form, and this fall accelerates as lignification proceeds. When the D-value is 65, ME content will be about 10.5 MJ kg^{-1}, and in fully mature grass with a D-value of 50–55 the ME content will have fallen to around 9.5 MJ kg^{-1}. Leafy autumn regrowth following flowering is usually of lower digestibility than spring grass, while winter pasture, with its high proportion of dead and senescent material, is generally of very low digestibility, with an ME content of 8–9 MJ kg^{-1}.

This marked trend for the digestibility of vegetation to decline with advancing maturity is a very important agricultural phenomenon and requires some explanation. Obviously, the sole source of energy in the plant is its dry matter, which can be classified loosely as cell contents and structural material. Structural material forms the bulk of the dry matter. The cell contents are, largely, digestible by animals and are thus ready sources of energy and of nutrients in general. The structural material, of which cellulose and hemicelluloses are the major components, presents serious problems that have major implications for plants as sources of energy and general nutrition.

No animals, other than certain species of a primitive and microbial nature, can themselves digest cellulose. This is exclusively the ability of a range of microbes. All non-microbial animals such as horses, rabbits, sheep and cattle which can digest cellulose have evolved digestive systems in some part of which live specialized microbes, chiefly bacteria and protozoa. It is these microbes, rather than any metabolic process of the animal itself, which degrade cellulose.

In agriculture, the most important group of animals possessing such digestive systems are the ruminants, including cattle, sheep, goats and deer. They house cellulose-degrading microbes in a specialized chamber of their digestive tract called the rumen, and are particularly efficient utilizers of the energy content of cellulose. Other major groups of animals living primarily on a diet of cellulose, including termites, have different arrangements but the end result is largely the same. For those animals such as pigs and humans which have no such symbiotic association with cellulose-degrading microbes, plants are a much poorer source of energy. As the breakdown products of cellulose provide molecular skeletons around which other metabolites necessary for the animal can be constructed, these differences extend also to more general areas of nutrition.

Aging of tissues in the plant sees the deposition of hemicelluloses and lignin on the cellulose of the cell walls. Since hemicelluloses are as digestible as cellulose, their pattern of deposition has little effect on digestibility. However, not only can no animal digest lignin, but the ability to do so is confined to an even more restricted range of microbes than in the case of cellulose. A few groups of animals, such as beetles with wood-boring larvae, have evolved associations with gut organisms able to digest lignin, but no large animals have done so. Thus, as the cellulose of cell walls becomes encrusted with lignin during aging of tissues, it becomes inaccessible to the digestive enzymes of rumen microbes, and the nutritional value of the feed, especially as a source of energy, declines drastically. This accounts for the indigestibility of wood, and for many of the common differences in digestibility between plant products. The low energy value of cereal straw, one of agriculture's main by-products, is largely caused by lignin deposition. We discuss the factors affecting the energy content of

plant materials in more detail in Section 12.1.4.

In cereals grown as green forage crops, the fall in digestibility with development and lignification of the flowering stem is less marked than in herbage grasses, because the declining digestibility of the stem is offset by the filling of the grains with highly digestible carbohydrate. The digestibility of cereal straw, however, continues to decline until harvest. Modern oat straw from a combine-harvested crop tends to be of lower digestibility and ME content than straw from a crop harvested in the old-fashioned way by binder at an earlier stage of maturity.

Fodder roots also become progressively lignified as they mature. Hence, they show a particularly rapid decline in digestibility at the end of the winter storage period when they enter the sequence of physiological changes leading to flowering.

Crop plant density. In fodder root crops, as was seen in Section 11.2.1, a high population density produces relatively small roots. These have a higher dry matter content but a lower digestibility than larger roots. For example, a swede with a fresh weight of 1.5 kg might contain 13.3 MJ kg^{-1} dry matter of ruminant ME, whereas a 0.5 kg swede from the same field might contain only 12.6 MJ kg^{-1}. High plant density in a kale crop, however, has the opposite effect, by increasing the proportion of leaf to stem, thereby increasing digestibility. A similar effect is obtained by late sowing of kale or forage rape.

Method of conservation. Artificially dried grass is similar in chemical composition, digestibility and ME content to the fresh grass from which it has been made. Well-made grass silage, although chemically very different from fresh grass, is also similar in ME content to the grass from which it has been made. Badly preserved silage, on the other hand, suffers major losses in digestible nutrients through microbial respiration and the release of ammonia. Grass which is not wilted before ensiling may lose much of the highly digestible soluble components in effluent. The result in either case is a product with a much lower ME content than at harvest.

Considerable losses in ME take place during haymaking. These losses, which are largely due to plant and microbial respiration and to leaching of soluble nutrients, are minimized by rapid drying, for example on tripods. In one study the ME content of tripod hay from mature grass containing at harvest 9.8 MJ kg^{-1} dry matter fell to 9.1 MJ kg^{-1} whereas traditional field drying reduced ME content to 8.3 MJ kg^{-1}.

Other treatments. A range of treatments are commonly employed in agriculture to improve the digestibility of feeds. Some treatments are designed to increase the physical accessibility of the cellulose and hemicelluloses, or other components, to digestive enzymes. These may be simple mechanical measures, like the rolling, crushing or bruising of cereal grain, which exposes the endosperm. Treating cereal straw with concentrated sodium hydroxide or ammonia to attack the lignin are examples of chemical treatments to expose cellulose to digestion. Other techniques, such as the addition of urea or protein to straw, are simply providing supplements to the diet in recognition of the principle that no component of the diet can be efficiently digested unless balanced amounts of other necessary components are also present.

12.1.5 Protein in plant foods

The dry matter of an animal body contains much more protein than that of the typical plant. Whereas carbohydrate (in the form of cellulose) is the main structural component of plant tissue, that of animal tissue is protein. Muscle, skin, hair, feathers – all consist very largely of protein, and of course protein in the form of enzymes and as components of cell membranes is as vital to animal as to plant metabolism. All this protein has to be built up from protein in the animal's diet.

Strictly speaking, what animals require in their food is not protein but amino acids. Animal proteins are constructed from some twenty-three different amino acids. Some of these are interconvertible in the animal body or can be manufactured from other substances, so that not all twenty-three are essential components of the diet. Only eight to ten amino acids are essential in this sense.

Even these amino acids are not essential in the food of the ruminant. The microbial inhabitants of the rumen (the first and largest chamber of the ruminant's digestive tract) make all the essential amino acids, which can then be ingested by the animal itself. Any source of nitrogen in the amino ($-NH_2$) form can be utilized by the rumen microbes; thus not only the true protein but virtually the entire digestible crude protein (DCP) content of the food is capable of being built, even-

tually, into the animal's own protein.

The DCP contents of some typical ruminant feeds have been listed in Table 12.2. As is the case with metabolizable energy, DCP feeding standards have been established for farm livestock of different classes, at different liveweights and to allow different levels of production (for example as milk or as liveweight gain). It should be remembered that DCP is a dual-purpose nutrient, supplying both nitrogen and energy to the ruminant. Recently a new protein feeding system for ruminants has been introduced, taking into account the differential degradability in the rumen of proteins from different sources.

An animal's requirement for amino acids depends not only on the species but to some extent on its stage of growth and on the growth rate and productivity expected of it. To demonstrate the point, in the adult human diet there are eight essential amino acids: isoleucine, leucine, lysine, methionine, phenylalanine, threonine, tryptophan and valine, but infants additionally require histidine. About three-quarters of the requirement for methionine can be met by cystine, and about three-quarters of the requirement for phenylalanine can be met by tyrosine. Laying hens require arginine and growing pigs require arginine and histidine in addition to the eight amino acids essential for humans.

The minimum protein requirement for non-ruminants such as pigs, chickens and humans is the quantity that will supply sufficient of all the essential amino acids. It will thus be determined by the requirement for the amino acid that is in shortest supply relative to that requirement. The other amino acids which are then consumed in excess of the requirement are available for interconversion and as a source of metabolizable energy. A problem in practice is that in plant proteins, certain amino acids, namely tryptophan, threonine, cystine, lysine and methionine, especially the last two, occur at fairly low levels relative to the needs of animals. The concentration of these in the diet of non-ruminants is thus often a limiting factor on their growth. Ruminants do not entirely evade this problem through their use of microbial proteins from the rumen, as proteins from these sources partially share the same deficiency.

By taking a varied diet, a human being can minimize the risk of amino acid deficiency. This is particularly important for people on vegetarian diets, since animal proteins are less likely to be deficient in particular amino acids. As we shall see, the importance of a varied diet is not purely a matter of protein nutrition but also involves vitamins and minerals.

Protein deficiency in children retards growth. In extreme cases, all too common in tropical Africa and America, it causes a wasting disease known as kwashiorkor. This condition can be relieved either by feeding protein-rich foods or by increasing the metabolizable energy intake so that the meagre protein content of the diet is not wasted by being used as an energy source.

Protein content of plant products varies widely; some of the main sources of this variation are listed below.

Part of plant. Actively growing or meristematic parts of plants tend to be relatively rich in protein – for example, the crude protein content of the curd of a cauliflower is about 300 g kg^{-1} dry matter and that of wheat germ about 270 g kg^{-1}. Seeds are richer than straw in protein; leaves are normally richer than stems.

Species of plant. The most striking difference is between legumes and non-legumes. The DCP content of European field beans (about 250 g kg^{-1}) compares very favourably with that of wheat grain (about 105 g kg^{-1}), while the DCP content of pre-flowering lucerne (alfalfa) herbage (about 213 g kg^{-1}) is greater than that of perennial ryegrass (about 146 g kg^{-1}) at the equivalent stage of growth.

Variety of plant. Cereal varieties differ considerably in the protein content of their grain. In wheat and barley there has been deliberate selection of varieties with high and low protein (strong and weak wheats, feeding and malting barley respectively). Breeders have put some effort into raising the lysine content of cereals, and have had some success with barley.

Stage of growth. As the crude fibre content of herbage rises with the onset of maturity, the crude protein content falls (Fig. 12.3). Thus grass harvested early for silage has a higher DCP content (typically around 150 g kg^{-1}) than more mature grass harvested later for hay (around 80 g kg^{-1}). Since leaves are richer in protein than stems, a leafy kale crop, produced by late or close sowing, contains more DCP than a stemmy crop.

Nitrogen fertilizer application. Almost all crops show a response in protein content to application of nitrogen fertilizer. The main exceptions are legumes, which are independent of externally applied nitrogen. Late or excessive nitrogen application increases the protein content of barley grain; this reduces its malting quality because it results in

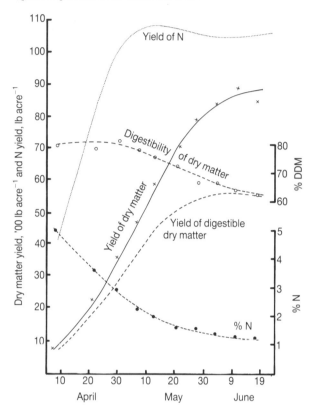

Fig. 12.3 Changes in yield and composition of perennial ryegrass (variety S24) with increasing maturity following the application of 60.4 kg ha^{-1} nitrogen fertilizer on 25.2.1965 and 78.4 kg ha^{-1} on 24.3.1965. From *NAAS Quarterly Review* 77 Autumn 1967.

unacceptable levels of nitrogenous compounds in the malt extract. Heavily fertilized grass shows a substantial increase in DCP content over less intensively managed grass. This is partly a result of the greater leafiness of the herbage and partly a result of protein enrichment of both leaves and stems. The increase in protein is mainly at the expense of non-structural carbohydrates such as starch, rather than of fibre. The influence of nitrogen fertilizer application on protein content and quality is discussed in more detail in Section 12.2.2.

Losses during conservation. As was seen in the foregoing section on energy, the losses of digestible nutrients in artificially dried grass and well-made silage are fairly small, but considerable losses of DCP can occur as effluent and ammonia from badly made silage. Some loss of DCP during hay-making is inevitable, but this can be minimized by rapid drying.

Food preparation. In the milling of wheat the protein content is increased by the removal of the bran, which contains little protein, and decreased by the removal of the protein-rich wheat germ. The net effect is that white flour has a somewhat lower protein content than the original grain and a considerably lower protein content than 'wholemeal' flour which contains the germ but only some of the bran. Most of the protein content of potatoes is often discarded in the peelings, since the tuber has a protein-rich layer just beneath the skin.

12.1.6 Lipids in plant foods

The animal body contains a lot of lipid, most of it in the form of fat. Animal fats and vegetable oils are very similar chemically, both being esters of glycerol with three fatty acids. A fat is solid at room temperature, whereas an oil is fluid. It is unfortunate that this minor difference has led to the use of two separate terms for these materials.

Only 10–20% of the fat in a normal healthy animal performs essential physiological functions. The remainder is an energy store which can be severely depleted without causing serious harm.

As was seen earlier, the lipid component of food (measured as the ether extract) is a concentrated source of metabolizable energy, contributing about two and a quarter times as much ME as an equal weight of carbohydrate or protein. Most of the foods consumed by farm animals are relatively low in lipids; seldom do animals receive more than about 5% of their ME from lipids. Oils and fats are much more important in the human diet, where they contribute up to 40% of the ME intake in the developed countries. The great bulk of this is animal fat, in meat, eggs and dairy products, but a growing proportion is vegetable oil.

Lipids are nutritionally important not only for the energy they contain but also for a small number of *essential fatty acids* which, if absent from the diet, cause dermatitis, cessation of growth and, ultimately, death. These acids, linoleic, linolenic and arachidonic, can partially substitute for one another in some animals. Humans, for example, can manufacture linolenic and arachidonic acids from linoleic acid and so have an absolute requirement for only one essential fatty acid. In practice, essential fatty acids are hardly ever deficient in human or animal diets.

A more important dietary function of lipids is that they act as a vehicle for certain lipid-soluble vitamins (see Section 12.1.8). These vitamins can be taken in by the animal only as part of the lipid component of food. Fats and oils also have an important effect on the palatability of food, as will be discussed later in this chapter.

Much has been said and written about the effects of saturated and unsaturated fatty acids in the human diet. Plant products tend to have a higher proportion of unsaturated fatty acids – those containing double carbon-carbon bonds – in their lipid content than do animal products. It appears that a high intake of certain saturated fatty acids encourages the deposition of large amounts of cholesterol and other lipids in the blood vessels, restricting blood circulation. The most dramatic effect is atherosclerosis (hardening) of the coronary artery, a heart disease which presently kills or permanently disables more people in western Europe and North America than any other disease. It must be emphasized that diet is only one of many factors said to be implicated in atherosclerosis; other predisposing factors include smoking and diabetes. However, there is evidence that an increased intake of unsaturated fatty acids, for example by replacing butter with certain forms of margarine or animal fats with particular vegetable oils, can reduce the level of cholesterol in the body and lower the risk of coronary atherosclerosis.

12.1.7 Fibre in plant foods

As we have seen, the main constituents of the crude fibre component of plants are the cell wall materials – mainly cellulose, hemicellulose and lignin. The digestibility of a feed, for both ruminants and non-ruminants, tends to decrease with increased crude fibre content. Typically a 1% increase in crude fibre brings a 1% decrease in digestibility for ruminants and a 2% decrease for pigs. Crude fibre adds bulk to the food of a non-ruminant without greatly adding energy. It may therefore be a useful component of a pig's diet where a lean carcase is required.

Digestibility is affected not only by the total amount of crude fibre present but also by the composition of the crude fibre. Specifically, the higher the lignin content the lower is the digestibility because, as explained in Section 12.1.3, lignin cannot be digested even by ruminants. We have already seen how the digestibility of herbage falls with maturity; this fall is associated with the progressive lignification of the flowering stems. Herbage species and varieties differ in their lignin content at any one growth stage; clearly the most desirable ones agriculturally are those that lignify most slowly and least thoroughly.

In human nutrition the term *dietary fibre* is often used. This is not synonymous with crude fibre, but refers to all the organic matter of plant foodstuff that is not capable of being digested in the human digestive system. Although virtually all the dietary fibre consumed is excreted in the faeces it is wrong to assume that it plays no useful role in human nutrition.

Fibre appears to be necessary for the normal functioning of the intestines. The natural laxative effect of fibre is partly due to its ability to absorb water and increase the bulk of faeces and partly to the production of volatile fatty acids (acetic, propionic and butyric) in the gut as a result of microbial breakdown of some of the cellulose. The greatest laxative effect is produced by fibre low in lignin. Cabbage, for example, containing 3% lignin, is more effective than wheat bran which contains 8% lignin.

Increasing affluence in western societies brought a marked reduction in fibre consumption, a trend which is now being reversed. Wholemeal bread is regaining popularity, and fruit and vegetable consumption is increasing. These changes, and the greatly increased popularity of bran-based breakfast cereals, are at least partly due to the publicity given to the linking of certain diseases such as diverticulitis and cancer of the colon to inadequate dietary fibre.

12.1.8 Vitamins in plant foods

The role of most, but not all, vitamins is to work along with certain enzymes, in which role they are known as *co-enzymes*. An example of a coenzyme is NAD, nicotinamide adenine dinucleotide, which, as we saw in Section 1.8.1, has an essential role in energy transference in respiration. Both plants and animals respire; both therefore require NAD. Plants make their own NAD but animals need a supply of nicotinamide or the chemically similar nicotinic acid from which to manufacture NAD. Since NAD is continuously regenerated in the process of oxidative phosphorylation, the animal, like the plant, needs only a small amount at a time. Some of the nicotinamide is metabolized or excreted and must be replaced, hence the requirement for a small daily intake of nicotinamide or nicotinic acid. This is the vitamin known sometimes as niacin or vitamin B_5.

Vitamins, some properties and sources of which

are listed in Table 12.4, are classified according to whether they are soluble in water or in lipid. The various B vitamins and vitamin C are water-soluble, while vitamins A, D, E and K are lipid-soluble. Since lipid-soluble vitamins can be stored in the body fat, it is less essential for the animal to have a regular daily intake than is the case with the water-soluble vitamins.

Certain lipid-soluble vitamins exist in the diet as *provitamins* – substances which can be converted to the active vitamin in the animal's body, for example, the various carotene pigments of plants, present in all green tissues and abundantly in the root of the carrot. Most carotenes are convertible to vitamin A by animals; the pigment with the greatest vitamin A activity is beta-carotene. By contrast, vitamin D in plants can be absorbed in the digestive system only when it is in the active form, ergocalciferol. This is produced in plants by the action of ultraviolet light on a related substance, ergosterol, which itself is of little nutritional value to animals. Hay is thus richer in vitamin D if made in sunny weather. Vitamin D is also made in the skin of humans and animals, again by the action of ultraviolet light. The dietary requirement is therefore greater for housed animals in winter than for animals grazing summer pastures.

Many vitamins are manufactured by microbes living in the digestive tract – principally in the rumen of ruminants or the intestine of non-ruminants. There is sufficient microbial synthesis of all the B vitamins and of vitamin K in the rumen to make a dietary supply of these vitamins unnecessary for the ruminant. Similarly, there is usually sufficient synthesis of the B vitamins biotin and folic acid and of vitamin K in the intestine, to meet normal human requirements.

The daily requirement in the adult human diet for those vitamins not, or inadequately, synthesized in the intestine ranges from the merest trace (e.g. vitamin B_{12}) to around 50 mg per day (e.g. vitamin C). It must be appreciated, however, that no hard and fast figures can be quoted for human or animal dietary requirements, since the adequacy or otherwise of a certain intake of a dietary component such as a vitamin or a mineral depends also on the other components of the diet and on the condition of the person or animal.

Shortage of a vitamin in the diet of a human or animal induces a set of symptoms characteristic of deficiency of that vitamin. Sometimes the symptoms bear an obvious relationship to the metabolic function of the deficient vitamin. For example, vitamin A deficiency in humans is characterized by night-blindness; the vitamin is essential to the regeneration of the eye pigments involved in seeing in dim light (and also has many other vital functions). In other cases, the relationship is less obvious. Nicotinamide deficiency, for example, produces inflammation of the skin in areas exposed to sunlight, as well as diarrhoea and nervous disorders. The combination of symptoms was well known, and had been given the name pellagra, long before it was realized that a dietary deficiency was responsible.

Nicotinamide is unusual in that the amino acid tryptophan can substitute for it in the diet, but it takes about 60 mg of tryptophan to make 1 mg of the vitamin. Pellagra is especially characteristic of people on a staple diet of maize because maize grain is deficient in both nicotinamide and tryptophan.

Vitamin C or ascorbic acid was one of the earliest vitamins to be identified and synthesized (it is a fairly simple derivative of glucose) but its basic function in the animal body remains unclear. It is known to be involved in the formation of collagen, a structural protein which binds cells together in animal tissues.

Virtually all animal species, including those domesticated as farm animals, possess the necessary enzymes for the conversion of glucose to ascorbic acid. Only man and a handful of other species require a dietary source of this vitamin. The characteristic vitamin C deficiency disease in humans is scurvy, the symptoms of which include the bursting of blood vessels, swollen and bleeding gums, the failure of wounds to heal and the reopening of previously healed wounds.

Scurvy was the most serious health problem in the British Royal Navy until the late eighteenth century, when it was discovered that a daily ration of citrus fruits (oranges, lemons or limes) or their juice would both cure and prevent the disease. It is now known that vitamin C, of which citrus fruits are a rich source, is the anti-scurvy factor. Scurvy also used to be prevalent in northern Europe in the late winter, when no fresh fruits or vegetables were available, but it disappeared with the introduction of potatoes, about the only energy-rich food which contains significant amounts of vitamin C.

Recently there has been controversy over the advocacy of very large doses of vitamin C (1–5 g per day or 20–100 times the recommended dietary intake) to prevent the common cold. Even higher doses have been suggested for treatment of colds. Both the effectiveness and, more important, the safety of these doses must be regarded at present as unproven.

Generally, as a vitamin becomes depleted, the efficiency of body chemistry begins to deteriorate long before obvious symptoms appear. This may heighten an animal's susceptibility to disease, as suggested in the vitamin C example, or have other effects.

Young animals generally show retardation of growth at vitamin intake levels which are low but not so low as to cause deficiency symptoms to appear. Thus calves require, for optimum growth, about twice the intake of vitamin A needed to prevent night-blindness. The recommended human daily intake of vitamin B_{12} is six to eight times the level necessary to prevent the characteristic deficiency disease for this vitamin – pernicious anaemia.

Foods derived from plants are, in general, much richer than animal products in vitamin C, E and K but poorer in biotin and vitamin D. Plants contain absolutely no vitamin B_{12}; vegetarians must be careful to supplement their diet with yeast or other preparations containing this vitamin. The value of a particular plant food as a vitamin source depends on the vitamin in question, as shown in Table 12.4.

12.1.9 Minerals in plant foods

All of the mineral elements which are essential for plants (see Chapter 3) are, with the single exception of boron, also essential for animals. In addition, animals require substantial quantities of sodium and trace amounts of at least nine other elements none of which are essential to plants; these include iodine, fluorine, cobalt and selenium.

Relative to human or animal requirements, certain minerals are often in short supply in foods of plant origin. Such deficiencies must be made up in the drinking water or by feeding minerals as supplements, such as common salt in the case of sodium deficiency, otherwise animal health will suffer.

Just as vitamin deficiencies produce characteristic symptoms, so do mineral deficiencies. For example, iodine deficiency results in impairment of the functioning of the thyroid gland, causing the disease known as goitre. Without iron, the body cannot make the red blood pigment haemoglobin, and so iron deficiency shows as anaemia. Bones are made largely of calcium and phosphorus, thus a deficiency in either of these elements causes rickets

Table 12.4 The major vitamins required in human or animal diets

| | Vitamin A | Vitamin B complex | | | |
		Vitamin B$_1$	Vitamin B$_2$	Niacin	Vitamin B$_{12}$
Chemical nature	retinol (can be made in animal tissues from carotenes)	thiamine	riboflavin	nicotinic acid, nicotinamide	various cobalamins
Functions	component of eye pigments necessary for night vision; also various more general functions	involved in enzyme systems responsible for release of energy from carbohydrates	involved in enzyme systems responsible for release of energy from carbohydrates	required to make NAD, essential for release of energy from carbohydrates	involved in synthesis of methionine and nucleic acids; essential for dividing cells as in bone marrow
Deficiency symptoms	night-blindness; other symptoms in severe cases; farm animals show reduced secretion of milk and other disorders	beriberi, a disease characterized by nerve and heart impairment	skin and tongue lesions	pellagra, a disease characterized by skin inflammation and nerve disorders	pernicious anaemia, caused by depressed red blood cell formation in bone marrow
Typical sources	fish, liver; no retinol in plants but all leafy vegetables are good sources of carotene; also carrots, tomatoes; potatoes contain only a trace of carotene but sweet potatoes are a rich source	most plant foods are good sources, especially cereal grains, but milling can remove much of the thiamine	leafy vegetables, legume seeds; also abundant in meat and dairy products	widely present in plant foods, but in cereals may be largely in a form unavailable for human metabolism	none in plants; synthesized by microbes (e.g. yeast); occurs in most animal products, especially liver

or osteomalacia, as in the case of vitamin D deficiency.

Minerals commonly deficient in the diet of a herbivore fall into three categories.

1. Minerals for which the plant has no requirement are often in very short supply in the plant, especially on soils with low contents of these minerals. Perhaps the best example is sodium, which fortunately is easily supplied as salt. Large areas of the world have soils deficient in iodine, so that human and animal populations in these areas show a high incidence of goitre. Similarly, selenium and cobalt deficiencies are widespread in farm livestock in many parts of the world. Microbes in the ruminant digestive system require cobalt for the manufacture of vitamin B$_{12}$; non-ruminants require their cobalt in a form already made into vitamin B$_{12}$. As we have seen, this vitamin cannot be obtained from foods of plant origin. The ruminant can derive sufficient cobalt for its needs from herbage, provided the available cobalt level in the soil is adequate.

2. Minerals, for which the requirement of animals is much greater than that of the plants on which they feed, are often, for that very reason, deficient in animal diets. A good example is copper. A copper content in grass as low as 2 mg kg^{-1} dry matter is quite consistent with good grass growth, but the recommended copper intake for cattle is six times that concentration in the diet. Maize has particularly low copper and manganese contents; the feeding of maize without supplementation often gives rise to deficiencies in these elements, particularly in poultry.

Others	Vitamin C	Vitamin D	Vitamin E	Vitamin K
pyridoxine (B_6), biotin, pantothenic acid, and others	ascorbic acid	cholecalciferol, ergocalciferol	tocopherol	naphthoquinone
involved as coenzymes in various enzyme systems	necessary for synthesis of collagen (a protein important in formation of healthy skin, bones, tendons)	enables animals to use calcium and phosphorus; essential for calcification of bones	essential for maintenance of muscle function and reproductive fertility	involved in blood clotting
seldom or never deficient because of adequate synthesis by intestinal microbes	scurvy, a disease characterized by skin lesions, failure of wound healing	osteomalacia, rickets, caused by retarded bone calcification	seldom seen in humans; farm animals show muscular wasting and reproductive disorders, eggs fail to hatch	internal bleeding; but seldom seen because of adequate synthesis by intestinal microbes
widely distributed in plant and animal foods but not normally required in diet	fruits, green vegetables, potatoes; virtually none in animal foods; required by humans but not in diet of farm animals	cholecalciferol produced in animal and human skin by action of sunlight; ergocalciferol formed by action of sunlight in field cured hay	most plant foods, especially cereals (located mainly in embryo, e.g. wheat germ) and vegetable oils; also in animal foods	widely distributed in plant and animal foods but not normally required in diet

3. The availability for assimilation by the animal of certain minerals in foods of plant origin is often restricted. This may be because the mineral forms an organic complex in the plant which the animal's digestion system cannot break down to release the mineral, or because there is an imbalance between two or more mineral elements which results in poor assimilation of one of these elements.

An example of an organic complex is phytic acid or any of its salts (phytates). As much as 80% of the phosphorus in plant foods may be in the form of phytates. Ruminants can degrade phytate to release the phosphorus, but for non-ruminants this phosphorus is largely unavailable. Foods such as cereals, with a high phytate content, also show reduced availability to the non-ruminant of calcium, magnesium and, perhaps most seriously, zinc. Most of the phytate content of wheat is localized in the bran. Bread made from wholewheat flour, including bran, is therefore very deficient in available zinc. The problem is particularly serious in societies where the staple diet is unleavened wholewheat bread. The leavening process breaks down some of the phytate, releasing zinc.

The best understood example of a mineral imbalance causing reduced availability of one of the interacting elements is the case of calcium and phosphorus. As mentioned, a straight deficiency of either element causes disorders of bone formation. The same deficiency symptoms, however, can appear when both elements are above the normal deficiency range, but when the calcium/phosphorus ratio in the diet is greater than about 2 or lower than about 1. A great excess of calcium over phosphorus causes reduced phosphorus retention by the animal; similarly an excess of phosphorus over calcium causes reduced calcium retention. Legumes have a high calcium/phosphorus ratio – 5 or more in the case of lucerne (alfalfa) – and should therefore be combined in the diet with products having a lower ratio. Cereals have a very low ratio, typically around 0.2.

Less well understood is the cause of grass staggers, or hypomagnesaemic tetany, seen in grazing cattle, most commonly in spring. The condition, which is often fatal if untreated, can rapidly be overcome by injection of magnesium, but very often the herbage being grazed is apparently adequate in magnesium content. It appears that there is some antagonism between magnesium and potassium, which is often present at high concentrations in heavily manured spring grass.

An example of a threefold interaction is the one between copper, molybdenum and sulphur. The presence of molybdenum and sulphur (in the form of sulphate) limits the retention of copper by the ruminant. The higher the molybdenum content the higher must be the copper content of the diet to prevent deficiency, but only in the presence of sulphate. Swayback, a disorder of the nervous system in lambs, is an example of a molybdenum-induced copper deficiency. It is impossible to draw a line between copper deficiency and molybdenum toxicity. The condition in cattle known as teart is associated with high molybdenum content in herbage and is normally thought of as a toxicity condition, but it can be treated as if it were a straightforward copper deficiency, by administration of supplementary copper.

Mineral toxicities most commonly arise with those elements for which the toxic dose is not much greater than the nutritional requirement. Elements with a narrow range of safety include molybdenum, selenium, fluorine and copper. More will be said about toxic constituents of plants in Section 12.5.

The three chief factors that cause variation in the mineral content of herbage are growing conditions, stage of maturity, and species composition. Variation due to growing conditions will be considered in Sections 12.2.2 and 12.2.3.

The content of most minerals in herbage declines with advancing maturity. Changes also take place through the growing season in the mineral content of regularly cut or grazed herbage which is not allowed to mature. There are, for example, rises in magnesium and copper and a fall in potassium.

Legumes are richer than grasses in calcium, magnesium, iron, copper and cobalt. There are also differences between species within each of these groups. Cocksfoot, for example, is often richer in minerals than perennial ryegrass, perhaps because its deeper rooting system draws minerals from greater depths in the soil. Certain non-leguminous broad-leaved herbs which occur commonly in pasture have been shown to be relatively rich in minerals. They include dock, sorrel, dandelion and ribwort plantain. Chicory, yarrow, sheep's parsley and forage burnet are very occasionally sown in pasture or in special strips to enrich the diet of grazing animals in minerals.

Trace element deficiency in grazing animals is often less severe than might be expected on the basis of the content of these minerals in the grazed herbage. This is because animals ingest some soil along with the herbage, boosting their mineral intake. As well as soil, they consume atmospheric dust trapped in the vegetation. This, though small in amount, can enormously increase the intake by livestock of certain pollutants, including toxic minerals such as lead and fluorine.

The mineral intake in ingested soil can be considerably greater. As much as 2–10% of a grazing herbivore's dry matter intake may be soil. Cattle tend to eat more soil than sheep because they tear up vegetation as they graze and do not bite it off cleanly as sheep do. This soil can be an essential source of minerals. Soil ingestion is a particularly important source of iron for ruminants and of copper for pigs. Pigs reared indoors and unable to root for food in soil, some of which they would ingest, must have their diet augmented with copper additives. It should be noted that certain soils, if ingested, have the opposite effect, reducing the availability of copper for absorption in the gut.

Wild herbivores in some parts of the world dig up and consume natural deposits of salts or soil heavily impregnated with salts. These salt-licks, as they are called, almost certainly act as an important supplement of certain minerals, especially sodium.

12.2 Effects of growing conditions on the nutritional quality of plants

The aspect of quality of prime concern in this section is nutritional value. In passing, mention will also be made of other aspects of quality, including taste, smell and colour, but these will be more fully dealt with in Section 12.3. Quality also covers a disparate range of characteristics including size of product, its resistance to damage, its suitability for certain markets, and its health. All of these are mentioned, where appropriate, in the text. In this section the term 'growing conditions' is meant to convey that combination of environmental influences and husbandry practices to which every crop is subjected.

Where growers have been affected directly by shortcomings in the quality of their produce, for example through refusal for certain markets, there has been considerable research into the effects of growing conditions on quality. This has included the nutritional value of grass, the malting quality of barley and the sugar content of sugar beet. Where growers have not been directly affected in this way, the study of crop quality has tended to be neglected. Curiously, this has included the quality of produce for human nutrition. Yet, although the nutritional quality of a plant is largely determined by its genetic make-up, the variations that occur in nutritional value due to the activities of growers are often large and important. To bring home the point, a recent Danish study of the protein content of barley over several seasons, sites and varieties and two rates of nitrogen fertilizer application revealed a range of 91–153 g kg^{-1} in crude protein content. Thus the highest value was nearly 70% greater than the lowest. Another study showed considerable variation in the concentration of certain vitamins in barley grain. Vitamin E varied by up to 100% between seasons, and sizeable variations were also found in the concentrations of vitamins of the B complex. The significance of such differences in nutritional value depends on how large the variation is, how limiting the affected nutrients are in any diet and how critical the aspect of quality affected is to the marketability of the product.

12.2.1 Main causes of variability in nutritional value

The largest variations and greatest diversity of differences found in nutritional value and in qualities such as scent and flavour are undoubtedly those revealed when products from different sites or years are compared. A comparison of vegetables grown in northern and southern Sweden made this point well. Those grown in the north were superior in crispness, flavour and aroma, and had higher vitamin C and sugar content, but a lower carotene content. The yearly variations in vitamin contents in barley, quoted earlier, also demonstrate this point. When nearby sites are compared, differences in quality are usually due primarily to differences in soil conditions arising from soil type and treatment, but differences between products from sites distant from one another are usually compounded by, and are more attributable to, climatic differences between the localities.

Unravelling the influence of the many individual factors covered by soil and weather conditions and their interactions is extremely difficult. Even where the impact of a single environmental factor is thought to be detected, its influence is not necessarily straightforward. The Swedish studies just mentioned included data on some flower crops. These data showed better scent and colour in flowers grown in the north. This was attributed to the effect of temperature, but not only to high or low temperature; rather it was thought to be due to larger fluctuations in temperature. A systematic American study of the composition of turnip foliage over a wide range of sites, climates and seasons, correlated a single factor, low soil moisture, with fifteen aspects of leaf composition, including content of dry matter, vitamin C and potassium. If the correlations reflected cause and effect, there are obviously numerous ways in which low soil moisture could have induced the observed changes.

Fortunately, the results of a few systematic studies, such as the American one above, allow some generalizations to be made. Since soils are the source of minerals, soil conditions, especially soil type and treatment, have a major influence on the mineral content of the plant. In contrast, they have little direct influence on the moisture content of plants, or on the contents of their organic fractions, including proteins, lipids and vitamins, with certain important exceptions. High levels of soil nitrogen lead to increased moisture content in plant parts. Further, as nitrogen is part of all amino acids and sulphur is part of some, the supply of nitrogen and sulphur directly affects the level and composition of proteins in plants. Potassium also has important influences on the organic matter composition of crop products, as will be explained below.

Weather and plant age are, on the other hand, the two chief determinants of variations in the organic constituents of plants. Weather not only

exercises a major influence on the organic constituents and moisture content of the plant, it also
influences soil conditions such as temperature and
moisture content, which then affect the availability
of minerals in soil. Weather thus also indirectly
influences, to a considerable extent, the mineral
content of plants.

The susceptibility of plant constituents to variation through the influence of these factors differs
greatly between constituents. At one extreme,
some constituents appear to be such integral parts
of plants that, as a percentage of dry weight, they
can vary only within narrow limits in response to
variations in climate or soil. Crude fibre content is
a good example. At the other extreme, the content
of some constituents such as lipids or certain
minerals, including sodium and iron, seem not to
have this close linkage to the integral structure of
plants, and can vary widely in response to growing
conditions. This pattern of variation can be seen in
the data quoted from the American study of turnip
foliage mentioned above (Table 12.5), in which the
highest recorded sodium content was nearly seven
times greater than the lowest.

The causes of such variations in plant composition are often complex, but some mechanisms can
be discerned. A number of important factors
influencing energy and protein content have been
mentioned in Sections 12.1.4 and 12.1.5. Here, a
number of more general mechanisms are discussed.

Table 12.5 Variation in the chemical composition of
turnip leaves from a range of sites in the southern USA;
each figure represents the mean of numerous samples
collected at one site; all samples were collected at one
time of year

Component	Units	Highest	Lowest
Dry matter (DM)	g kg^{-1}FW	121	107
Crude protein	g kg^{-1}DM	395	328
Ether extract	g kg^{-1}DM	55	28
Crude fibre	g kg^{-1}DM	78	63
Vitamin C	g kg^{-1}DM	12.2	9.1
Thiamine	mg kg^{-1}DM	17.3	11.6
Riboflavin	mg kg^{-1}DM	30.4	23.3
Potassium	g kg^{-1}DM	4.8	3.2
Calcium	g kg^{-1}DM	3.6	2.2
Magnesium	g kg^{-1}DM	0.54	0.25
Sodium	g kg^{-1}DM	0.81	0.12
Iron	g kg^{-1}DM	0.49	0.20
Phosphorus	g kg^{-1}DM	6.8	4.9

From H L Lucas et al. (1959) Southern Coop. Ser. Bull.
52.

In the simplest situation, plants are induced to
take up more of a mineral, as happens through
potassium fertilizer application, or to manufacture
more of an organic constituent. For example,
carotene production in leaves is stimulated by high
light intensity. Since animals convert carotene to
vitamin A, the effect of high light intensity is to
increase the vitamin A value of the leaves.

More commonly, interactions between molecules or groups of compounds are involved, with
growing conditions altering the balance between
them. Low nitrogen supply to a crop restricts
protein synthesis, but leaves the energy that
would have been utilized in that process available
for the synthesis of carbohydrates, which therefore
increase in concentration in the plants. Constituents can interact in other ways. As described in
Section 3.5.2, mineral ions may be antagonistic to
one another, as when increased uptake of
potassium decreases that of calcium and some
others, an example of a competitive interaction
between constituents. Alternatively, they may
interact in a positive way, as when increased nitrate
stimulates sodium uptake. Through interactions
like these, fertilizer usage affects plant composition, and thus nutritional value.

Influences on other aspects of the plant's physiology are also causes of variation in nutritional
value. Perhaps the best known example is where
increased potassium uptake leads to more rapid
and complete filling of seeds, fruits and vegetative
storage organs through its stimulus to the transport of organic and inorganic substances within
the plant. Here both quality and yield are altered.
The accumulation of storage carbohydrates also
depends greatly on the balance between their production by photosynthesis and their utilization in
respiration and as structural materials. Increased
temperature, at a given light intensity, usually
leads to increased growth rate without increased
photosynthesis, and hence to increased utilization
and depletion of storage carbohydrates. In contrast, increased light intensity, at a steady
temperature, leads to increased photosynthesis and
therefore a rise in storage carbohydrates.

Since the plant's composition changes markedly
with age and stage of development, especially in
the case of leafy material such as herbage or green
vegetables, it is inevitable that any environmental
influences or husbandry practices that alter the
pattern or rate of growth and development will
markedly influence plant composition. A simple
and common situation in which this happens is

where high rates of nitrogen or potassium application, or both together, continue to stimulate growth but without any increase in total dry matter production. Any added yield is water only, which inevitably boosts moisture content and dilutes the concentration of all other components in the plant, including organic materials and minerals. Another common and simultaneous effect of such high rates of nitrogen and potassium application is to influence a competitive relationship by stimulating shoot growth at the expense of the development of the organs of yield, where these are storage organs. This shoot growth forms an enlarged sink within the plant, which then competes for resources with fruit, seed or vegetative storage organs that are being formed, and slows down and depletes their filling.

Certain of the major variations that occur in the nutritional value of crops are of sufficient dietary importance to merit special attention. These relate to protein content and quality, and to mineral composition.

12.2.2 Variation in protein content and quality and its dietary significance

The substantial variation in protein content that occurs within a single crop species may be less important than expected in developed countries, where the population has a high food intake, consumes more protein than it needs, and obtains much of its protein from animal products. In less developed countries, where the general population is on a lower plane of nutrition and depends very largely on plant protein, this variation is much more significant for the human diet. It is still often not clear what factors cause these variations in protein content.

The most prominent single factor known to influence protein content is the rate of nitrogen fertilizer application. Increases in nitrogen can substantially increase the protein content of all parts of plants, including foliage, vegetative storage organs and seeds, as mentioned in Section 12.1.5. The protein content of barley grain has, in some experiments, been doubled by increased nitrogen fertilizer application, though the relationship between the two factors is not directly proportional. For example, a doubling of nitrogen application does not produce a doubling of protein content. In experiments with rice the protein content as a proportion of dry matter has continued to rise even when the total dry matter

yield has passed its maximum and is declining with successive further increases in nitrogen application. The maximum protein yield, then, occurs at a higher nitrogen rate than that which gives the maximum total yield. It is probable that this occurs with other crops also. However, the chief effect of nitrogen applications on protein yield is in fact to increase greatly the total protein harvested by greatly increasing the total amount of crop growth.

More subtle but equally important variations can also occur in the amino acid composition of proteins. Certain amino acids essential for animals occur only at low levels in plant proteins (see Section 12.1.5). This is the chief drawback of plant proteins from a nutritional point of view. In cereals, lysine is the essential amino acid in shortest supply, though some others are also low. Legume seeds such as beans are low in methionine and cystine.

As the protein content of many cereals increases in response to rising applications of nitrogen, the content of certain essential amino acids as a percentage of total protein declines. In barley, this occurs with lysine and threonine, and in wheat with lysine, cystine and possibly some others. The response of crude protein and lysine content and yield in wheat grain in response to increasing rates of nitrogen fertilizer application is shown in Fig. 12.4. The explanation for these changes appears to lie in the fact that in some seeds the increase in protein content is mainly due to rises in certain storage proteins – the prolamins – which are low in lysine and certain other amino acids. Wheat, barley, maize, rye and millet are in this category and the same effect appears to extend to beans. In oats and rice, the storage protein which increases is glutelin, which has a moderately high content of lysine and the other limiting essential amino acids, and the effect of increased nitrogen rate on protein quality is therefore less marked.

Similar problems can be induced by a low sulphur supply, which can lead to lower levels of amino acids containing sulphur, such as cystine and methionine. The balance between nitrogen and sulphur in plant nutrition can therefore be important. Sulphur deficiency could be of increased significance as the need for applications of sulphur becomes more apparent in intensive agriculture (see Section 3.2.3), more crops being at risk of receiving a suboptimal sulphur supply.

Increased protein content is almost always desirable in cereal grains and in crops in general but there are exceptions. With herbage for ruminants,

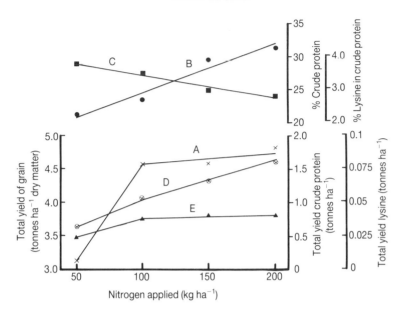

Fig. 12.4 Influence of nitrogen fertilizer application rate on crude protein and lysine content and yield in wheat. With increasing nitrogen application, total dry matter yield of wheat grain (A) increases, as does the crude protein content of the grain (B). The content of lysine as a percentage of crude protein (C) declines with increasing nitrogen fertilizer rate. Thus, while crude protein yield (D) rises steeply with increasing nitrogen, lysine yield (E) shows only a modest increase. From M A Kirkman, P R Scheury and B J Miflin (1982) *J. Sci. Food Agric.* **33**, 115–27.

for example, because of their relative independence of plant proteins, little or no improvement in animal performance is obtained when the crude protein content of the crop rises above about 10%, and the excess protein is used largely as a source of energy. Sometimes high protein content in a crop is a severe disadvantage. Maltsters reject high protein barley for two reasons. It produces a variety of compounds such as nitrosamines which can complicate the fermentation and give rise to off-flavours in the final product. Also, the high protein content is accompanied by a relatively low carbohydrate content, and this gives a reduced alcohol yield. In potatoes, high nitrogen fertilizer applications produce a higher protein content, but also a high content of free amino acids, which result in foaming and darkening of the tubers during cooking, consequently reducing consumer appeal. High amino acid content also damages various aspects of quality in other vegetables such as tomatoes and cucumbers.

12.2.3 Variation in mineral composition and its dietary significance

Applications of nitrogen, phosphorus and potassium usually, but not invariably, increase the concentration of the applied mineral within the plant. In herbage at least, this is not always beneficial. As has been seen in Section 12.1.9, additional potassium may interfere with magnesium utilization by the animal, while extra phosphorus lowers the already low calcium/phosphorus ratio in grass. It is important to bear in mind these effects of fertilizers on the nutritional quality of herbage when applying them for the purpose of increasing herbage yield.

Applications of micronutrients to soils have no consistent effects on micronutrient uptake by plants. Mainly, this is because micronutrient uptake by plants is much more strongly influenced by inherent soil properties, especially soil type, and to a lesser extent by the impact of weather on soil conditions, than by any direct attempt to change the mineral supply in the soil by applying fertilizers. As was pointed out in Sections 3.2.3 and 3.5.2, influences on the availability of a micronutrient in the soil are more important than the total soil content of it, but total soil content should not be dismissed as being an insignificant factor in mineral uptake and the nutritional value of plants. Mapping of areas deficient in minerals such as cobalt may well indicate the likelihood of the corresponding deficiencies occurring in farm livestock there (Fig. 12.5). Endemic goitre in humans and animals is closely associated with soils deficient in iodine. Copper deficiency in ruminants may reflect soils with an absolute copper deficiency or soils of high pH in which availability of copper for plant intake is reduced.

Variations in plant mineral content induced by growing conditions can have a considerable dietary significance. Plants growing at reduced soil temperatures or moisture contents often have a low phosphorus content as a result of the effects of these conditions on phosphate availability in the soil. On low phosphate soils especially, this can be important for stock. They may receive inadequate

Fig. 12.5 Mapping of cobalt levels in soils as indicators of where deficiency of cobalt in animal diets is liable to occur. L = low, M = medium, H = high level of cobalt in the soil. From *The probable cobalt status of soils.* The Soil Survey of Scotland. November 1983.

phosphorus, particularly if they are young and growing, carrying young, or producing milk, all conditions in which animals have a high phosphorus requirement. Changes in soil conditions may induce more complex chains of effects. Thus, in the poorly drained soils that often underlie the rough grazings of cool temperate regions, the waterlogged, acid conditions favour the availability of most trace elements for plant uptake, including nickel, iron, manganese, copper and cobalt. Where these grazings are made more productive by liming and drainage, the availability of these minerals is decreased, and copper and cobalt deficiency may be induced in sheep.

Where and when variations in mineral content due to variations in growing conditions are most likely to be nutritionally significant is a matter of some debate. Grazing animals tend to live off a restricted variety of plant species, especially on intensive pasture based on a few species of grass,

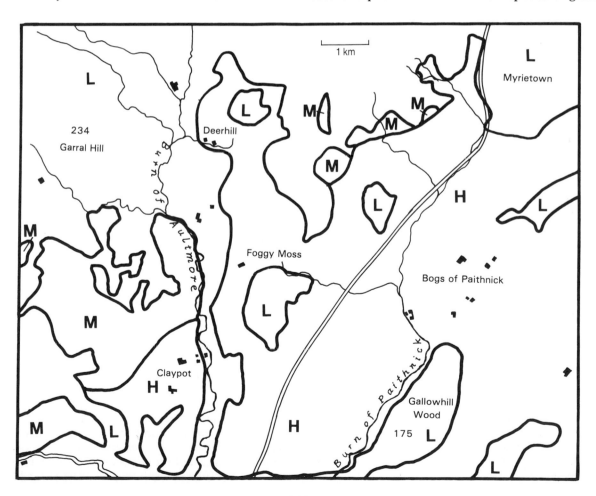

and growing on one or a very few soil types within a restricted locality. Where local climatic conditions, soil types or interactions between the two induce a low supply of one or more minerals in the herbage, such animals are vulnerable, especially if heavy demands are being made on their productivity. Patterns of geographical occurrence of mineral deficiency in stock can, indeed, often be correlated closely with the distribution of soil types.

The likelihood of feedlot animals experiencing mineral deficiencies depends on the feeding system. They are most at risk when fed diets based on a few components grown at one or a few localities. Increasing either the diversity of the diet, or drawing components of it from a greater number of localities with a wider range of weather and soil conditions reduces the risk.

Most humans in rural areas of third world countries eat a restricted variety of plant produce, grown on local soil types under local climatic conditions. They live on a generally low plane of nutrition, with a low intake of animal products in their diet as an alternative source of minerals. They therefore share the vulnerability of grazing animals to mineral deficiencies, and for much the same reasons. It is often held that people in developed countries, consuming many different foods from widely different sources, including a considerable intake of animal products, have a high and varied mineral intake and are thus not vulnerable to deficiencies in the diet. Nonetheless, a recent study of Finnish diets revealed a very low intake of selenium, and other authors have asserted that there is evidence of deficiencies of iron, zinc and some other minerals in the typical diet in the USA. There is also a growing number of people in developed countries living on a strictly vegetarian diet. Thus, even in developed countries, variations in the mineral content of plants in the human diet may merit closer attention than they have been given in the past.

Finally, other aspects of quality can be affected by mineral content. For example, increasing the potassium content of tomatoes improves their flavour and keeping quality. Increasing trace element content can have similar effects.

12.3 Acceptability of foods

We have so far discussed the quality of plants as food from only one viewpoint – that of nutritional value. But no matter how nutritious a food is, it is of no use if the animal will not eat it. Or, in the case of produce such as fruits and vegetables marketed for direct human consumption, it is of no value if the consumer will not buy it. Those aspects of quality which have a bearing on the readiness or otherwise with which humans will buy, or animals will eat, a food, can be collectively termed the *acceptability* of the food.

Acceptability to the animal is reflected in two quite different ways. The first is the *intake* of the food when given ad lib and when no choice is available. The second is the degree of *preference* exhibited by the animal for the food when given a free choice. Acceptability to the human consumer is related to a wide range of characteristics which may be thought of as constituting 'consumer appeal'.

12.3.1 Food intake by animals

Animal species differ greatly in the range of foods they will eat. Most farm livestock can be persuaded to eat a wide range of plant materials; the goat has the reputation of being particularly catholic in its tastes. At the other extreme, there are animals which derive virtually all their sustenance from a single plant species. Think, for example, of the giant panda and its bamboo shoots, or the koala and its eucalyptus leaves. For any given animal species, an *edible* food can be defined as one that will be eaten by a hungry animal, without harm to itself, if no choice is available.

Virtually all herbage, fodder, grain and pulse crops are edible to the domesticated ruminant. But the dry matter intake is not the same for all edible foods. The most valuable foods are those that combine high nutritional value, as discussed in the foregoing sections of this chapter, with high intake. The importance of intake as an aspect of food quality is frequently understated. It should be realized that, say, a 10% increase in intake under ad lib feeding conditions is as valuable as a 10% increase in ME content, as, for example, between grasses of D-values 67 and 74.

The non-ruminant, or the ruminant on a diet of concentrate feeds, has an appetite controlled largely by the levels of certain products of food metabolism in the blood. The animal, if fed ad lib, stops eating when these blood metabolites reach a threshold level of concentration. The intake of a ruminant on a diet of roughages, however, is determined by the fullness of the rumen. The animal in this case stops eating when the rumen is filled to capacity.

It follows that if the food is slow to pass through the rumen, then intake per unit time will be reduced. A number of food characteristics can lead to slow digestion and therefore reduced intake.

Low digestibility. In roughages as a whole, there is a general tendency for intake to increase linearly with digestibility (Fig. 12.6). Many instances, however, are known of differing intakes of materials of similar digestibility. Forage legumes such as lucerne (alfalfa) and red clover show greater intakes than grasses at the same level of digestibility; considerable differences also exist from one grass species to another in intake at equal digestibility. Furthermore, species differ in the relationship they show between digestibility and intake. In the study illustrated in Fig. 12.6, intake of meadow fescue fell off more sharply than that of Italian ryegrass with declining digestibility.

Digestible crude fibre content. Many of these

Fig. 12.6 The relationship between intake and digestibility of organic matter for six grasses and two legumes. Voluntary intake is measured here as material consumed per kg of 'metabolic live weight' of animal per day. This is a measure of an animal's energy requirements, obtained by raising its liveweight to the power 0.75.

instances of modified intake-digestibility relationships can be explained by differences in the composition of the digestible dry matter. Digestion in the rumen is slower if a high proportion of the material being digested is crude fibre. Digestible crude fibre content is greater in grasses than in red clover at the same level of total digestible dry matter content; similarly, timothy has a higher digestible crude fibre content than perennial ryegrass of equal total digestibility. The digestible crude fibre content of timothy increases less markedly with maturity than that of perennial ryegrass.

Very low crude protein content. The intake of roughages containing less crude protein than about 60 g kg^{-1} dry matter is often lower than would be expected from their digestibility. This is probably because the rate of digestion is further slowed down by the inadequate nitrogen supply for the rumen microbes. Increased use of nitrogen fertilizers in the growing of such forages, or the addition to the forage of nitrogen in some form such as urea, often improves intake. There is, however, no evidence of stimulation of intake by the addition of nitrogen to roughages whose crude protein content is greater than 60 g kg^{-1} dry matter.

Rumen pH. Digestion proceeds most rapidly when the pH of the rumen contents is about 6.8. A diet rich in legumes such as lucerne or red clover

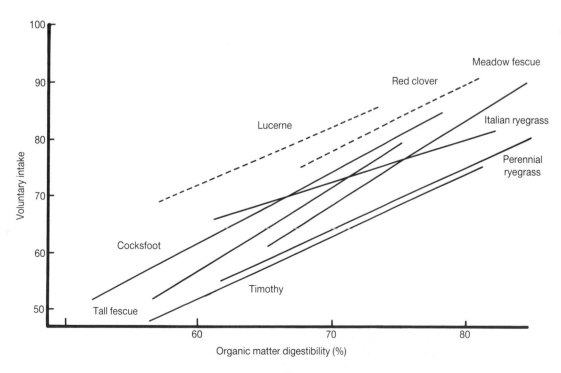

gives a rumen pH around this optimum level, but grasses tend to give a lower rumen pH and consequently are digested more slowly.

Intake is influenced not only by the rate of digestion but also by palatability factors. These are dealt with in Section 12.3.2. It must be emphasized, however, that differences in palatability as revealed by animal preference do not necessarily indicate that intake will differ when no choice is offered.

12.3.2 Animal preference for foods

An animal given a free choice among two or more foods or forage species will almost always eat more of some than of others. It will, in other words, exhibit preference for one or more of the materials on offer. The most important practical situation in which preference is exercised is in grazed pasture, where it leads to *selective grazing*.

On the whole, sheep are more selective grazers than cattle, but there are big differences in selectivity between breeds and even between individuals of the same breed. Grazing behaviour also varies with time; a particular species may be selected at one time of year and rejected at another. This may reflect changes in the plant or in the animal, for example as it goes through the cycle of reproduction and lactation.

As a rule, both sheep and cattle show strong preference for legumes, and their diet may be twice as rich in legumes as is the herbage on offer. Among non-leguminous herbs, sheep discriminate in favour of broad-leaved species, whereas cattle tend to discriminate against them. Selective grazing is an important factor leading to changes in the botanical composition of grassland swards.

The preference of an animal for one species or food over another is a measure of the relative *palatability* of the two materials. The senses of sight and smell may aid the animal in selection, but it is the sense of taste and to some extent that of touch that determine preference.

Many plant factors, both physical and chemical, influence palatability. Physical factors causing reduced palatability include resistance to breaking (closely correlated with lignin content and therefore inversely with digestibility), hardness (usually due to high silica content) and hairy texture. This last factor is often blamed for the low palatability of the grass Yorkshire fog.

Among chemical factors high crude protein content is one that is frequently said to confer palatability. Certainly legumes are more palatable than grasses but whether or not this is because of their higher protein content is not known for sure. The application of nitrogen fertilizer tends to depress, not to increase, palatability of grasses, in spite of an increase in protein content.

Perhaps more important than crude protein is water-soluble sugar content, which confers a sweet taste on the food. Sweetness appears to be an attractive taste attribute for almost all animals. Big increases in palatability (but, it must be stressed again, not necessarily in intake if no choice is given) can be obtained by spraying molasses on to roughage feeds.

Bitter-tasting substances have an adverse effect on palatability. Strains of reed canary-grass with a low content of alkaloids (see Section 12.5 on toxic constituents of foods) are more palatable than strains with a high alkaloid content. In this instance the difference in palatability does cause a difference in intake in the absence of choice. Another bitter-tasting substance is coumarin, thought to be responsible for the low palatability of sweet vernal grass and sweet clover. 'Sweet' in the names of these species refers to the smell of the coumarin, not its taste.

12.3.3 Consumer preference for plant foods

The acceptability of a food to an animal depends, as has been seen, on characteristics of the animal as well as of the food. The same is true of acceptability to the human consumer. Many perfectly nutritious materials are not eaten because of strong religious or social taboos, as apply, for example, to certain animal products such as pig-meat in Muslim and Jewish societies, beef in India or horse-meat in most of the English-speaking world.

Plant foods generally arouse less passionate feelings for or against, but there exist marked national and regional preferences for different kinds of bread and other cereal products, vegetables and fruit. For example, the Scottish consumer's taste is for a drier, more 'floury' variety of potato than that of his English counterpart. Likewise the Scottish consumer eats less green vegetables and more sugar. Very often, however, regional tastes simply reflect the crops that can be grown locally, hence the southern European emphasis on olive oil and spaghetti, made from durum wheat. Samuel Johnson's eighteenth-century dictionary definition of oats as 'a grain, which in England is generally given to horses, but in Scotland supports the people' says as much about the relative inability of most of Scotland to support crops of wheat, as about the English preference for wheat over oat products.

Of less importance to the food producer but of great importance to the diet of any individual is the latter's own particular preferences. All five senses (taste, smell, sight, touch and sound) are involved in consumer appreciation of plant foods, and it should not be forgotten how individual and subjective these various responses are.

We are here concerned with the principal plant factors affecting consumer acceptability, but other factors such as price, advertising, presentation and packaging may be just as important in determining choice. We shall not consider highly processed foods; processing is the subject of the next section of this chapter.

Taste. The importance of *flavour* as an attribute of food can be measured by the commercial significance of two major world crops, sugar cane and sugar beet. Sugar, although of nutritional value as an energy supplier, is included in or added to foods primarily for the sweet taste it imparts. Of lesser, but by no means negligible commercial importance, are mustard, spices and flavouring herbs, all of which add characteristically to the taste of food.

As sweetness is contributed by sugars, primarily sucrose, so bitterness is contributed by acids such as malic and citric acid. In small concentrations these acids give a pleasant 'tang' to fruits and some vegetables, but in greater concentrations, as in unripe fruit or raw rhubarb, they are definitely off-putting.

Superimposed on sweetness or bitterness, each fruit or vegetable has its characteristic flavour, determined by a vast range of organic compounds present only in trace amounts. No fewer than three hundred different compounds contributing to the flavour of apples have been isolated. Differing proportions of these give rise to the differences in flavour between varieties of apple, or in one variety in different seasons, or at different stages of ripeness, or with different methods or times of storage. The possibilities for breeding new varieties with new flavours are virtually limitless. No one variety will ever be universally regarded as the 'most delicious' because of variation in individual consumer preferences, as illustrated in Fig. 12.7.

Variety, stage of maturity and storage conditions are important factors affecting the flavour not only of apples but of all plant foods. Another set of

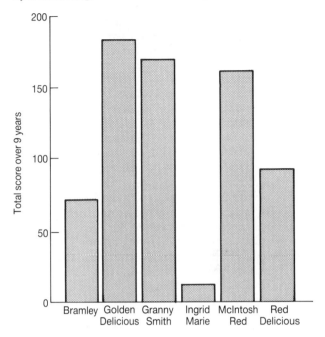

Fig. 12.7 Scoring of six apple varieties as 'most delicious' by a wide range of tasters over nine years.

factors can be grouped together under the heading of growing conditions. In general, any environmental conditions or husbandry treatments which increase yield will do so at the expense of flavour. This is especially true of high rates of nitrogen fertilizer. In a very competitive market it may be better for the grower to aim for a moderate yield of a saleable product rather than a high yield of a product that has to be sold at a loss or fed to pigs.

Smell. The odour of a food often gives the consumer a clue to its flavour and may be quite important in determining choice of purchase at the greengrocer or supermarket. Smell is also a guide to freshness.

Sight. Fruits and vegetables must be visually appealing to the consumer. In particular, produce must be free of blemishes resulting from disease or careless handling during harvest, storage and transport. Fruits and vegetables of high water content are prone to bruising, caused by rupture of cells near the surface followed by necrosis and the invasion of rot-causing microbes, and this can seriously disfigure the produce without significantly affecting its nutritional quality.

Size is also important. To the consumer, small means excessive waste and preparation time (for example in potatoes, swedes or carrots) while large means tasteless. A happy medium has to be sought – say 3–5 cm diameter for Brussels sprouts for sale fresh, or 6–10 cm for ware potatoes. A uniform crop is necessary to give the maximum yield within the desired size range.

Other aspects of appearance which influence consumer appeal include regularity of shape, not only for ease of peeling, as in potatoes, but as an indication of health and internal homogeneity. Colour, too, is important, as an indication of ripeness and, again, of health, but there is also much unfounded colour prejudice which, in particular markets, may adversely affect the saleability, for example, of purple-skinned potatoes or unblanched celery. (*Blanching*, in this context, is a husbandry technique whereby part or all of a plant is made to complete its maturation in darkness, to inhibit chlorophyll synthesis and give a pale colour to the produce. In addition to celery, asparagus and Belgian endive are often managed in this way. Most commonly, blanching is done by heaping soil around the plant, but black plastic sheeting may be used to give the same effect. The term 'blanching' is also used for a totally different process, as described in Section 12.4.3.)

Environmental and husbandry factors affecting appearance include: choice of variety, sowing or planting density (influencing size of produce); pest and disease control (to promote regularity of shape and freedom from blemishes); weed control (weed competition will result in smaller produce and may influence shape – for example, lettuces may fail to heart properly); use of growth regulators to promote uniformity of produce or prevent premature fruit drop leading to bruising; mineral nutrition (boron deficiency, for example, causes necrosis and rotting in the centre of many brassicas); blanching; harvesting date; and handling during harvest, storage and transport.

Touch. Under this heading we can include such varied attributes of food as surface texture (in most cases this should be smooth and hairless), internal firmness and homogeneity ('soggy' apples and 'gritty' pears are not in demand), degree of lignification (when over-mature, carrots become woody and celery and runner beans become 'stringy'), and the presence or absence of seeds (seedless fruits have obvious attractions). Choice of variety and harvesting date are the most important ways in which the grower can influence these qualities.

Sound. Perhaps the most important way in which the sense of hearing is involved in consumer acceptability of plant foods is in the 'crunchiness' of commodities such as apples or celery. This is closely connected with internal firmness and depends on the turgidity of the tissue, that is the degree to which the cells are inflated with water. This in turn depends on variety, stage of maturity and storage conditions.

12.4 Processing and preserving of plant products

Processing can encompass a wide range of operations, including cleaning, upgrading of quality by the removal of substandard or inedible parts such as shrivelled grain, tenderizing, milling into flours, blending with other products, extraction of constituents, preserving, and cooking. Here we can consider only the more important of these operations.

The aims of processing are also various. The aim may be to extract one or more constituents in a more or less pure form, for instance oil from oilseeds such as rape, linseed or sunflower, or sugar from beet or cane. Sometimes the aim is to convert the food into a different, often more edible form. For example vegetable oil is made into margarine, wheat into bread, maize into corn flakes, and soya beans into TVP (textured vegetable protein). An increasingly important aspect of processing is the manufacture of 'convenience' and 'fun' foods such as instant mashed potato and potato crisps respectively. Large quantities of food are used to make alcoholic beverages by fermentation, for example beer and whisky from malted barley, grain spirit from maize and wine from grapes.

Some plant foods such as fruits and salad vegetables are most commonly eaten raw, but most are cooked in one way or another. Cooking is done for a variety of reasons – to make the food digestible (an example being potatoes), more tender (beans, cabbage), or more palatable (onions) or to destroy toxins (cassava, red kidney beans – see Section 12.5). Sometimes the most important role of cooking is to sterilize the food, in situations where it may be contaminated with human or animal parasites or microbial infectious agents. Cooking is a form of processing which is still largely carried out by the end consumer, after marketing.

The proportion of plant foods which are sold

direct to the public in the form in which they are harvested is decreasing all the time. Commercial processors of one kind or another therefore represent the most important outlet for crop produce in market economies, and the grower must take account of the processor's demands in terms of quality and regularity of supply.

The most important aim of food processing is to preserve the food in a palatable and nutritious condition so that it can be stored and transported without deterioration. Frequently, the aim is not to prevent deterioration completely, but merely to delay it sufficiently to allow the product to be consumed in good condition.

The very fact that food supports human and animal life means that it also supports microbial life and is therefore subject to spoilage through microbial activity. One important objective of preservation is thus to limit or prevent such activity. Stored plant foods, unprocessed, are less subject to microbial decay than animal products because they are normally alive and able to maintain defences against bacteria and fungi as well as repair damage. They are not, however, immune from the effects of microbes, particularly if stored at temperatures above about 10°C and with inadequate ventilation.

Spoilage of foods is not only a result of microbial activity. Insects and mites in grains, flours and other products, as well as larger animals such as rats and mice, consume, or through contamination render unfit for human consumption, some 10–15% of the world's food supply each year. Enzyme activity in the living plant tissue and larger-scale developmental changes such as sprouting of potato tubers or lignification of roots in store also contribute to deterioration in food quality, as discussed in Section 11.8. Preservation therefore usually involves the killing of plant material, while protecting it from attack by pests and microbial decay. The terms preservation and conservation are both used rather loosely to mean similar things; indeed the dictionary makes no clear distinction between the two words. However, the term conservation is most commonly used to cover storage and protection from decay and destruction of live material, while preservation generally relates to similar protection of killed or processed material.

12.4.1 Preservation by asepsis

All systems of preservation centre round a very few basic and simple principles. The simplest in concept is to exclude spoilage organisms, either partly or wholly, a method known as *asepsis*. This is often one objective of packaging, by using coverings to prevent gross contamination of products during handling and transport. Canning is perhaps the best example of asepsis, for the product to be canned is first sterilized by heat and then sealed in the can to prevent its recontamination by spoilage organisms. Irradiation after sealing in clear plastic is an increasingly common process, more as for animal products such as bacon than for plant products.

All other methods of preservation work on the principle of removing one or more of the conditions that spoilage organisms need for growth, these being water, adequate temperature, suitable pH, oxygen for most species, and freedom from substances toxic to microbes. Obviously, spoilage organisms also need organic and inorganic nutrients but these are supplied by the product itself and cannot be excluded. In practice, nearly all techniques of preservation apply a combination of methods.

12.4.2 Preservation by lowering water potential

One group of techniques acts by making water in the product less available to spoilage organisms. It is not necessary to remove the spoilage organisms to achieve this. All that is necessary is to lower the water potential of the plant material to a point at which the organisms can no longer absorb water from the material. One way of doing this is to immerse foods in fairly concentrated solutions which, as discussed in Section 2.2.2, have a large negative solute potential. For example, fruits are commonly preserved in syrups, which contain a high concentration of sucrose or other sugars, and certain vegetables are preserved in brine, which is a solution of common salt.

A more widely used approach using the same principle is to dry the product, so that the concentration of solutes in the residual water in the plant cells increases as more and more moisture is removed, thereby lowering water potential. The method can be used for dead materials such as dried forage (hay) or straw, or for live seeds such as dried cereal grains, leguminous seeds and nuts. Much confusion about this technique persists in agriculture.

The first point that should be recognized is that a stored, dried product constantly exchanges moisture with the air within the bulk. This exchange is brought about by differences between the product and the air in water potential, and tends to even out these differences. Thus the moisture content of the product is strongly affected by any changes in the water potential of the ambient air within the bulk of stored material, and vice versa. The water potential of the air at any given temperature is, in turn, dependent on its relative humidity. The lower the relative humidity of the air, the greater is its tendency to absorb water from stored products, and the higher the moisture content of the product, the greater is its tendency to lose water to the air. The relative humidity at which the water potential of the air equals that of the product is the *equilibrium relative humidity* (ERH) of the product. At this ERH there is no net exchange of water between the air and the product.

Thus to maintain a stored product at or below a given moisture content, the relative humidity of the ambient air must be at or below the ERH corresponding to that moisture content. Figure 12.8 gives the ERH of different products at different moisture contents. It shows that the higher the moisture content the higher is the ERH of the product, but that at any one level of moisture content the ERH of wheat is much lower than that of peanuts. This is typical of the difference between starchy and oily seeds, starchy seeds having a stronger tendency than oily seeds to draw water out of the air. Both kinds of seeds, however, do have a strong tendency to absorb water because of the high matric forces in seeds which cause imbibition (Section 2.2.1).

Now what limits the development of spoilage organisms is the relative humidity of the atmosphere within the bulk, and not the moisture content of the stored material as such, which only influences spoilage organisms indirectly through its strong influence over that relative humidity. Hence stored products must be kept at a moisture content at which their ERH is lower than the relative humidity at which spoilage organisms can grow.

Further, the ERH of a product at a given moisture content varies with temperature. To be more specific, the water potential of the product rises with temperature, and the moisture within the product thus becomes more available to spoilage organisms at higher temperatures. For any product, therefore, there is a corresponding ERH for every combination of moisture content and temperature. From this follows the second point that has been the source of much confusion in practice. For drying to be successful as a method of preservation, both

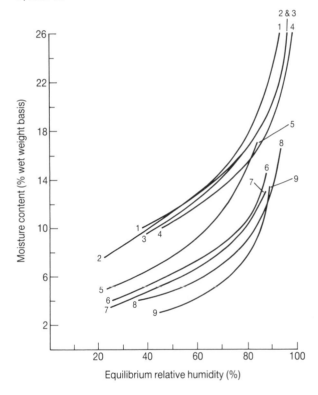

Fig. 12.8 The equilibrium relative humidity of various products, showing the difference between non-oily and oily grains, at 25°C. 1. English maize; 2. wheat variety Cappelle; 3. barley; 4. American maize; 5. soya beans; 6. linseed; 7. sunflower seed; 8. peanuts; 9. coconut copra. Note that the curve that develops when materials reach ERH by drying is different from that produced by absorption of moisture. From S W Pixton and S Warburton (1971) *Journal of Stored Products Research* 7, 261–9.

moisture content and temperature must be controlled.

Certain microbes, notably fungal species of the genera *Aspergillus* and *Penicillium*, can grow at lower water potentials, and therefore on products at lower moisture contents, than can other spoilage organisms. The complex of species belonging to the *Aspergillus glaucus* group are prominent examples. Such microbes, called *osmophiles*, will grow on grain at moisture contents of 18–20%, and indeed are often troublesome on other dried products of which the water potential is marginal for preservation.

The abilities of different groups of spoilage organisms to thrive at different relative humidities and temperatures lead to a proliferation of restrictions as to the conditions under which dried products can be stored. Thus barley at 18% moisture content, stored at 2°C for 32 weeks, is safe from all forms of spoilage, but at 5°C is vulnerable to invasion by mites, and at 10°C to invasion by mites and moulds. Some storage insect pests can survive at relatively low water potentials, and grain is increasingly vulnerable to invasion above 12% moisture content and 15°C. An increase in either of these parameters encourages invasion. The manner in which threats from these diverse groups of organisms confines the safe storage of dried grain to certain restricted combinations of moisture content and temperature is illustrated in Fig. 12.9, although the combinations vary slightly among different cereals, pulses and oilseeds.

Inadequate cooling of grain that has been heated to dry it is a common cause of loss. In areas or seasons when harvesting conditions are not cool, grain frequently comes in from the field too warm to store, even at very low moisture contents, without artificial cooling. Part of the problem is that convection currents arising in the warm grain carry warm air to cooler parts of the bulk where moisture condenses, thus encouraging the growth of spoilage organisms.

Fig. 12.9 Safe storage conditions for grain for a 32-week storage period. From N J Burrell (1982) *Storage of Cereals and their Products*, 3rd ed. American Association of Cereal Chemists.

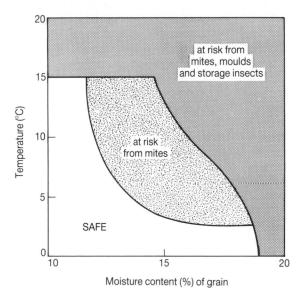

12.4.3 Cold storage

Chilling to temperatures just above 0°C will give only short-term preservation unless it is combined with drying, as in the case of barley at 18% moisture content and 2°C described above. Both enzymic activity within the plant and the activity of many spoilage microbes continue below 0°C, though at reduced rates. The fungus *Botrytis cinerea* is a well-known example and can cause the disease grey mould on soft fruit at temperatures as low as −4°C. Some other fungi can grow even at −9°C.

In order to stop virtually all metabolic activity, both in plant material and in spoilage organisms, it is necessary to store at temperatures around −20°C or lower. This *deep freezing* technique is now so familiar for both domestic and commercial food storage that little needs to be said about it here.

It should be noted, however, that not all plant products are equally suitable for freezing. Many show undesirable effects such as loss of flavour. During the chilling before the temperature falls to freezing point, there may be substantial conversion of starch to sucrose. Below freezing point, ice crystals form in the tissue, rupturing the delicate cell membranes. As a result, water comes out of the food when it is thawed for use. The damaged tissue is then very prone to microbial decay and enzymic spoilage. To minimize this, vegetables are generally scalded before freezing to destroy the activity of enzymes and to kill microbes. This technique is known as *blanching*.

12.4.4 Sealed storage

Practically no preservation technique depends solely on the removal of oxygen, not least because certain spoilage organisms derive their energy by anaerobic respiration and can therefore grow in the absence of oxygen. However, if anaerobic conditions are accompanied by accumulation of carbon dioxide, metabolic activity is suppressed even in these organisms. Undried grain, at moisture contents up to 30%, is preserved by placing it in tower silos which are then sealed. The respiration of microbes on the grain, and of the grain itself, is left to exhaust the supply of oxygen trapped within the tower. Near-anaerobic conditions develop, resulting in the death of the grain and inhibition of spoilage organisms, including insects and mites, but it is the concomitant build-up of carbon dioxide from respiration that is most effective in preserving the product.

12.4.5 Adjustment of pH

Different groups of microbes differ markedly in their pH requirements for growth. One group of bacteria, the actinomycetes, grow readily at pH 7.5 but are somewhat inhibited at 6.5, a pH that favours most other groups of bacteria. These, in turn, are greatly inhibited by a pH of 5.5, which nevertheless favours the growth of most fungi. The many common techniques of preservation which rely on lowering the pH below the tolerance range of spoilage organisms must therefore aim well below 5.5. Preservatives which act by lowering pH include formic, acetic, lactic and propionic acids. Butyric acid is equally effective but is strongly malodorous and is therefore avoided. These acids, especially formic acid, also have a direct toxic effect that augments their preservative qualities. One method is simply to immerse material in an acid solution, as in the pickling of vegetables in vinegar, which is a solution of acetic acid. A variation on this theme is the preservation of grain and hay by spraying it with mixtures of propionic, formic and acetic acids or with one of these preservatives on its own.

The alternative approach to pickling is to use a controlled fermentation of the material so that lactic acid is produced from the plant sugars naturally present, mainly by a group of microbes called lactic acid bacteria. In practice, more than one technique is usually combined. For example, undried grass is converted to silage by compacting it in a silo, so that exclusion of the air inhibits fungi but encourages fermentation by lactic acid bacteria. The resulting accumulation of lactic acid in the bulk lowers pH and eventually inhibits bacterial growth. Additives containing formic and propionic acids are often applied to aid preservation of grass silage.

Cabbage is fermented by a similar method in special containers to give sauerkraut for human consumption, but salt is added to lower the water potential and aid the inhibition of undesirable microbes. Onions, gherkins, olives and other products are similarly fermented, although with the addition of brine rather than dry salt. The tartness imparted to the taste by the acid is an important part of the attraction of these foods.

12.4.6 Quality of crop produce for processing

The aspects of quality of crop produce that are important to the processor fall into two groups: those aspects that influence the quality of the final processed product (that is, its suitability for the target market), and those that affect the efficiency of the process.

Let us first consider some effects on the quality of the final product. Protein content of wheat determines whether or not a risen, light-textured loaf can be baked from the flour. A series of proteins collectively known as gluten occurs in the endosperm. The more gluten that is present the better are the bread-making qualities of the wheat. Gluten content depends on variety and on growing conditions; thus, wheat matured in hot, dry, sunny weather contains more gluten than wheat which is allowed to ripen more slowly in cooler, damper weather. For this reason a variety of good bread-making quality when grown in continental Europe will often be unsuitable for bread-making when grown in Great Britain. Increased rates of nitrogen fertilizer raise the protein, and hence the gluten, content of wheat grain.

Other examples of chemical composition, as influenced by crop variety, affecting the quality of processed products include fructose content in potatoes, high levels of which make the potatoes unsuitable for crisp making because the product is unacceptably dark in colour, and the content of phenolic compounds in cider apples, high levels of which are essential for flavour in the cider. Beetroot for boiling and pickling must be fully mature, with the cambium rings in the root no longer active, otherwise the sliced roots tend to split along the rings after boiling, reducing consumer appeal.

Turning now to some effects of crop quality on the efficiency of crop processing, a high protein content in malting barley is undesirable because it lowers the efficiency of fermentation as well as giving unacceptably high levels of nitrogenous substances in the resulting brew. As in wheat, rapid ripening and heavy nitrogen fertilizer application raise protein content, thus in the case of malting barley they lower quality. A good malting variety is one that can stand fairly high fertilizer rates to boost yield without raising protein content to an unacceptably high level.

Wheat varieties are classified as 'hard' or 'soft' depending on the physical consistency of the endosperm. A hard wheat is easier and cheaper to mill for flour than a soft wheat because it is less likely to clog the sieves.

The efficiency of any extraction process is enhanced by a high content of the desired material in the crop produce. Thus high sugar content in sugar beet or high oil content in oilseeds improve acceptability to the processors and therefore the price commanded by the produce.

Size of produce affects efficiency of processing for a number of reasons. For example, Brussels sprouts for freezing should be small (2–4 cm in diameter) so that they freeze through to the centre quickly. Carrots and 'new' potatoes for canning should be small so that a greater weight can be accommodated in the can.

12.5 Toxic constituents of plants and plant products

The presence of toxic substances clearly has a major effect on the value of a plant material as food for human or animal consumption. In a few cases the presence at low levels of a substance which is toxic at higher levels has some desirable effect. An example is the stimulant alkaloid caffeine, which is largely responsible for the commercial importance of two major world crops, coffee and tea. Toxins from poisonous plants, such as digitalis from foxglove and morphine from opium poppy, used as drugs, have played a major role in medicine. In agriculture, poisonous plants are important both when they form part of an intended diet and in situations where they are accessible to grazing animals.

Poisoning, like nutritive value and acceptability, depends on animal as well as plant factors. Not all animals are equally affected by a particular plant species or food, or are affected at all. For example, wheat, the staple diet of much of the human race, can be lethal if fed in quantity to horses. Acorns are poisonous to ruminants but not to pigs. Sheep and goats tolerate much greater amounts of ragwort than can safely be eaten by cattle or horses.

12.5.1 Toxic and other injurious effects of plants and plant products

Almost anything, if eaten in sufficient quantity, will impair health, but the term toxin is usually confined to substances of plant, animal or microbial origin which cause poisoning when ingested in quite small quantities. Many common food plants, such as potatoes, contain toxins at such low concentrations that they seldom create problems at normal levels of intake. The plants we would normally consider toxic, in contrast, have deleterious effects on human or animal health when ingested in fairly small amounts. These effects are of many different kinds.

It is helpful to consider first the range of injurious effects that plants can have on animals. Some effects are purely physical, not involving toxicity in a true sense. For example, cattle may choke on potatoes and horses have been known to choke to death on sugar beet pulp, which has formed into a solid mass in the gullet. Large masses of indigestible material in the stomach can also cause severe illness or death; cases have included chickweed in lambs and ash leaves in cattle.

Certain foods, not in themselves toxic, may lead to poisoning through the inability of the animal's digestive system to cope with them or because toxic by-products are produced when they are partially digested. The poisoning of horses with wheat or barley, or the gorging of cattle on almost any concentrate feed in excess of their ability to digest it, come into this category.

Mineral excess or imbalance can lead to toxicity problems. We have seen in Section 12.1.9, for example, that a high molybdenum level in herbage causes an induced copper deficiency in the grazing animal. Some plant species, notably certain North American milkweeds, accumulate toxic levels of selenium. Plants growing on soils rich in toxic heavy metals such as lead or nickel, for example on mine spoil heaps, cannot entirely control their uptake of these minerals, and the accumulation of heavy metals in the leaves can lead to poisoning of grazing animals. However, the commonest form of mineral toxicity is caused by excessive amounts of nitrate in plants, almost always associated with very high levels of nitrogen fertilizer application to crops. Ruminants convert nitrate to ammonia by way of nitrite. Nitrite is toxic, but normally does not accumulate in the rumen. If the nitrate content of a plant food is very high, nitrite is produced in the rumen faster than it can be converted to ammonia, and poisoning results. Some plant species are more likely to contain dangerous levels of nitrate than others; mangels, rape and oats are among them. Nitrate poisoning of humans has been recorded, resulting from high levels of nitrate in spinach.

This is not the only method by which plants can

be rendered toxic. Many species, when subjected to stress by factors such as drought, high temperature, or even mechanical damage during handling at harvest, accumulate compounds, previously present only in trace amounts, to much higher concentrations. These *stress metabolites* often inhibit the growth of invading organisms and some have been shown to be toxic to animals. By implication, some at least are toxic to humans. As mentioned in Section 4.10.2, invasion by pathogenic bacteria and fungi can induce the accumulation not only of stress metabolites naturally present at low concentrations in healthy plants, but also the synthesis and build-up of substances absent from healthy plants. These substances are toxic to invading organisms and possibly also to other life forms.

The significance of these metabolites for human and animal nutrition has received some study, but the questions raised have yet to be answered. Certainly, in one well-established case, invasion of sweet potato tubers by the fungus *Fusarium solani* induces the accumulation of a range of plant toxins, the best known of which is called ipomeamarone. Cattle fed sweet potato tubers infected with this pathogen develop characteristic lung disorders associated with the toxin.

A better known and commoner situation is where the invading bacterium or fungus produces the toxin itself. Fungi are particularly prone to do so, producing toxins called *mycotoxins*. The range of known mycotoxins is wide and diverse, and they have undoubtedly caused many outbreaks of serious poisoning of humans and livestock over the world. They may be produced by fungi on the plant in the growing crop or after harvest, usually in store. An example is the fungus *Claviceps purpurea*, which invades the flowering parts of some cereals and grasses and there produces small, hard masses of fungal tissue called ergots. These ergots contain toxic alkaloids and may be consumed by stock when harvested with the grain or by humans when milled with grain to produce flour. *Aspergillus flavus*, growing on peanuts and some other products in store, produces the group of extremely virulent carcinogens known as aflatoxins.

Almost any plant food, or contact with any plant material, can cause an allergic reaction in some people, but it is uncommon to develop such reactions to a wide range of products. It is a highly individual response, and may take many different forms, including skin rashes and migraine headaches.

Finally, many plant species are intrinsically toxic, as a result of organic substances present in them. These species occur in a very wide range of families, but certain agriculturally important plant families, including the Solanaceae (the potato family) and the Fabaceae (the legume family), contain an especially large proportion of toxic species. In the Solanaceae the toxins are mainly alkaloidal glycosides, and are seldom present in the edible parts in quantities that cause injury. In the Fabaceae the problem is much greater and a much wider range of toxins is involved. Thus a number of leguminous seed crops have to be detoxified by various processes before they are consumed. Soya beans, for example, (one of the most important animal feeds and widely consumed by humans) contain up to fifteen or more different toxic factors. They must be heat-treated before consumption to destroy these factors. There are some crops in other families where generalized toxicity problems arise if intake is not regulated, such as kale and other forage brassicas. They contain oestrogenic agents that can upset the breeding cycle of stock, and other agents that can cause internal haemorrhage by reducing blood clotting. More generally, plant poisoning is a problem that arises in grazing stock, through their inadvertent consumption of toxic wild species, often growing as weeds. This most readily occurs with inexperienced young animals, or those with depraved appetites developing from inadequate or imbalanced nutrition, causing the animals to consume species they would normally ignore.

12.5.2 Plant poisons and their effects on humans and animals

Some of the more important classes of toxic substances found in plants are listed below.

Alkaloids. An alkaloid is one of a large number of compounds containing nitrogen, present in plants in small amounts, and not involved in the mainstream of plant metabolism. Most, but not all, alkaloids are toxic to animals. Many have been exploited as drugs, for their characteristic and, at low doses, useful effects on the human body, but most alkaloidal drugs are now synthesized chemically rather than extracted from plants.

Alkaloids are as diverse in their toxic effects as in their chemical structure. The tryptamine alkaloids of canary-grass and the yew alkaloid taxine cause sudden death by heart failure. Aconitine in monkshood and atropine in deadly nightshade are narcotic, affecting the central nervous system, while cytisine in laburnum and nicotine in tobacco paralyse the nerves which control breathing. The pyrrolizidine alkaloids of ragwort cause progressive damage to the liver. Perloline in tall fescue kills rumen microbes. In addition to their acute toxicity, many alkaloids are teratogenic (causing malformations to the foetus when consumed by the pregnant mother) and carcinogenic (causing cancer).

Glycosides.   These consist of a monosaccharide sugar (commonly glucose) attached to another fraction, which is usually toxic. The toxic fraction is released by enzyme action either in the plant in response to damage, or in the digestive system of the animal.

The *cyanogenic glycosides* release cyanide, a poison which interferes with oxidative phosphorylation (see Section 1.8.1) and causes heart failure. Ruminants can degrade cyanide more effectively than non-ruminants, but in doing so produce thiocyanates which bind iodine and induce iodine deficiency. Cyanogenic glycosides are therefore goitrogenic in ruminants. Examples of these substances include amygdalin and prunasin in almonds, plum kernels, apple seeds and the leaves of cherry laurel, linamarin in linseed and cassava, lotaustralin in certain strains of white clover, and dhurrin in the leaves of sorghum.

Other glycosides, known as *glucosinolates*, release isothiocyanates, which are irritant poisons and goitrogens in both ruminants and non-ruminants. An example is allyl glucosinolate, found in mustard and horseradish. It releases the strong-tasting allyl isothiocyanate. Closely related glucosinolates are responsible for the slightly pungent taste of cabbage, kale and turnip, and for the goitrogenic activity of these foods if eaten in excess.

The *cardiac glycosides* release toxins which affect the heart. They include digitoxin, the main toxic principle of foxglove, and a variety of glycosides found in lily of the valley, hellebores and some North American milkweeds. *Alkaloidal glycosides* release alkaloids; they include solanin from all green parts of the potato.

Saponins.   These are substances which cause frothing by lowering the surface tension of water; they contribute to the various causes of bloat in ruminants. They also promote the bursting of red blood cells. Saponins occur in corncockle, chickweed, lucerne (alfalfa) and many other plants.

Acrid irritant poisons.   In this category we could include the glucosinolates, mentioned above, and protoanemonin, the toxic principle of buttercups and related plants. This group also includes tannins, which occur in many woody plants, and are particularly abundant in acorns (oak fruits). In addition to being irritant, tannins also bind to proteins, lowering their digestibility.

Substances interfering with vitamin or mineral nutrition.   Here we might include the goitrogens, which, as we have seen, interfere with iodine supply mangels. Oxalic acid and oxalates, found for example in beets, mangels and the leaves of sorrel and rhubarb, withdraw calcium as the insoluble calcium oxalate, sometimes causing calcium deficiency. The solid crystals of calcium oxalate may cause further problems when deposited in the urinary system. Bracken and horsetail contain thiaminase enzymes which destroy thiamine (vitamin B_1). This accounts for their toxicity to horses, but not their toxicity to ruminants which do not have a dietary requirement for thiamine. Sweet clover disease in ruminants is caused by dicoumarol, a vitamin K antagonist, formed in sweet clover as it begins to decay following a wet harvest.

Photosensitizing substances.   Certain substances found in plants cause the skin of animals to react to light as in acute sunburn, the skin erupting and peeling to leave painful, slow-healing wounds. Areas of skin not heavily pigmented or not covered with dark hair are most seriously affected. One of the commonest causes of photosensitization is phylloerythrin, formed in a malfunctioning liver as a result of the incomplete breakdown of chlorophyll. A mycotoxin produced by a fungus growing on dead leaves of ryegrass in New Zealand causes liver damage, which in turn leads to phylloerythrin photosensitization of sheep – a condition known as 'facial eczema'. Other photosensitizing substances, which may act either by contact or by ingestion, are found in buckwheat, St John's wort and in many umbelliferous plants (members of the carrot family, Apiaceae). The umbelliferous photosensitizers are furocoumarins. Some plants contain large amounts at all times – for example, giant hogweed, which often seriously affects children who use its hollow stems as blowpipes. In other cases the furocoumarins are produced by the plant only in response to attack by a fungal disease. Such is the case with celery, which sometimes causes outbreaks of photosensitization in pickers.

Other contact toxins. A variety of plant toxins are known which cause inflammation of the skin without the action of light. A notable example is urushiol, the toxic principle of poison ivy. Casual contact with this common North American plant or especially its sap gives rise to a painful dermatitis in many people. Even smoke from burning poison ivy plants can carry enough urushiol to produce dermatitis in people who do not go near the plants themselves.

Miscellaneous toxins. Some of the most potent of all poisons are proteins, such as ricin from castor beans and abrin from certain tropical legumes, which cause the blood to coagulate. A toxic amino acid, dopa, has been extracted from broad beans. Essential oils, such as those of mint, inhibit many microbes from performing their normal function in the rumen. Oestrogens, for example in red clover, upset the hormone balance of pregnant and lactating animals. The full list would be very long – there is, for instance, a substance in kale which can cause anaemia, a toxic pigment in cotton seeds (gossypol) and a series of resinous substances which are toxic principles of cowbane and hemlock water dropwort. The chemical nature of some toxins has not yet been elucidated – for example, we do not know the nature of the poison in North American locoweeds, nor do we understand why bracken is poisonous to ruminants. A condition resembling grass staggers has been attributed to chronic poisoning with canary-grass but it appears not to be caused by the tryptamine alkaloids of that plant.

12.5.3 Effects of toxins on animal products

When a plant product is consumed by stock, it is the beginning of a chain of events, the next stage of which is usually the consumption of the meat or dairy products by humans. Such a series of organisms each living off the previous one is called a *food chain*, and in the example just outlined there are three stages – plant, to animal, to man. Both in agriculture and in the wild such chains can consist of much more than the simple three-stage process outlined above. For example, cows' milk may be fed to calves which in turn may be eaten by humans. Both poultry manure and poultry offal are sometimes incorporated into animal feeds. In both these cases, the chain is extended to four stages – plant, animal, animal, man.

Toxins which enter food chains may move from stage to stage. It is important to realize that this includes not only plant toxins, but toxins which can contaminate human and animal foods. Such toxins arise from many different sources, including pesticides applied to crops in the field or in store, atmospheric pollution such as fluoride deposited on grass from nearby aluminium smelters, and additives such as antibiotics, supplementary minerals or animal growth stimulants placed in animal feeds. Mishandling of agricultural systems can permit many of these, or plant toxins, to make plant products toxic or otherwise affect their quality.

To consider plant toxins in particular, most of these are bitter-tasting, and this taste can persist as a taint in the flesh or, more commonly, the milk. Tannins from acorns, protoanemonin from buttercups, and essential oils from mint and other plants all impart a characteristic taste to milk and butter. Certain non-poisonous but strong-tasting substances, for example those occurring in wild onion and yarrow, also taint milk. Some toxins, such as the essential oils of mint, interfere with the ability of milk to clot during butter-making. Others impart an undesirable colour to milk, such as buttercups which give a yellowish-red tinge, and dog's mercury which gives a bluish tinge. To demonstrate that such effects are not confined to food chains involving large animals, honey tainted with the toxins of ragwort or some species of rhododendron, through bees gathering from these plants, becomes unsaleable.

The more deadly effect, however, is where the animal product is rendered toxic by the animal consuming a toxic plant product. It is not unknown for this to happen with a plant toxin. For example, the quantity of ragwort alkaloids passing into goats' milk may be sufficient to be a human health hazard. The problem more often occurs with toxins contaminating the plant material. For instance, dangerous quantities of aflatoxins, quoted earlier as examples of mycotoxins, can appear in milk. Other mycotoxins, and toxic heavy metals, can similarly appear in meat.

The fate of toxins entering food chains varies with the toxin, the species eating the plant food, and various other factors. Frequently, toxins are broken down by the animal's metabolism, or passed out in their faeces or urine, and thus leave the food chain. Note, however, that if poultry manure is included in animal feed, any toxins excreted by poultry will continue in the food chain. It is where toxins are passed on in significant quantities to the next stage in the chain that problems arise. They may be altered by the animal's metabolism to more toxic or less toxic forms. A particularly dangerous situation is where a toxin is persistent and is not excreted by animals, but instead accumulates in them, building up to higher levels at each stage in the food chain. The insecticide DDT, widely used for many years, built up in this manner in food chains.

In reality, many cross-connections exist between food chains. A single group of farm animals for example, may draw its feed from a variety of sources, and likewise most humans draw their diet from a range of plant and animal foods. Such networks of connected food chains are called *food webs*, and movement of toxins within them can be complex.

12.5.4 Consumption by animals of poisonous plants

It has been suggested that plants evolved toxins as a defence mechanism, not primarily against large grazing animals, but against insect pests. It is interesting that the quinolizidine alkaloids of lupins provide an effective defence against attack by aphids, strains low in alkaloids being much more susceptible to aphid attack. It should be remembered, however, that flowering plants and insects have evolved together as food sources and pollinators. It should come as no surprise, therefore, to learn that certain insects have become specialized to feed on certain toxic plants. They either detoxify the poisonous substances or sequester them in some part of the body where they cannot exercise their toxic effect. A good example is the caterpillar of the cinnabar moth, which feeds almost exclusively on ragwort. The caterpillar accumulates the ragwort's pyrrolizidine alkaloids to the extent that it becomes toxic itself. The insect thereby defends itself against predation by birds. The caterpillar's bright warning colours have been mimicked by a number of non-poisonous insects, giving them some degree of protection.

Mammals carry in the liver a battery of enzymes which can oxidize a large number of potentially toxic materials. They are general detoxifiers, whereas insects tend to be specialists. The toxins categorized in this section, however, are those which the livers of particular species of mammal cannot detoxify, or which they detoxify too slowly to give protection.

Mammals have evolved to meet the danger of poisonous plants, not principally through improved enzyme systems (although this has occurred to some extent in highly selective feeders), but through reaction to the taste or smell of toxins. Alkaloids, glycosides or their breakdown products, saponins and other toxins all taste bitter. Indeed, we make use of the bitter taste of allyl glucosinolate in mustard and quinine in tonic water. The palatability of canary-grass to cattle is closely and inversely related to its content of tryptamine alkaloids; the palatability of lucerne (alfalfa) is similarly related to its saponin content. Common poisonous plants in grazed pasture in Great Britain, but seldom eaten fresh because, we assume, of their bitter taste, include buttercup, ragwort and bracken. If it were not for the bitterness of these plants, poisoning would not be the rarity that it is.

Under what circumstances, then, does plant poisoning of livestock occur?

1. The plant may be the only food available. Hunger lowers any palatability barriers to feeding, and so hungry animals may not refuse alkaloid-rich canary-grass or saponin-rich lucerne hay. In pasture during winter or in time of drought, the only green plants available may be toxic. Poisoning of cattle by ragwort, a plant which they normally avoid, has been recorded in this situation.

2. Some poisonous plants are not unpalatable. This is true of locoweeds and green sorghum, as well as of nitrate- and oxalate-rich plants. The attractive red fruits of yew trees are sweet-tasting to children but contain a deadly poisonous seed.

3. A poisonous plant may become more palatable as a result of some treatment. Perhaps the commonest of such treatments is cutting, as for hay. Wilting tends to be accompanied by a rise in the sugar content of leaves, and the increased sweetness may overcome the bitterness of any toxic materials present. It is therefore important not to make hay containing a lot of ragwort, nor to cut the weed and leave the stems lying in the field where cattle can have access to them. Some toxins do not persist in a wilted plant. For example, in living buttercups, protoanemonin is continuously being created and remetabolized to other non-toxic products; during wilting, however, no more protoanemonin is formed, and the existing toxin is metabolized. Buttercups are therefore perfectly safe in stored herbage. Certain herbicides, including those used to control buttercups and ragwort, raise the sugar content without lowering the toxicity. Cattle should therefore be kept out of sprayed

fields for a few weeks if poisonous plants such as these are present.

4. The animal may be unable to differentiate between poisonous and non-poisonous plants. This is especially so with animals feeding on silage, in which the strong flavours resulting from fermentation mask those of alkaloids or other toxins. It is therefore in silage that the main danger of ragwort poisoning exists, the problem being aggravated by the tendency of the pyrrolizidine alkaloids of ragwort to diffuse through the entire pit or silo, contaminating parts which were ragwort-free.

5. Young animals may not have learnt to discriminate between plant species, and are more likely to select unusual items. Compare this with the behaviour of children who are attracted by poisonous fruits such as those of nightshades, spurge laurel, ivy, privet, mistletoe and laburnum.

6. The normal behaviour pattern of an animal may be upset so that it eats things it would normally avoid. Freely ranging animals, when penned, become less selective. Cases have occurred, for example, of red deer being poisoned by foxgloves when confined in paddocks. Animals with a nutrient imbalance, especially a mineral one, also become less selective. Mineral deficiencies have been associated with increased incidence of ragwort poisoning in cattle; similarly, cattle on a diet low in fibre have been known to eat bracken. Some poisonous plants, notably hard rush, are addictive, so that once an animal has developed a taste for the plant it returns to it again and again.

7. Finally, underground organs of poisonous plants may be exposed by excavation, as in drainage work, and may then be eaten by animals. This is the commonest cause of poisoning by cowbane, hemlock water dropwort, and yellow flag.

12.5.5 Factors affecting the toxicity of plants

Part of plant. The toxicity of all parts of a plant is seldom uniform. In lupins, for example, the greatest concentration of alkaloids is in the seeds, in ragwort in the leaves and in comfrey in the roots. Sorghum has no glycoside in the seeds but toxic levels are present in the leaves and stems. In darnel, by contrast, the seeds contain narcotic alkaloids but the leaves and stems are harmless. In the potato, solanin is found only in green parts of the plant.

Growth stage. Very often a plant is at its most toxic when very young or when growing again after cutting. Examples include sorghum and ragwort.

Soil and fertilizer. Where the toxicity of a plant is due to mineral excess or imbalance, high levels of the offending mineral (e.g. selenium, molybdenum, nitrate) in the soil will clearly increase toxicity. The mineral content of leaves is more strongly responsive to soil and fertilizer than that of other parts such as seeds or fruits. Thus grazed or conserved herbage is the most common source of mineral toxicity in animals, and likewise green vegetables such as spinach in humans. Nitrogen fertilizer increases not only the nitrate content of plants in general, but also the alkaloid content of tall fescue and canary-grass, and the goitrogen content of kale.

Variety. Canary-grass varieties show a 60-fold difference between the highest and lowest alkaloid concentration. Variation in the content of cyanogenic glycosides in white clover varieties is as much as 125-fold. There is therefore much scope for breeding activity to reduce the toxicity of these and other crops. As has been seen, however, it is possible that varieties of low toxicity may be more susceptible to insect pests. On the other hand, the presence of the toxin is in some cases a 'signal' for specific pests. Reduced glucosinolate levels in brassica crops is associated with reduced susceptibility to the cabbage white butterfly, and also, interestingly, to the fungus which causes club root disease.

Cooking. Where toxicity of a raw foodstuff is due to cyanogenic glycosides, as in cassava and red kidney beans, heat causes the toxic component to be released as hydrogen cyanide. This is volatile and, provided cooking is done in an open vessel, is driven off leaving a perfectly safe and nutritious food. Uncooked tapioca, which is made from cassava meal, is non-poisonous because sufficient heat is used in the drying of the meal to destroy all the toxin.

Summary

1. Plant foods are sources of energy, protein, vitamins, a few other essential organic nutrients, minerals and water for animals. Energy is needed for body maintenance and for production. It is supplied by four of the five proximate constituents of the dry matter of plants: crude protein, ether extract (mainly lipids), crude fibre (mainly cell wall materials) and nitrogen-free extract (mainly non-structural carbohydrates), but not by the fifth, ash (inorganic matter).

Not all the organic matter is digestible. Digestibility depends on the animal's digestive system – ruminants, for example, digest much more of the crude fibre content of plants than non-ruminants. Metabolizable energy (ME) is that portion of the gross energy content of a food that is digestible (i.e. not lost in faeces) and is not lost in urine or as methane. ME content varies with the part of the plant harvested, crop species and variety, stage of growth when harvested, crop plant density, method of conservation and other post-harvest treatments.

Only 8–10 amino acids are essential in the diet of non-ruminants; the others which go to make up animal proteins can be synthesized from these in the animal body. Ruminants have no requirement for particular amino acids but must consume enough digestible crude protein to supply their requirements for amino-nitrogen, any form of which can be built into body protein. The protein content of plant foods depends on the part of the plant harvested, crop species and variety, stage of growth, growing conditions and post-harvest treatment.

Lipids are nutritionally valuable as a concentrated energy source, as a source of a small number of essential fatty acids and as a vehicle for certain lipid-soluble vitamins. Among plant foods only oil-rich seeds contain significant amounts of lipids.

Vitamins are organic substances required in small amounts for the functioning of some enzymes. Not all are essential in the diet because they can be synthesized in the digestive tract. Ruminants require vitamins A, D and E in their diet, while non-ruminants also require several B vitamins. In addition, poultry require vitamin K and humans need vitamin C. Plant foods are in general richer than animal products in vitamins C, E and K but are poorer in the B vitamin biotin and vitamin D. Plant materials, with the exception of seeds, also contain abundant carotene, a pigment which can be converted in the animal to vitamin A, but there is no vitamin B_{12} in any green plant.

Minerals which are commonly deficient in plant foods in the quantities required by animals fall into three categories: those such as sodium, iodine, selenium and cobalt for which the plant has no requirement; those such as copper for which animal requirements are much greater than plant requirements; and those whose availability for assimilation by the animal is restricted in certain foods.

Examples of this third category include phosphorus, calcium, magnesium and zinc, rendered unavailable by phytates in the plant; and mineral imbalances as between calcium and phosphorus, magnesium and potassium, or copper and molybdenum. The mineral content of plants depends on growing conditions, stage of growth and species.

2. The largest and most varied differences in plant nutritional value due to growing conditions are found when sites and seasons are compared. Soil conditions have a major influence on the mineral content of plants but not on their organic matter composition, which is determined chiefly by weather and plant age. Growing conditions may influence the concentration of a substance in a plant by affecting its uptake from the soil or its manufacture in the plant, or by altering its interactions with other substances, or by altering the plant's physiology or pattern of growth and development.

Increased nitrogen applications increase the protein content of crops, but concentrations of certain essential amino acids, already at a low level in plant proteins, may decline as a proportion of the protein. Variations in plant protein content may be important for humans on a low plane of nutrition and lacking significant sources of animal protein to augment their diet. Variations in plant mineral content may have considerable dietary significance for humans on a largely vegetarian diet and for grazing stock, especially where they are on a low plane of nutrition relative to their needs, and subsisting on a restricted variety of plant species grown in one or a few localities and on one or a few soil types.

3. The quality of plants as food depends not only on their nutritional value but also on their acceptability to the animal or human consumer. Acceptability to the animal is reflected in voluntary intake, which is influenced by digestibility and other chemical parameters, and also in preference when the animal is given a choice. The preference of an animal for one food over another is a measure of the relative palatability of the foods; this depends on the protein and soluble sugar content and on other chemical and physical parameters. Consumer appeal, which has a major bearing on the price commanded by crop produce, depends on flavour, odour, appearance, texture and even the sound made when it is bitten.

4. An ever-increasing proportion of plant foods is processed before sale to the consumer. To be suitable for processing the quality of the produce

must be such as to give a marketable and attractive end-product and not to interfere with the efficiency of the process itself.

Processing covers a wide range of operations including retexturing and extraction, but the most important processes are preservation and cooking. The aims of cooking are sterilization, improved digestibility and improved flavour or texture. Food is subject to decay by microbes and attack by insects, mites, rodents and other animals. Preservation is achieved by asepsis, by lowering the water potential through the use of solutions of strong solute potential or removal of moisture by drying, cooling, removal of oxygen, build-up of carbon dioxide, lowering of pH, or introduction of antimicrobial agents.

5. Injury to animals from plants may arise from purely physical effects such as choking, from the inability of the digestive system to cope with them, from allergic responses or from toxic substances. These may arise from pollution or contamination entering food chains, from mineral excess or imbalance, from microbial toxins, from pesticides or other materials retained by the plant, or from toxins of the plant itself, either induced by stress or normally present in the plant. Plant toxins include alkaloids, glycosides, saponins, acrid irritant poisons, substances interfering with vitamin or mineral nutrition and photosensitizing substances. They not only cause poisoning but may taint meat or milk.

Animals normally avoid poisonous plants when grazing but poisoning may occur if the toxic plant is the only food available, if it is not unpalatable, if it has been rendered palatable by wilting (as in hay) or by herbicide spraying, if it is present in silage when the animal's ability to differentiate between species is minimal, if it is available to young animals which have not fully learned to discriminate between one species and another or to other animals whose normal selective behaviour is upset, or if the toxic underground organs of the plant have been exposed by excavation.

Toxicity of a species can depend on the part of the plant eaten, on growth stage, on soil and fertilizer use, on variety and on whether it is eaten raw or cooked.

Index

abrasion 48, 94, 96, 97, 106, 115, 130
abscisic acid (ABA) 195, 198–201, 205
abscission 97, 155, 156, 166, 179, 180, 182, 195, 198, 203, 258
achene 112, 181
adventitious bud 241, 242, 245
adventitious roots
  effects of auxins 198, 203
  in vegetative propagation 203, 241–3, 245, 246
  growth in soils 139
  origin and formation 117, 127, 133, 134, 136, 150, 198
  specialized 148, 271
aerobic respiration 25, 119
aflatoxin 330
after-ripening 115
air layering 246
aleurone layer 111, 118, 180
alfalfa *see* lucerne
algae 5–7, 53
alginate 5
alkaloid 322, 329–34
alkaloidal glycoside 107, 330, 331
allele 224–30
allelopathy 137, 271
allergy 154, 175, 330
almond (*Prunus amygdalus*) 272, 331
aluminium (Al) 76, 79, 80, 140
amidochlor 206
amino acid 12, 36, 119, 214–17, 222, 307, 315, 317, 318
amylase 118
anaerobic respiration 25, 51, 52, 119, 327; *see also* fermentation
analysis of variance 211
anemone (*Anemone* sp.) 115
angiosperm 6, 8, 111, 159, 160, 168
annual 132, 133, 156, 244
annual meadow grass (*Poa annua*) 162, 164
anther 159, 160, 169, 170, 174, 176
anthocyanin 75, 182
antiauxin 203
antigibberellin 203
aphids 18, 105, 106, 277, 333
apical bud 96, 146, 150, 151, 156, 241, 242, 245, 261
apical dominance 146, 197, 201, 203, 205, 206, 241, 242, 291, 292

apical meristem 9, 87, 96, 97, 107, 117, 143–6, 160, 164, 191, 196
apomixis 180
apoplast 39–41, 70–3, 96
apple (*Malus domestica*)
  breeding and varieties 178, 254
  chilling injury 102
  cider making 328
  consumer preference 323, 324
  control of growth and development 189, 190, 195
  cuticle of fruit 94, 95
  flower structure 170
  flowering initiation 164
  fruit development and structure 180, 181, 280
  grafting 247–9
  incompatibility 178
  mineral nutrition 76, 290
  nutritive value 305
  ripening 182, 195
  seed dormancy 113, 115
  storage and post-harvest losses 95, 287, 288, 290
  toxic substances in 331
ascorbic acid *see* vitamin C
asepsis 325
asexual reproduction 159, 163, 180; *see also* apomixis, vegetative propagation
ash (*Fraxinus excelsior*) 329
ash in analysis of food 302, 303
asparagus (*Asparagus officinalis*) 176, 293, 324
Aspergillus sp. 326, 330
ATP (adenosine triphosphate) 13–16, 23–5, 62
aubergine (*Solanum melongena*) 291
autonomous development 198
auxin 166, 167, 192–4, 196–8, 201, 203–5, 246
avocado (*Persea americana*) 182, 290
awn 127, 174, 280
axil 9, 10, 164
axillary bud 9, 97, 107, 145, 146, 148, 150, 151, 156, 164, 197, 203, 205, 242, 245, 251, 252

bacteria
  as plant pathogens 94, 105, 106, 251, 252, 276, 281, 289, 290, 293, 330